The Handbook of Environmental Chemistry

Volume 121

Founding Editor: Otto Hutzinger

Series Editors: Damià Barceló • Andrey G. Kostianoy

In over four decades, *The Handbook of Environmental Chemistry* has established itself as the premier reference source, providing sound and solid knowledge about environmental topics from a chemical perspective. Written by leading experts with practical experience in the field, the series continues to be essential reading for environmental scientists as well as for environmental managers and decision-makers in industry, government, agencies and public-interest groups.

Two distinguished Series Editors, internationally renowned volume editors as well as a prestigious Editorial Board safeguard publication of volumes according to high scientific standards.

Presenting a wide spectrum of viewpoints and approaches in topical volumes, the scope of the series covers topics such as

- local and global changes of natural environment and climate
- anthropogenic impact on the environment
- water, air and soil pollution
- remediation and waste characterization
- environmental contaminants
- biogeochemistry and geoecology
- chemical reactions and processes
- chemical and biological transformations as well as physical transport of chemicals in the environment
- environmental modeling

A particular focus of the series lies on methodological advances in environmental analytical chemistry.

The Handbook of Environmental Chemistry is available both in print and online via http://link.springer.com/bookseries/698. Articles are published online as soon as they have been reviewed and approved for publication.

Meeting the needs of the scientific community, publication of volumes in subseries has been discontinued to achieve a broader scope for the series as a whole.

Series Preface

With remarkable vision, Prof. Otto Hutzinger initiated *The Handbook of Environmental Chemistry* in 1980 and became the founding Editor-in-Chief. At that time, environmental chemistry was an emerging field, aiming at a complete description of the Earth's environment, encompassing the physical, chemical, biological, and geological transformations of chemical substances occurring on a local as well as a global scale. Environmental chemistry was intended to provide an account of the impact of man's activities on the natural environment by describing observed changes.

While a considerable amount of knowledge has been accumulated over the last four decades, as reflected in the more than 150 volumes of *The Handbook of Environmental Chemistry*, there are still many scientific and policy challenges ahead due to the complexity and interdisciplinary nature of the field. The series will therefore continue to provide compilations of current knowledge. Contributions are written by leading experts with practical experience in their fields. *The Handbook of Environmental Chemistry* grows with the increases in our scientific understanding, and provides a valuable source not only for scientists but also for environmental managers and decision-makers. Today, the series covers a broad range of environmental topics from a chemical perspective, including methodological advances in environmental analytical chemistry.

In recent years, there has been a growing tendency to include subject matter of societal relevance in the broad view of environmental chemistry. Topics include life cycle analysis, environmental management, sustainable development, and socio-economic, legal and even political problems, among others. While these topics are of great importance for the development and acceptance of *The Handbook of Environmental Chemistry*, the publisher and Editors-in-Chief have decided to keep the handbook essentially a source of information on "hard sciences" with a particular emphasis on chemistry, but also covering biology, geology, hydrology and engineering as applied to environmental sciences.

The volumes of the series are written at an advanced level, addressing the needs of both researchers and graduate students, as well as of people outside the field of "pure" chemistry, including those in industry, business, government, research establishments, and public interest groups. It would be very satisfying to see these volumes used as a basis for graduate courses in environmental chemistry. With its high standards of scientific quality and clarity, *The Handbook of Environmental Chemistry* provides a solid basis from which scientists can share their knowledge on the different aspects of environmental problems, presenting a wide spectrum of viewpoints and approaches.

The Handbook of Environmental Chemistry is available both in print and online via https://link.springer.com/bookseries/698. Articles are published online as soon as they have been approved for publication. Authors, Volume Editors and Editors-in-Chief are rewarded by the broad acceptance of *The Handbook of Environmental Chemistry* by the scientific community, from whom suggestions for new topics to the Editors-in-Chief are always very welcome.

Damià Barceló
Andrey G. Kostianoy
Series Editors

Preface

This second volume of the global agricultural soil degradation project is dedicated to Europe. It is well known in Europe that agricultural soil degradation severely impacts the environment. For instance, the extension of soil degradation in the Mediterranean area is still being determined [1]. Soil intensive exploitation due to intensive agriculture practices is a source of the food chain and water reserves contamination and species extinction. Approximately 40% of European soils are subjected to industrial agriculture. This represents tremendous pressure in these areas. Approximately 36 million ha in European Union countries are affected by compaction.[1] Water erosion is also one of Europe's most common forms of soil degradation. In 2016, about 43 million ha were affected by water erosion. 80% of these areas were agricultural or grasslands. This is especially serious in southern European countries.[2] On the other hand, agricultural areas are more vulnerable in northeast Europe [2]. In the EU, mean soil erosion is 1.6 times higher than the mean soil formation [3]. Secondary salinization (e.g., saltwater intrusion, wastewater use, groundwater table fluctuation, water logging, fertilizers, or saline water irrigation) is estimated to affect 3.8 Mha in Europe [4]. The costs of soil salinization in Europe are enormous. They range between €577 and €610 million per year.[3] Soil pollution due to agriculture management in Europe is severe due to agrochemicals use [3]. In European Union countries about 65–75% of agricultural soils, the use agrochemicals and other products s more than the necessary.[4] There are different dynamics among member states. For instance, Italy, Germany, Spain, and France consumed

[1]https://esdac.jrc.ec.europa.eu/themes/soil-compaction.

[2]https://ec.europa.eu/eurostat/statistics-explained/index.php?title=Agri-environmental_indicator_-_soil_erosion&oldid=581847#Introduction.

[3]https://pr.euractiv.com/pr/economic-impacts-salinization-exceeds-half-billion-euro-year-europe-alone-and-cost-projected-rise.

[4]https://www.eea.europa.eu/themes/soil.

two-thirds of pesticides in all of Europe.[5] The usage of nitrogen and phosphorus is increasing in Europe. Compared to 2010, in 2020, the European Union countries used more than 10 million tons of nitrogen and 1.2 million tons of phosphorus, respectively.[6] Finally, the concentration of metals such as cadmium or copper is very high in some areas. The use and abuse of agrochemicals substantially increased the presence of nutrients in water bodies threatening flora and fauna in these ecosystems. It is recognized that the agrochemicals regulatory regime in agriculture is rather weak, which is an important barrier to halting and reversing their use [3]. There is much evidence that agricultural intensification is an ongoing process in Europe. Several strategies have been developed to reduce the impacts of intensive agriculture on the ecosystems and soil degradation, such as the European Green Deal.[7] One of the aims is to ensure "A healthy food system for people and the planet." The EU goals are (1) *to ensure food security in the face of climate change and biodiversity loss,* (2) *to* ***reduce the environmental and climate footprint of the EU food system***, (3) *to strengthen the EU food system's resilience, and* (4) *to lead a global transition towards competitive sustainability from farm to fork.*[8] The objectives are ambitious. However, there is a lot to do to make them a reality. In volume II of the book, global agricultural soil degradation contributions from Croatia, Czech Republic, Estonia, Germany, Greece, Hungary, Iceland, Italy, Latvia, Lithuania, Portugal, Slovenia, Spain, Sweden, and Ukraine were published.

Vilnius, Lithuania
Seville, Spain
Zagreb, Croatia
Beijing, China

Paulo Pereira
Miriam Muñoz-Rojas
Igor Bogunovic
Wenwu Zhao

References

1. Ferreira CSS, Seifollahi-Aghmiuni S, Destouni G, Ghajarnia N, Kalantari Z (2022) Soil degradation in the European Mediterranean region: processes, status and consequences. Sci Total Environ 805:150106
2. Borrelli P, Ballabio C, Panagos P, Montaranella L (2014) Wind erosion susceptibility of European soils. Geoderma 232–234:471–478

[5]https://ec.europa.eu/eurostat/statistics-explained/index.php?title=Agri-environmental_indicator_-_consumption_of_pesticides.

[6]https://ec.europa.eu/eurostat/statistics-explained/index.php?title=Agri-environmental_indicator_-_mineral_fertiliser_consumption.

[7]https://commission.europa.eu/strategy-and-policy/priorities-2019-2024/european-green-deal_en.

[8]https://commission.europa.eu/strategy-and-policy/priorities-2019-2024/european-green-deal/agriculture-and-green-deal_en.

3. EEA (2019) Land and soil in Europe. Why we need to use these vital and finite resources sustainably. Publications Office of the European Union, Luxembourg, p 59
4. Daliakopoulos IN, Tsanis IK, Koutroulis A, Kourgialas NN, Varouchakis AE, Karatzas GP, Ritsema CJ (2016) The threat of soil salinity: a European scale review. Sci Total Environ 573:727–739

Acknowledgments

The editors appreciate the support of Prof. Damia Barcelo and Sofia Costa in developing this book. Their encouragement was crucial to bring this project to light. We would also like to acknowledge the support of the project No. 42271292, funded by the National Natural Science Foundation of China.

Contents

Agricultural Soil Degradation in Croatia 1
Igor Bogunovic, Lana Filipovic, Vilim Filipovic, and Ivica Kisic

Agricultural Soil Degradation in the Czech Republic 35
David Zumr

Agricultural Soil Degradation in Estonia, Latvia and Lithuania 59
Paulo Pereira, Miguel Inacio, Igor Bogunovic, Lyudmyla Symochko, Damia Barcelo, and Wenwu Zhao

Agricultural Soil Degradation in Germany 87
Manuel Seeger

Agricultural Soil Degradation in Portugal and Greece 105
Carla S. S. Ferreira, António C. Duarte, Anne K. Boulet, Adélcia Veiga, Giorgos Maneas, and Zahra Kalantari

Agricultural Soil Degradation in Hungary 139
M. Birkás and I. Dekemati

Agricultural Soil Degradation in Iceland 159
Isabel C. Barrio and Ólafur Arnalds

Agricultural Soil Degradation in Italy 179
Demetrio Antonio Zema, Giuseppe Bombino, and Santo Marcello Zimbone

Agricultural Soil Degradation in Slovenia 223
Matija Zorn

Agricultural Soil Degradation in Spain 263
Natalia Rodríguez-Berbel, Rocío Soria, Raúl Ortega, Manuel Esteban Lucas-Borja, and Isabel Miralles

Agricultural Soil Degradation in Sweden 299
Ana Barreiro and Linda-Maria Dimitrova Mårtensson

Agricultural Soil Degradation in Ukraine 325
Oleksandr Menshov and Oleksandr Kruglov

Correction to: Agricultural Land Degradation in the Czech Republic . . . C1
David Zumr

Agricultural Soil Degradation in Croatia

Igor Bogunovic, Lana Filipovic, Vilim Filipovic, and Ivica Kisic

Contents

1 Introduction 2
2 Soil Compaction and Erosion 2
3 Overgrazing 9
4 Soil Salinity 13
5 Soil Contamination with Chemicals 18
6 Soil Contamination with Microplastics 27
7 Conclusion 28
References 28

Abstract Croatian agriculture has a small impact on the country's economy (around 4% of the Croatian Gross Domestic Product), but agricultural activities affect agricultural land and reduce quality and productivity. The types of land degradation are not distributed equally in the territory, considering an environmental diversity between the Mediterranean, Mountain and Continental parts of Croatia, different pedological and geomorphological conditions, and cropping and management systems. In Croatia, the most widespread degradation problems are soil erosion by water and soil compaction. Together with soil salinity and soil contamination, these processes decrease soil ability to provide ecosystem services in quantity and quality. Soil compaction and erosion by water present the greatest challenge for soil protection, considering the impact of future climate change conditions. Croatia still has not established comprehensive monitoring of the soils. Thus, currently, the less visible degradation processes in Croatia (microplastics, metal contamination) are not recognized as an important threat to soil quality and productivity. This chapter summarizes the occurrence, distribution, and causes of various physical and chemical soil degradation processes and national reclamation and conservation practices.

I. Bogunovic (✉), L. Filipovic, V. Filipovic, and I. Kisic
Faculty of Agriculture, University of Zagreb, Zagreb, Croatia
e-mail: ibogunovic@agr.hr; lfilipovic@agr.hr; vfilipovic@agr.hr; ikisic@agr.hr

Paulo Pereira, Miriam Muñoz-Rojas, Igor Bogunovic, and Wenwu Zhao (eds.),
Impact of Agriculture on Soil Degradation II: A European Perspective,
Hdb Env Chem (2023) 121: 1–34, DOI 10.1007/698_2022_919,

Published online: 14 December 2022

Keywords Land abandonment, Machinery traffic, Sediment detachment, Soil contamination

1 Introduction

Soils support various ecosystems on the Earth through their functions. Soils regulate climate and filter pollutants [1]. Soils stores water, filtrates it, and through colloidal complex binds various substances [2]. This is also true for various pollutants emitted from various sources such as industrial activities, inadequate waste disposal and mining or by accident into the soil. In this role, soil also protects groundwater and aquatic ecosystems from contamination. Furthermore, soil provides the basis for the urban areas, roads, recreational areas, or landfills. Unsustainable soil management triggers soil degradation. These processes occur on a World level, and Croatia is not an exception. Decades of ploughing, chemical fertilizers, and plant protection agents make Croatia face land degradation problems. Nowadays, the degradation of the Croatian agroecosystems is when detailed studies and organized activities to prevent degradation processes are needed. Lack of soil protection measures and inappropriate soil management in Croatia increase the soil vulnerability to climate changes. Different types of degradation confirmed in Croatian agricultural lands will be evaluated in the next subchapters. This work objective is to analyze the causes and prevalence of land and soil degradation associated with agricultural practices in Croatia.

2 Soil Compaction and Erosion

Soil compaction and erosion are recognized as the most threatening degradation processes in Croatia [1, 3, 4]. Nowadays, between 25% and 35% of all agricultural land in Croatia are affected by compaction [5]. Although landowners are aware that soil compaction is closely related to structure deterioration (reduced soil permeability and macroporosity), this type of degradation is not monitored correctly in Croatia since subsoil compaction is usually hard to identify from the surface. In Croatia, most soil degradation is anthropogenically induced by tillage or machinery traffic [6]. Compaction occurs in several soil types (Cabisols, Chernozems, Stagnosols, Gleysoils) and different climatic conditions (Mediterranean, Continental), generally everywhere in Croatia where conventional management is applied [7–11]. In Croatia continental region, tillage-induced compaction affects a large area, mainly because of less favourable economic circumstances [12].

Tillage-induced soil compaction varies according to the tillage depth, mainly depending on the tools used (e.g., disk pan or plough pan). Compaction is a consequence of multiple tillage interventions performed at the same depth for several

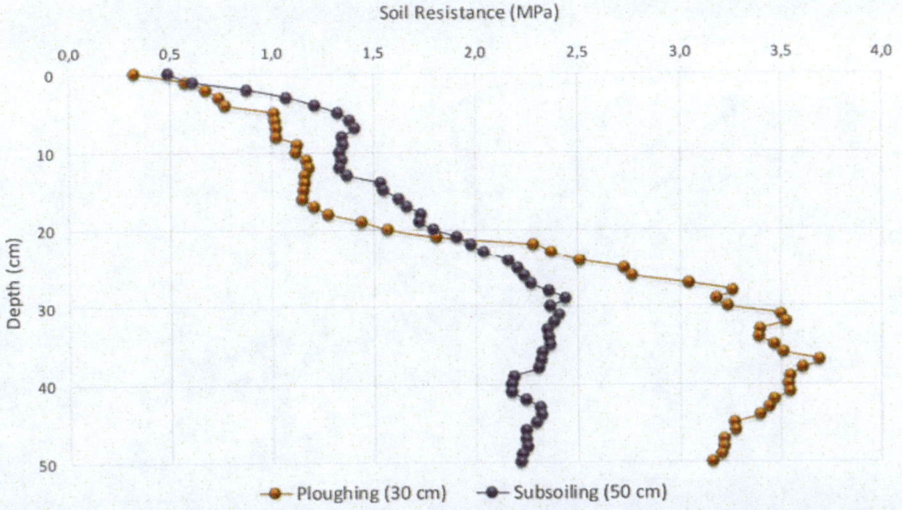

Fig. 1 Soil compaction in ploughed and subsoiled cropland in a Croatian Stagnosol (June second 2012, Blagorodovac). Compaction was measured with a handheld penetrometer (Eijkelkamp, Netherland) using a cone with a 2 cm^2 base area, 60° included angle, and 80 cm driving shaft

years. Soil compaction is vulnerable when the soil is in wet conditions. In this context, Fig. 1 shows the example of soil compaction (30–35 cm) induced by tillage in Central Croatia. The high compaction is a consequence of ploughing the soil at the same depth in the last 40 years. Although subsoiling was applied to alleviate compaction, a permeable subsoil layer is still identifiable.

Plough-pans occur in Croatian croplands mostly between 25 and 35 cm depth, whereas deeper (up to 45 cm) occurrence is recorded mostly for clay and silty clay soils in the western part of Croatian Pannonia. In Eastern Croatia on loam, sandy clay loam, or silty loam soils plough pan mainly occurs at a low depth (i.e., 25 cm). Here farmers usually perform shallower primary tillage. Moreover, a small number of farmers perform a reduced tillage system, where the disk is used as a tool. These soils usually suffer from disc-pans underneath the most frequent disking depth, between 12 and 18 cm [12, 13].

Compaction induced by trafficking is also widespread in Croatian soils. The agricultural mechanization in Croatia resulted in an increase in the weight of farm vehicles during the last decades [14]. Several researchers identified severely compacted soils in croplands [15–18] or permanent plantations [10, 19–21]. As shown in Fig. 2, the compaction changes in depth in trafficked and not trafficked silty soils in Central Croatia. On arable cropland under winter wheat, after 1, 3, 6, and 7 passes of eight-tonne axle load, soil strength increased more than 300%, exceeding the threshold limit of 2.5 MPa for optimal crop root growth [22]. Such increase in soil resistance has negative impacts on soil since it changes the structure (destructs the macro-aggregates, reduces the macropores and water infiltration, decreases air movement), reducing the yields of tobacco [8], maize [23], winter

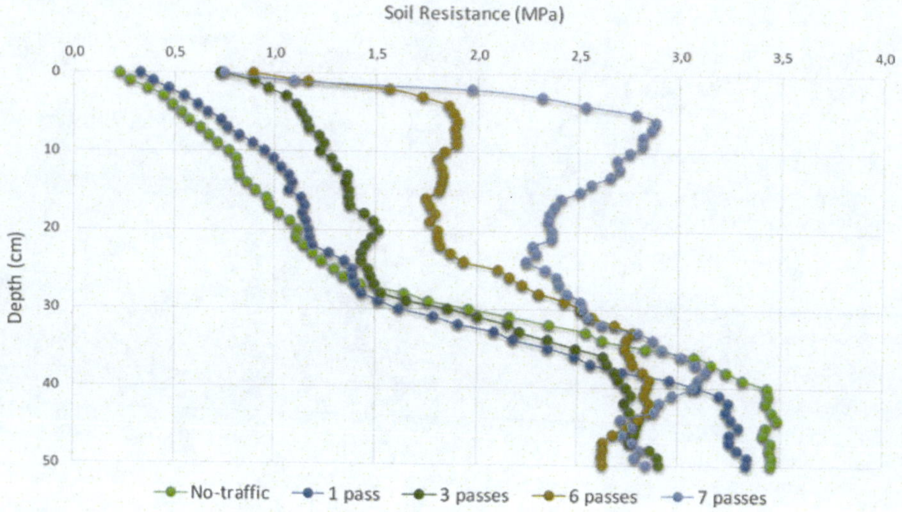

Fig. 2 Soil compaction in a no-trafficked and trafficked soil after 1, 3, 6, and 7 passes of the tractor (eight-tonnes axle load weight) during 2014 in Croatian Stagnosol near Potok. Compaction was measured with a handheld penetrometer (Eijkelkamp, Netherland) using a cone with a 2 cm^2 base area, 60° included angle, and 80 cm driving shaft

Fig. 3 Examples of poor agro-technical practice and waterlogging near Potok (Photo taken by Igor Bogunovic, March, 2015)

wheat [7], oat [11] and Sudan grass [24]. Moreover, the negative effect of distorted structure and poor water-air relation in compacted soil is evident after heavy rains (Fig. 3).

The compaction shown in Fig. 2 is result from "good agricultural practices", where farmers follow the permanent tracks in the field. However, these examples are only observed in a small percent of arable lands, where producers implemented GPS technology. The majority of farmers do not apply precision farming technology. According to the available data, we estimate that the percentage of wheeling on soil surface exceeds 90% of the track area of the field area only during tillage operations.

In comparison, for all agricultural machinery interventions during the cereal production season, the track area in 1 year exceeds 300% [25].

Coupled with degradation induced by heavy tractors, traffic-induced compaction by light tractors (< 4 t axle load) is also standard in Croatia. Such degradation is mainly related to permanent plantations: vineyards and orchards, where this type of mechanization is frequently used. Only in one season, more than 15 passes occur for tillage, fertilizer application, mowing, pomo-technical or ample-technical operations and chemical protection [10, 18]. Figure 4 shows the temporal increase of soil compaction in a Croatian vineyard. Regardless of the soil management, it is visible that soil tillage (Fig. 4a) has a positive impact only at the beginning of the season, while after 12 passes, soil compaction increases above limits for normal root growth (>2.5 MPa). In a grass-covered vineyard, the initial compaction is high at the beginning of the vegetation season. However, the additional number of tractor passes also dramatically increases the penetration resistance (Fig. 4b). Croatia has more than 22,000 ha of vineyards and more than 41,000 ha of orchards [26]. The majority of them are conventionally managed. Therefore, it is very likely that these areas also suffer from severe compaction because subsoil ripping is very limited and is applied only by few grape and fruit producers.

Water and wind erosion occur because of inappropriate agricultural practices causing soil structural degradation [13]. As confirmed in several studies, soil structural degradation in Croatian agroecosystems is mainly associated with poor soil organic matter content [27–30].

In Croatia, the conversion from forest to agricultural land use implies a loss of more than half of soil organic matter [31]. Moreover, different soil management practices affect organic matter decomposition rate and increase CO_2 emissions. As proof of the predominantly impact of management on soil organic matter loss, we present the part of the results from the "Soil erosion and degradation in Croatia" – SEDCRO project. This project is focused on soil and land use management impact on soil properties and hydrological response on a national scale. Some of the obtained results are presented in Fig. 5, which shows the harmful effects of conventional soil management on soil organic matter and aggregate stability. However, it is necessary to highlight that variable differences between soil management occur due to initial soil organic matter concentrations, duration of particular soil management and different soil characteristics. The tillage and no-tillage management were compared considering different land uses (croplands, vineyards, and orchards). A reduction of soil organic matter from 7% to 71% was observed in tilled soils compared to no-tilled soils. Similarly, a reduction of water-stable aggregates percentage on tilled soil ranged from 8% to 40%, depending on the location. Such results confirm that tillage management in Croatia severely impacts soil organic matter content and the structural stability in various soil types, land uses, and environmental conditions. Moreover, ploughing and other tillage intensive operations were confirmed as management which has a detrimental impact on soil structure and soil organic matter content. Since more than 95% of Croatian farmers still use conventional ploughing to manage croplands [28, 32], it can be assumed that these areas suffer from severe erosion. In this context, it is necessary to adjust tillage management to become an

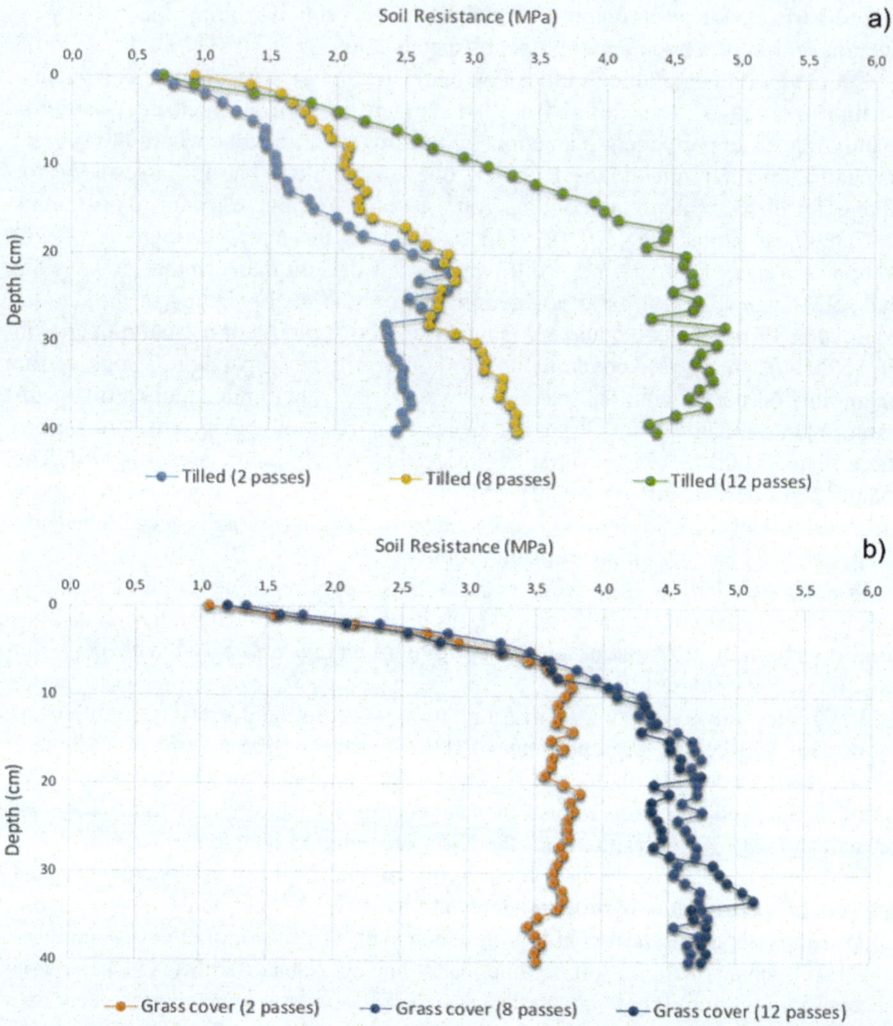

Fig. 4 Soil compaction on tilled (**a**) and grass-covered (**b**) inter-row in a vineyard after 2, 8 and 12 tractor pass (three-tonnes axle load weight) in Jazbina (2017). Compaction was measured with a handheld penetrometer (Eijkelkamp, Netherland) using a cone with a 2 cm^2 base area, 60° included angle and 80 cm driving shaft

environmentally friendly practice by using already proved conservation practices in the Croatian pedosphere like no-invertive tillage [18], tillage across the slope [33], no-tillage [35] or mulching [20]. However, there is an opinion that Croatian farmers and some agricultural experts are often not educated enough in the area of conservation practices. This is important to be considered for the national map of erosion prevalence. Husnjak [36, 37] mapped the potential and actual risk of water erosion in Croatia. There is a potentially high risk of erosion on 1,800,265 ha and an actual high

Fig. 5 Comparison of tilled and no-tilled soil organic matter (SOM) content and water-stable aggregates (WSA) percentage in vineyard, orchard and cropland soils in Croatia. Data were derived from "Soil erosion and degradation in Croatia" national project. Data present reduction of SOM and WSA as directly connected to tilled and no-tilled soil at each location

risk of soil erosion on 746,475 ha. The dominant mapped areas are the plains with less than 3.5% slope, where soil erosion is not observed. Milder forms of erosion are mapped on 27.2% of the surface with a slope of up to 8.8%. Severe erosion areas are visible in 48.3% of the area. Figure 6 shows the risk of occurrence of erosion processes. Therefore, appropriate soil tillage management is crucial for agroecosystem sustainability.

In most cases, tillage is not adjusted to the slope of the terrain in Croatia [18, 20, 35]. In some northwest and mountain parts of Croatia, the croplands often occur on slopes, where this land should not be tilled or should be under different land use (Fig. 7). A long-term study (20 years) conducted in Central Croatia [38] revealed that the most effective cropland management to decrease erosion by water is the use of no-tillage (Fig. 7). Compared to the conventional ploughing performed up and down the slope direction, no-tillage reduced soil water erosion by 85% (from 13.9 t ha^{-1} to 2.1 t ha^{-1} annually). The differences are even more pronounced when observing the crops grown in Croatia. In this context, no-tillage management reduced soil loss by 81%, 84%, 91%, 92%, and 97% under maize, soybean, wheat, oilseed rape, barley, and soybean double-crop, respectively (Fig. 8).

No-tillage management reduced nutrient losses and water resources pollution from agrochemicals [33, 34]. Long-term research conducted on permanent

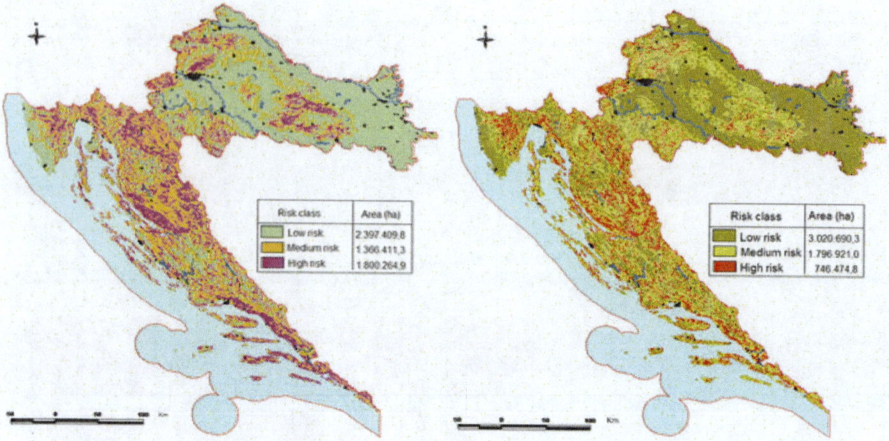

Fig. 6 Map of potential (left) and actual (right) risk from soil erosion by water in the Republic of Croatia. Source: Ministry of Agriculture [39]

Fig. 7 General morphology of northwest Croatia. Croplands were annually tilled on high slopes near, regardless of the potential soil erosion. Location: Gornja Pušća, Croatia (Photo taken by Ivan Dugan, February 2021)

plantations has been not performed yet for the Croatian agroecosystems. However, some rainfall simulation experiments were conducted at different locations in Croatia. The results are presented in Fig. 9, where the detrimental impact of inadequate tillage management in vineyards and orchards on soil loss and hydrological response can be seen. The soil loss under simulated rainfall (60 mm h^{-1} for 30 min duration) was higher than 9.6 t ha^{-1} [40] on tilled soils than grass-covered soils in one rainstorm event. Soil loss reduction under conservation management in permanent plantations vary from 21% to 99%, confirming the beneficial impact of permanent grass cover.

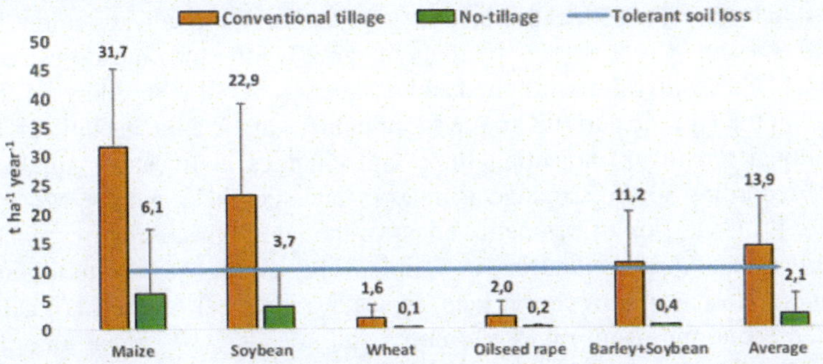

Fig. 8 Conventional and no-tillage management impact on soil loss in a 20-year experiment, Daruvar, Croatia. Each crop appears on four occasions in crop rotation. Hanging bars represent standard deviation. Soil loss tolerance level was based according to Kisić [3]

Fig. 9 Comparison between tilled and no-tilled vineyards and orchards and their impact on soil loss obtained by rainfall simulation experiments in Croatia. Data were derived from "Soil erosion and degradation in Croatia" national project. Data present a reduction of soil loss under grass-covered permanent plantations compared to tilled soils under the same land uses

3 Overgrazing

It is well known that overgrazing triggers multiple degradation processes such as compaction, erosion, acidification, depletion of organic matter or structural instability [13]. Croatia has large landscape diversity, where lowland is mixed with hills and mountains rich in forests, meadows, and grasslands. From all agricultural land (1,556,000 ha) in Croatia, the permanent grasslands and meadows occupy 597,000 ha [41]. Besides the mentioned areas, a great part of the meadows is already abandoned in the Dinaric part of Croatia (dominantly karstic area, occupying 48.9% of the national territory), indicating important changes in livestock number and diversity. According to the Croatian Bureau of statistics [41] and the Ministry of Agriculture [42], the lack of competitiveness in the agricultural sector compared to average European Union countries is visible. The average agricultural farm in Croatia uses 11.6 ha of agricultural land, raises 5.6 livestock units (LU), achieves an expected economic result of 15,134 euros and spends 1.19 annual units for its work. Compared to the average agricultural farm at the level of the European Union, the average Croatian agricultural holding uses 30% less agricultural land, raises 56% less LU, and achieves 57% lower standard economic result. Data presented here give a clear context for why national research on overgrazing in Croatia does not exist. Nowadays, Croatian meadows and grasslands inhabit only 25,000, 419,000, 945,000, 100,000, and 10,389,000 of Equidae, cattle, pigs, goats, and poultry, respectively [41]. Although livestock had the most important role in the survival of the population in this area, nowadays, the socioeconomic changes and poor economic results of agro-pastoralism made this profession undesirable for a great part of the public and especially abandoned by younger generations. However, remained producers have intensive livestock production and good economic results, indicating that certain areas suffer from possible degradation processes. Overgrazing in Croatia was noted in several places. One example was cattle overgrazing ("Angus" cattle) in Velika Jasenovača in the continental part of Croatia. The comparison of heavily grazed and not grazed plots ($n = 10$) revealed a drastic decrease of vegetation cover (Fig. 10) by more than 75%. Similar was also observed in the Mediterranean part of Croatia under sheep grazing [43]. Ljubičić et al. [43] reported the negative impact of overgrazing on vegetation community at northern Adriatic islands rocky pastures, where plant diversity was reduced. Such trends of vegetation biomass reduction have a detrimental impact on soil organic matter. Before the mentioned project, "Soil erosion and degradation in Croatia" includes the research on animal farms in the Continental part of Croatia (Velika Jasenovača), where overgrazing by cattle increased the soil compaction (measured by bulk density from 1.46 to 1.57 g cm^{-3}), while also modifying the pore characteristics measured by water holding capacity. Overgrazing of plots reduced the capacity of soil to hold water by 15%. Figure 10 shows the condition of soil surface at overgrazed and not grazed plots, which undoubtedly affected the hydrological processes in soil and caused sediment and nutrient losses in overgrazed areas. Simulated rainstorm events with the intensity of 60 mm h^{-1} for 30 min showed 14 times higher soil losses on overgrazed plots compared to the non-grazed. Similarly, in the Croatian continental area, nutrient losses on overgrazed plots were 27, 35, 39 times higher for phosphorus, nitrogen, and carbon, indicating a great threat from overgrazing on soil sustainability and possible ground and surface water pollution.

Although the detailed research in the Continental part of Croatia still missing, reported results indicate the possible degradation issues by overgrazing only in small

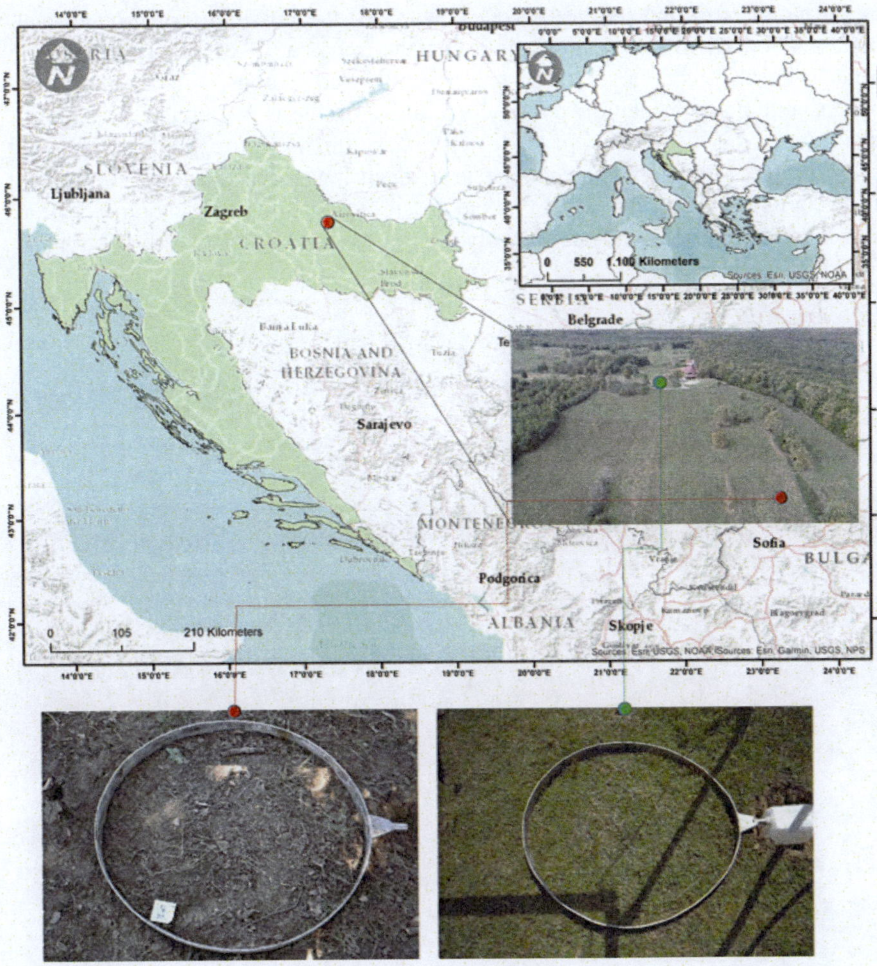

Fig. 10 Vegetation cover and surface soil state on overgrazed (left) and non-grazed plots (right) in Mala Jasenovača, Croatia (photo taken by Igor Bogunović, July 2019)

local areas. The absence of cattle reduces the possibility of spreading the overgrazing negative effect on the soil environment to a broader extent. However, several modern farms located in the Northern and Eastern parts of Continental Croatia are causing different degradation processes. On such farms, farmers were taught to collect the animal slurry and then use it in an area a few kilometres around the farm, which undoubtedly affected the soils [44]. Bašić [44] reported that the application of animal slurry through agro-technical operation, including high pressure, impacts soil surface, increasing compaction, while slurry acidifies soils and increases nitrates loss and pollutes the groundwater.

Literature indicates that livestock modifies pH and chemical fluxes in soil by ingesting nutrients and returning them via excreta [13]. As stated above, the lack of animal numbers during the last decades decreased the possibility of groundwater contamination by animal urine, which contains nitric acid. Moreover, 48.9% of Croatian territory is a karstic area [45], generally very porous, with limited water sources and poor vegetation cover. Regardless of these mentioned limitations, karstic land was dominantly used for livestock grazing throughout history. On karstic pastures, the animal stocking rate is much lower due to sparse vegetation. Lower animal stocking rate decreases the possibility for soil chemical modification from excreta and urine, while dominantly rocky area enables the generation of runoff after vegetation removal (Fig. 11). This is more expected on higher altitude pastures

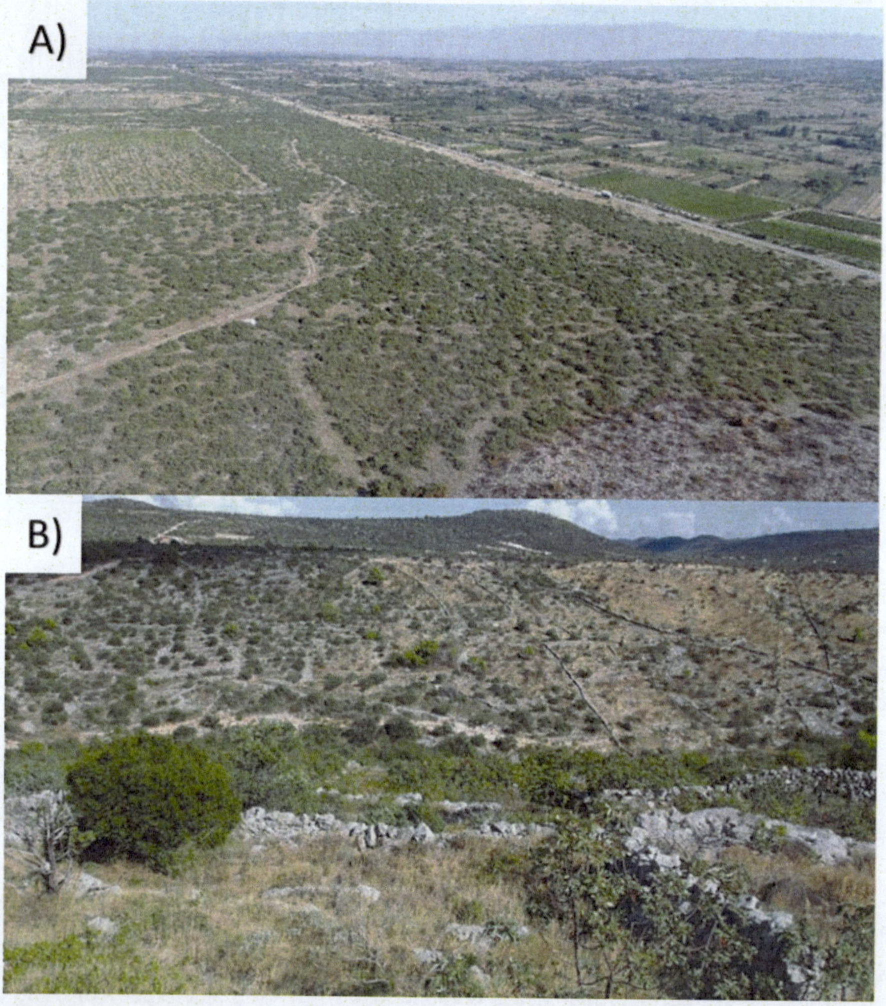

Fig. 11 Common karstic pasture in the Mediterranean part of Croatia (A: Nadin, Zadar County, B: Podglavica, Šibenik-Knin County) with the dominant stoniness limitation (photos were taken by Igor Bogunovic, September 2018)

in karstic areas, where environmental conditions enable the grazing during only about three to four summer months. Low possibility for biomass production (around 0.65 t ha^{-1} of hay on the mountain, and in the range from 1.0 to 2.5 t ha^{-1} on hilly pasture [46]) decreased the LU number in this area and reduced the possible degradation issue.

Besides overgrazing, in the karst environment of Mediterranean Croatia, several ongoing processes like littoralization, socioeconomic restructuring, and agricultural and livestock abandonment negatively impact the management of karst pastures. It is well known that agro-pastoralism is the optimal management of this area. With the absence of goats and sheep, land abandonment processes were speeded up with partial afforestation and dominance of scrublands (maquis), which accumulates organic matter. Afforestation and politics of ultimately fire suppression lead to excessive accumulation of burning fuel. Several mega-fires occurred in Croatia during past decades, with a highly negative degradation effect on abandoned pastures through modified soil chemistry properties [47, 48].

4 Soil Salinity

Soil salinity is not recognized as a national scale problem in Croatia, very likely because this type of soil degradation affects only a small number of farmers, although its importance can be high on a local scale. Within Croatian institutions focusing on agricultural research, soil salinization is considered an essential environmental constraint mainly regarding the productivity of agricultural soils, i.e., in the context of the negative effect of high concentration of salts in the rhizosphere on crop yield amount and quality. In this context, several Croatian researchers reported a decrease of above shoot groundmass or yields of several crops such as oat [49], olives [50, 51], Brassica crops [52], watermelons [53] or strawberries [54–56]. All of these examples were induced by soluble salts accumulation in the rhizosphere and affecting plant physiological processes.

As is well known, soil salinization is usually more expressed in arid and semi-arid regions where evaporation is high, and irrigation is a prerequisite for profitable agricultural production. Although Europe has the lowest quantity of soils degraded by salinity (about 31 million ha) compared to other continents [57], the importance of salinization (and alkalization) processes is increasing for the Croatian agricultural land as well, possibly becoming even more pronounced because of the climate change effects, and intensification in land use and irrigation. Similarly to the natural prevalence of saline soils on a world level, Croatia has the same sequence of saline soils – most of them are located in dry environmental areas in the uppermost eastern part of Croatia or close vicinity of the Adriatic Sea. Soil salinity occurs as a result of natural processes and/or anthropogenic activity, with the most common causes as described by Shahid et al. [58] as follows: inherent soil salinity (weathering of rocks, parent material), irrigation with brackish and saline water, seawater intrusion into coastal lands and aquifers due to overuse of freshwater, restricted drainage and a

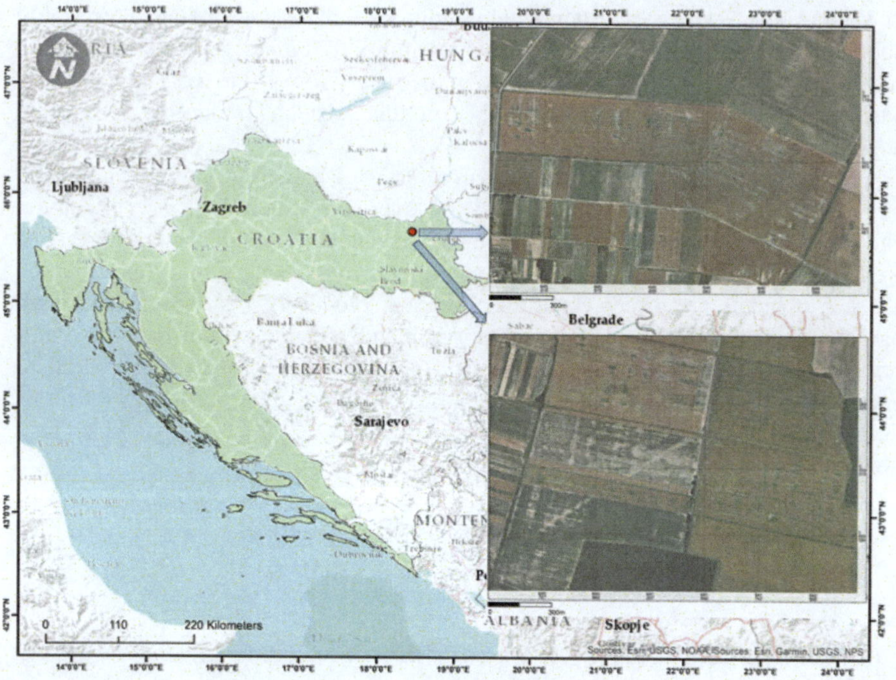

Fig. 12 Examples of saline soil patchy and scattered distribution near Marijanci (Eastern Croatia)

rising water-table, surface evaporation and plant transpiration, seawater sprays and condensed vapour which fall onto the soil as rainfall (near coast soils), wind-borne salts yielding saline fields, overuse of fertilizers (chemical and farm manures), use of soil amendments (lime and gypsum), use of sewage sludge or treated sewage effluent, and dumping of industrial brine onto the soil. In Croatia, the processes mentioned above have been identified, either with permanent or periodical character. According to the General Soil Map of Croatia, permanently Halomorphic soils (Solonchak and Solonetz) are distributed on <600 ha or 0.01% of total land [59]. In this context, it is important to mention that mapping and diversity of Halomorphic soils' order are minimal, similar to the assumptions about their total prevalence in the Croatian pedosphere. The reason for this is a generally small percentage of saline soils in Croatia and their point and clustered distribution (Fig. 12).

Geomorphological and microrelief diversity in eastern Croatia generates the occasional and sporadic groundwater close to the soil surface where it rises under the influence of evaporation and salinizes scattered point locations of agricultural areas (Fig. 12). As an example, Tomić et al. [60] stated that in Croatia, the order of Halomorphic soils covers an area of 410.5 ha, while Adam [61] estimates their prevalence on about 1,500 ha in Eastern Croatia only (mainly in Slavonia and Baranja regions). In the Eastern part of Croatia, saline soils were reported near

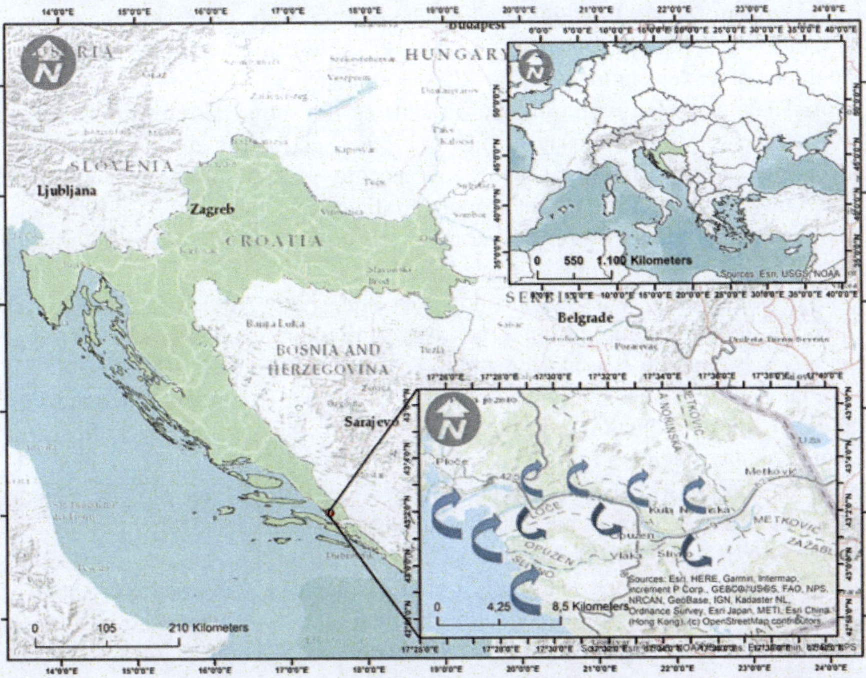

Fig. 13 Location of Neretva River estuary in Croatia, with blue arrows representing the intrusion of sea water

Tovarnik and Bizovac [62, 63] or on locations such as Poljic and Marijanci [64]. Saline soils in Eastern Croatia are mostly developed by weathering of saline parent material and high groundwater levels. Such a type of salinization is recognized as the typical sodium-type salinity.

Besides the mentioned salinity issue in the Eastern part of Croatia, the issue of soil salinity is mainly identified in Mediterranean areas of Croatia, e.g., in the karstic carbonate Neretva River valley [53], rivers Mirna and Raša in Istria [1, 65], Adriatic islands and the area of Ravni Kotari [66]. All of these areas are affected mainly by chloride-type salinization. Due to the pronounced climate change effects, seawater penetration into river valleys and inland accelerates and endangers the arable land in Mediterranean Croatia [65]. Salt accumulation in the rhizosphere modifies soil physical and chemical properties and microbiological processes, inhibiting plant growth and crop yields [49].

Neretva River delta is an area representative of many arable fields and alluvial estuaries with similar properties and exposure to salinization processes commonly found along the coastline of the Adriatic Sea (Fig. 13). Here, the main reason for soil salinization is the natural intrusion of seawater through porous media into calcareous aquifers and rivers, salinizing both soil and water (ground and surface). Also, surface and groundwater resources are used as irrigation water supply to achieve a profitable agricultural production [67]. Thus, farmers are frequently using water of poor quality for crop irrigation [54]. In addition, climatic conditions are increasing the demand for irrigation water, further intensifying the soil salinization issue [68], and raising the number of water engineering facilities, i.e., construction of power plants and industrial facilities [69] within the basin, causing the salinization of groundwater. However, the soil salinity issue in this area has a periodical character primarily, showing temporal variability because of specific geo(pedo)logical and climate interactions [57], with a significant spatial variability also identified [70]. Generally, the issue of salinity in the Neretva river aquifer is pronounced during summer when seawater intrusion reaches maximum values [71].

The Neretva River valley salinization was extensively studied by Romic et al. [72] through long-term monitoring focused on agricultural soil and water salinity in Neretva valley. Such focus on salinization processes at this area is because it is an area of intensive farming, significant for total Croatian agricultural production with a traditionally high production of fruits and vegetables, where salinity may affect crop yield and quality. Neretva River valley covers 4,500 ha of the cultivated area [73], where the majority of grown vegetables (such as cauliflowers, watermelons, melons, and potatoes) and fruits (e.g., high-quality mandarins) have a relatively low tolerance to salinity [67]. Although the average salinity of soil in the Neretva River valley (average values for measurements up to 1 m of soil depth) was defined as 2.0 dS m^{-1} [70], there is still the possibility of salt accumulation in the rhizosphere of deeper-rooted crops traditionally grown there (i.e., *Citrus* spp.). Therefore, as a part of the agricultural management in this, although not large, but agriculturally important area, identification, and a detailed description of soil salinization processes, as well as the long-term monitoring of soil and water salinity, with the implementation of traditional and new technologies, may prove to be necessary. Another river valley (Raša) in Istria was also investigated. Bogunovic et al. [65] reported a significant part of the lowland as saline because of seawater intrusion (Fig. 14). The great part of the Raša river valley was under sea level, and the embankment protects the valley from the sea, while two pumping stations control the water level in the channels surrounding the arable fields. This area was abandoned at least three decades before re-cultivation started again during 2014, which affected the functionality of small channels and their water level. Seawater intrusion in shallow groundwater significantly increased salt accumulation in these lands.

Bogunovic et al. [74] studied the impact of gypsum, farmyard manure and sulphur in different combinations on possible reclamation of saline soils. After 4 years of research, the addition of 6 t ha^{-1} of gypsum and 40 t ha^{-1} of farmyard manure increased soil aggregation by 14%, while exchangeable sodium percentage and electrical conductivity were decreased by ninefold (from 21.38% to 2.42%) and fourfold (from 8.03 to 2.27 dS m^{-1}), respectively. Moreover, at the same investigated field, adding 6 t ha^{-1} of gypsum and 2 t ha^{-1} of elemental sulphur decreased the exchangeable sodium percentage by eightfold and electrical conductivity by threefold. Such a reclamation strategy was adopted because the current plant production is under organic farming. Other amendments were not allowed because organic farming was the current system of crop production. Otherwise, other soil amendments should be tested as well. Besides such traditional measures of soil

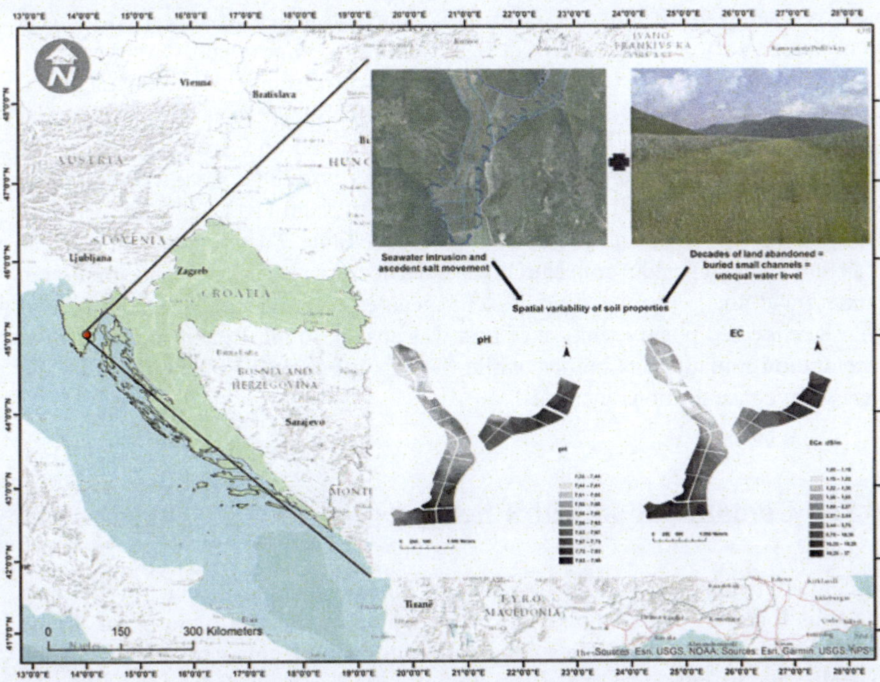

Fig. 14 The spatial variability of soil salinity (measured by electrical conductivity) and soil pH in river Raša valley. Black areas represent the most saline and alkaline soils

reclamation, various approaches and tools are applicable here to avoid future yield losses caused by seasonal increase salinity and achieve profitable agricultural production while maintaining soil fertility. New technologies such as, e.g., omics and nanotechnologies, or specific cost-effective traditional ameliorative measures such as conservation agriculture or the use of natural conditioners, can help against the adverse effects of increased soil salinity on crop production and quality [75]. This may prove to be especially important in the context of global climate change predicted to intensify short-term soil salinization likely. Therefore, technologies such as remote sensing for detecting and assessing soil salinity, with the combination of in situ and laboratory observations, could prove to be good approaches for mitigating difficulties in agricultural production caused by increased soil salinity [76]. Also, particularly beneficial would be if the implementation and combining of different strategies and measures for preventing and resisting soil salinity issues also maintains flexibility in fine-tuning the approaches according to their relevance for a specific area, i.e., preserving the site-specific focus.

Despite the new technologies that could help identify the endangered soils, well-known traditional measures should be regularly used. Inversion tillage like ploughing should be strictly avoided to reduce the possibility of raising extremely saline layers from the depth to the surface [77]. Only shallow tillage or tillage which aerates the soil should be used regularly. Here, we present the results obtained for

Raša river valley by Bogunovic et al. [74], where research of 4-years discing at 10 cm compared to discing at 10 cm with ripping to 30 cm depth. In 4-year crop rotation oat – triticale – barley – oat, the treatment with discing and ripping increased the grain yield by 13, 25, 32, and 15%, respectively. There is no other similar research in Mediterranean Croatia besides the presented one. However, the reclamation strategies in the Continental part of Croatia (Eastern Croatia) were tested on alkaline soils. Galović [78] studied the impact of gypsum (10 t ha^{-1}) and gypsum (10 t ha^{-1}) mixed with peat (40 t ha^{-1}) on soil properties. They reported an increase of calcium and magnesium concentrations and a decrease of sodium concentration in treatment with the gypsum addition (exchangeable sodium percentage from 27% to 5%). The expected positive impact of peat was absent. In this treatment, the sodium concentration remained as before, while the organic matter content, calcium and magnesium concentrations increased.

5 Soil Contamination with Chemicals

Soil contamination with various inorganic or organic chemicals is one of the most dangerous forms of degradation with consequences on the entire biosphere. In the context of food production, contamination of agricultural soils with metals may be especially significant, with the consumption of contaminated food as the most significant pathway of the exposure of the general population to metals [79], and with other possible ways of exposure, such as inhalation and dermal exposure, presumably more significant for people producing food on metal contaminated soils, i.e., occupational exposure. Additionally, the latter's significance may increase in the future as the path that any contaminant takes from the soil to the plate is very short, and food safety is a crucial and increasingly important requirement that is becoming more difficult and complex to implement. In order to define the contaminated soil, it can be said that it is the soil in which human activity or natural phenomena has increased the content of harmful substances whose concentrations may be harmful to human health, that is, by the consumption of contaminated food [80, 81], or by inhalation of contaminated soil particles and dermal exposure. In Croatia, soil metal contamination is primarily identified sporadically, including agricultural soils, therefore mainly of local and small-scale significance for the food production, and, were found, maybe originating from anthropogenic activities (Figs. 15 and 16) and/or natural sources (released into the soil from parent material, rocks, and sediments by weathering processes).

The Government of the Republic of Croatia published a Report on the state of the environment for the period from 2013 to 2016, in which they identified fuel combustion in the industrial sector as the primary source of cadmium, mercury, and lead air emission into the environment, followed by the sector of small stokehole, then traffic and industrial processes [45]. They further explained that the metal emission mainly depends on the type of the used fuel, with higher

Fig. 15 Example of anthropogenically induced soil contamination. Ruptured gas and oil pipeline as a source of soil contamination near Garčin (Photo taken by Ivica Kisic, July 2019)

Fig. 16 Example of plant production in zones of high soil contamination risks between railway and road near Breznica Našička (photo taken by Ivica Kisic, September 2008)

cadmium emission if fuel oil was used and higher mercury emission if the fuel used was natural gas.

Common metal(loid) present at contaminated locations in Croatia are arsenic, copper, zinc, cadmium, cobalt, chromium, mercury, nickel, and lead [45]. Therefore these elements may also be of interest for screening and monitoring in agricultural soils. Generally, high concentrations of toxic metals in Croatian soils (i.e., arsenic, cadmium, mercury, and lead) are confirmed at certain locations by the Report on the state of the environment published by the Government of the Republic of Croatia [45], where is summarized that: (1) soils in Coastal Croatia have the highest arsenic concentration (2.5–105 mg kg^{-1}, with a median value of 18 mg kg^{-1}), with the highest concentrations recorded (> 25 mg kg^{-1}) in the North and Middle Dalmatia (e.g., Obrovac city – Ervenik village and near Drniš city area), probably because of bauxite clumps found there; followed by Raša municipality in Istria, South Dalmatia and Lastovo Island; (2) soils in Coastal Croatia have the highest cadmium concentration (0.2–9.5 mg kg^{-1}, with a median value of 1.1 mg kg^{-1}), possibly originating from the lead and zinc mining industry, use of mineral fertilizers and pesticides: with Istria having the lowest cadmium concentration in soil, while the rest of the region has much higher concentrations (up to 3.5 mg kg^{-1}), although here was also recognized that at certain locations much higher cadmium concentrations are present in soil, probably resulting from certain local anthropogenic activities; (3) soils in Mountainous Croatia have the highest mercury concentration, especially at Ivanščica and Kalnik Mountain regions, followed by Gorski Kotar county, which was connected to rocks of Paleozoic complex (cinnabarit), especially in uttermost Northwest part of the county; (4) soils at the Drava and Mura River valleys had the highest measured lead concentration, with the highest concentration in topsoil because of complexation processes with soil organic matter, and probably resulting from anthropogenic activities from upstream, such as mining and industrial activities, similarly as for the cases of elevated cadmium and zinc concentrations; however the highest average lead concentrations in soil were recorded in Coastal Croatia (46–60 mg kg^{-1}, with a median value for the whole region of 48.7 mg kg^{-1}), with soils below the Velebit Mountain, at Dalmatian hinterland, and at Brač and Hvar Islands identified as the leading regarding lead concentration; additionally, high lead concentrations were also recorded at the mountainous area of Gorski Kotar county (i.e., Risnjak Mountain), as well as Lika county (Velebit Mountain), probably because of the composition of their present red soil and the atmospheric deposition. As it is seen from these data, agricultural use of mineral fertilizers and pesticides is so far identified as the possible source of metal pollution only at certain locations (e.g., cadmium contamination of soils in Coastal Croatia). In contrast, data from other parts of the country suggest that natural processes and other anthropogenic activities are releasing metals into soils. According to statistical databases published by the Croatian Bureau of Statistics [82], mineral fertilizer consumption in Croatia has been stable in the past few years, i.e., around 98.5 t of the active substance (nitrogen or phosphorus) per year since 2017 (Fig. 17). The majority of sold pesticides during the period from 2013 until 2019 were (1) herbicides, haulm destructors and moss killers, and (2) fungicides and bactericides, showing an apparent trend of a sale decrease

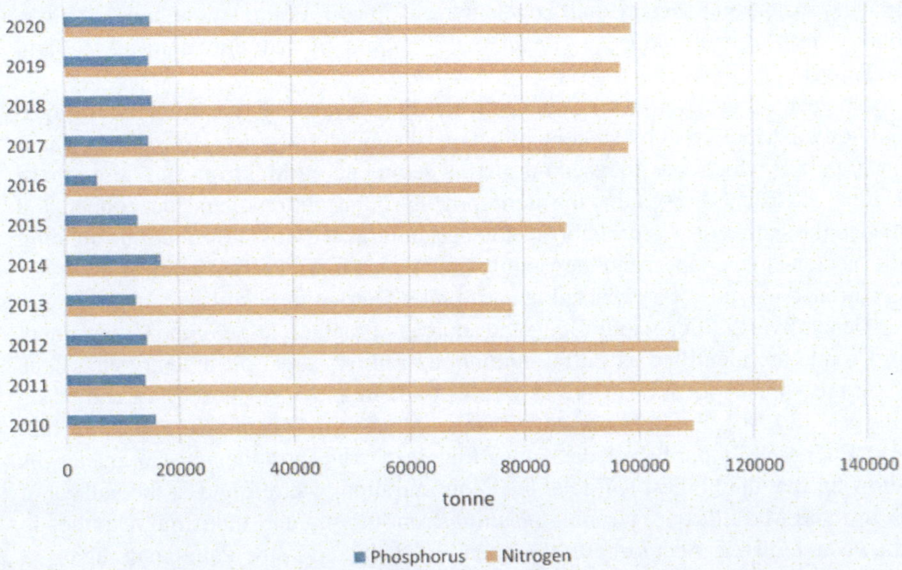

Fig. 17 Mineral fertilizer consumption in Croatia from 2010 until 2020 presented in tonnes of active substance (i.e., nitrogen or phosphorus), according to the statistical databases available from Croatian Bureau of Statistics [82]

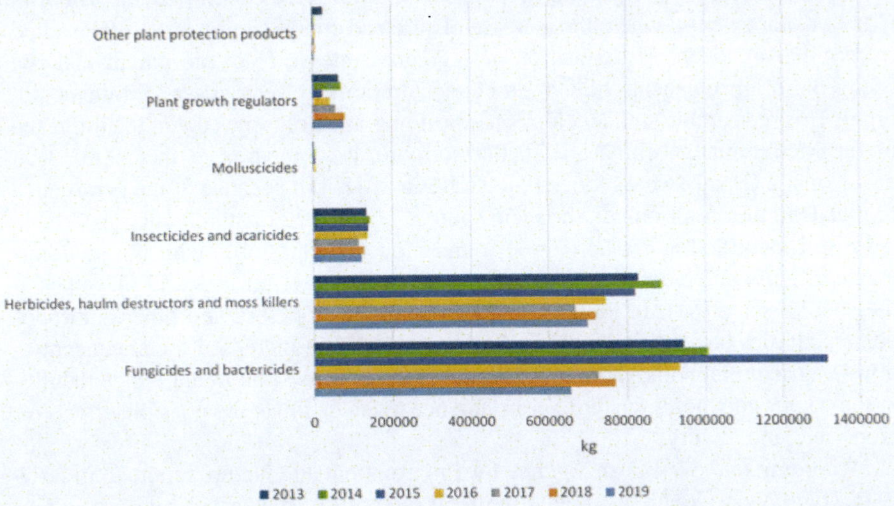

Fig. 18 Pesticide sales in Croatia from 2013 until 2019 presented by chemical classes and in kilograms, according to the statistical databases available from Croatian Bureau of Statistics [83]

since 2017, but, as presented in Fig. 18, still well over 650,000 kg for each chemical class [83]. Although metals are not the main active ingredient for the majority of here mentioned substances (except for, e.g., copper-based inorganic fungicides), metal traces or impurities present in agrochemicals are confirmed to be responsible for a gradual increase of metal (total) concentrations in soil. For example, mineral phosphorus fertilizers introduce a considerable amount of cadmium into agricultural soils [84].

However, there is a lack of research focusing on precise quantification of agricultural contribution to metal contamination of agricultural soils. However, research such as [85] which was focused on marine sediments in Makirina Bay in northern Dalmatia, clearly showed that the anthropogenic contribution of arsenic, copper, and lead to the sediment was probably due to agricultural activity in the area surrounding the bay (arsenic, copper) and the vicinity of the main road (lead). Therefore, a similar occurrence may also be expected in surrounding agricultural soils.

Generally, significant anthropogenic sources of metal release in the environment in Croatia are identified as traffic, mining, ex-military area, oil industry emissions [86], brownfields locations [87], as well as tourism and agriculture [88], and waste disposal [89, 90]. Furthermore, the Croatian Ministry of Environmental Protection and Energy [87] published data regarding lead concentrations in Croatian soils showing that the highest soil total lead concentrations were found in the valleys of Drava and Mura Rivers, possibly resulting from mining and industrial activities in the upstream area. According to the research conducted in the Pannonian region of Croatia, the total arsenic concentration varies from 0.9 mg kg^{-1} soil up to 490 mg kg^{-1} found in the topsoil of agricultural land, with an average value of 14.4 mg kg^{-1} soil [91]. Kisic et al. [92] explained that the increased arsenic concentration is flooding and the change of the Drava River flow in the recent past during the Holocene. Also, high lead concentrations were reported in Primorje-Gorski Kotar County, where the increase of soil lead concentration probably resulted from either atmospheric deposition or a gradual release from the parent material (geogenic origin), causing higher lead concentrations in Terra rossa (dominant soil type). This red cambic soil develops on limestones and dolomites under the influence of the Mediterranean climate [93]. Furthermore, as compared to the 1990s, lead emissions in Croatia were reduced by 98.5% in 2018, but because of the persistence of metals in the environment, the main sources of soil lead contamination in Croatia were still identified as traffic (Fig. 16) and industry emissions (with the predominance of glass and steel manufacturing), and various fuel burners and machines at the smaller scale [87]. In these cases, when natural processes and other anthropogenic activities besides agriculture are the sources of an increased metal concentration in agricultural soils, more site-specific studies focused on metal bioavailability in these soils and their transfer to crops are needed to estimate the possible effects on crop yield and quality.

According to the Croatian Agency for Environment and Nature report from 2018 [94], counties in Croatia estimated the total quantity of dangerous and potentially dangerous substances. The most polluted county is Primorje-Gorski Kotar (with 2,347,384.21 t), followed by Sisak-Moslavina county (1,367,393.39 t) and Zagreb, the capital of Croatia (176,622.44 t). In these areas, soils are used to produce food, even the capital of Croatia, Zagreb, where urban agricultural activities are increasingly popular. Since 2013, the City of Zagreb has designated 14 urban gardening areas (a total of 22.06 ha) with more than 2,152 urban gardens (allotments) conceded to the city residents for the production of vegetables, strawberries and other berry fruit, herbs, and flowers for their own needs [95]. Furthermore, Bašić and Kisić [96] concluded that the main metal polluters in the Primorje-Gorski Kotar county were cadmium, copper, and lead, with especially high concentrations in the soil where pollution deposition occurs, i.e., in the radius of 5 km from the source of pollution, explained by strong winds present in the area. Tisma [97] investigated soil quality in the lagoon environment in the area of an alumina factory near Obrovac (northern Dalmatia). The results showed the highest concentrations of the following elements: arsenic, chromium, nickel, zinc, manganese, vanadium, niobium, zirconium, strontium, yttrium, and titanium, which were determined in the direction of the dominant wind, suggesting that metal deposition may occur even in areas which are remote from the source of pollution, including agricultural soils, and especially considering strong, bora wind present in the area.

Also, Croatian soils, especially vulnerable and at the same time particularly susceptible to environmental pollution, are rather shallow soils developed over porous karst materials, commonly found at the Adriatic coastal area where agriculture is traditionally present. In Croatia, karst is found over 27,265 km^2, 48.9% of the country [45]. The connection between these soils and other environmental compartments is strengthened by the porosity of underlying karst material, enabling a permanent two-way link for transfer of metals (and other pollutants) between the soil and surface, ground and even seawater, suggesting more possible sources of metal contamination. For example, Bakić [98] presented mean values of total trace element concentrations in topsoil samples ($n = 312$) taken from the Adriatic area (in a regular sampling square grid) of Dinaric karst as follows: 1.02 mg cadmium kg^{-1}, 14.0 mg cobalt kg^{-1}, 73.3 mg chromium kg^{-1}, 78.3 mg copper kg^{-1}, 54.5 mg nickel kg^{-1}, 28.7 mg lead kg^{-1}, 86.9 mg zinc kg^{-1}, 49.4 g aluminium kg^{-1}, 34.0 g iron kg^{-1}. Also, the Neretva River delta was a subject of investigations of metal concentrations because of the importance of this part of the country for agricultural production, but also because river deltas are generally more susceptible to various contaminants because of pollutant translocation by water stream from their upstream source, followed by their accumulation in soils at river delta [99].

Halamić and Miko [100] published the Geochemical Atlas of the Republic of Croatia with maps showing the distribution of element concentrations in Croatian topsoil. Authors have taken topsoil samples (0–25 cm) in a regular square grid of 5×5 km^2 (one sample per 25 km^2) covering the whole country ($n = 2{,}521$). Mean, median, minimum, and maximum values with the standard deviation of selected metal and metalloid concentration in Croatian topsoil are adapted from maps provided by Halamić and Miko [100] and summarized in Table 1.

The Croatian region where the topsoil metals concentration is high is the Mediterranean one (Table 1). Here, agriculture is the main economic activity in many coastal parts [71], except for mercury. The region with the highest measured mean topsoil concentration was Mountainous Croatia, suggesting that mercury levels in Croatian soils are under atmospheric deposition. Also, maximum values for specific metal and metalloid concentrations measured at certain locations in Croatian topsoils (i.e., arsenic, cadmium, cobalt, chromium, copper, mercury, nickel, lead, and zinc;

Table 1 Mean, median, minimum, and maximum values for metal and metalloid concentrations in the topsoil of Croatia presented with standard deviation ($n = 2{,}521$) according to the Geochemical Atlas of the Republic of Croatia presented by Halamić and Miko [100]

	Metal or metalloid concentration [mg kg^{-1}]				
Metal/metalloid	Mean	Median	Minimum	Maximum	Standard deviation
Arsenic	13	12	0.5	105	8.34
Cadmium	0.7	0.4	0.2	15.5	0.97
Cobalt	14	13	3	120	5.82
Chromium	97	88	18	524	41.28
Copper	30	25	3	429	24.64
Mercury	0.09	0.06	0.005	4.54	0.15
Nickel	55	48	9	427	33.47
Lead	38	33	10	699	26.41
Zinc	99	88	23	1,432	61.81

Table 1), presented by Halamić and Miko [100], seemed to be very high in terms of soil metal contamination levels. However, although such locations with high topsoil metal concentrations were identified during this survey, it is also evident that these locations were relatively rare than a rule. This confirms that highly elevated metal (and arsenic) total concentrations in Croatian agricultural topsoils occur sporadically, possibly also suggesting their origin from a point source of pollution (either natural or anthropogenic). Additionally, even if such highly contaminated soils are not used for agricultural production, they still require special attention. Firstly, their effect on natural ecosystems and a possible cascade effect on environmental compartments other than soil (e.g., leaching to groundwater or affecting the wildlife health). Also, the probability of affecting local population health through, e.g., inhalation of soil particles or dermal exposure should be assessed. Therefore, implementing particular remediation, monitoring, risk assessment, or containment strategies could be appropriate measures, even if the issue of metal and metalloid raised concentration is occurring on a very small scale. Still, no law would impose the obligation of identifying contaminated or potentially contaminated locations in Croatia [45].

The high metal concentrations in Croatian soils are also attributed to the Croatian War of Independence, which lasted from 1991 to 1995. Vidosavljević et al. [101] investigated this particular topic while focusing on the concentration of metals and metalloids in the soils of Eastern Croatia, Croatian most developed agricultural area, representing about a quarter of Croatian territory and inhabited by nearly one million people (Fig. 19) which were affected by war. Authors compared soil samples from areas exposed to low-intensity combat activities with soil samples from areas that were under high-intensity combat activities and presented that the latter had higher concentrations of arsenic, mercury, and lead than allowed by the Croatian legislation for ecologic farming, with mercury concentrations higher than the allowed values for agriculture in general.

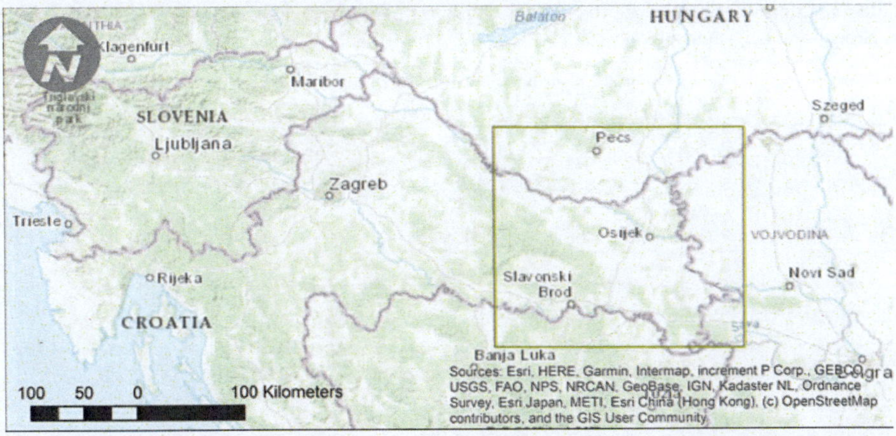

Fig. 19 The Croatian regions of Slavonija and Baranja affected by the Croatian War of Independence (1991–1995) studied by Vidosavljević et al. [101]

The additional issue here is that Eastern Croatia is rural, mainly with the population mostly engaged in agriculture and a long tradition of food production (plant and animal), which is still considered one of the critical activities for the development of eastern Croatia. Therefore, considering the persistence of metals once released into the environment, there might be a possibility of metal contamination of at least certain parts of eastern Croatian soils, which may be presumably used for agricultural production, and further suggesting the need for the scientific evaluation of this issue, as well as for the connected risks assessment (i.e., for the environmental and human health). Still, such presumably irregular and rare occurrence of elevated metal concentrations in soils of Eastern Croatia resulting from previous war activities may prove to be highly difficult to capture using regular soil sampling grids, especially after decades of metal ageing and relocation among environmental compartments (i.e., water-soil-plant-atmosphere). Nevertheless, Eastern Croatia is generally considered an unpolluted area with low total metal concentrations [102, 103], and therefore the anthropogenic influence on metal concentrations was mainly investigated in the Northwestern part of Croatia [104].

The responsibility for soil protection in Croatia is mainly under the Ministry of Environment and Energy's jurisdiction and the Ministry of Agriculture. Two key acts, Law on Environmental Protection [105] and Law on Agricultural Land [106], are focused on soil protection [107]. However, when estimating the concentrations of certain soil pollutants in Croatia, commonly used reference values for the allowable concentrations in soil are those found in NN 71/19 [108], a legislative document published by the Croatian Ministry of Agriculture in 2019, but only for agricultural soils (Government of the Republic of Croatia [45]; Table 2).

Although the importance of agricultural soils in terms of food production and the possible transfer of various pollutants from the soil into the food chain is evidently higher than that of non-agricultural soil, still a major role of soils in general

Table 2 Limit values presented in NN 71/19 [108] for metal concentrations and organic pollutants in agricultural soils as related to soil pH (issued by the Croatian Ministry of Agriculture, 2019)

Metal	Concentration in regard to soil pH in 1 M KCl [mg kg^{-1}]		
	<5	5–6	>6
Cd	1	1.5	2
Cr	40	80	120
Cu	60	90	120
Hg	0.5	1	1.5
Ni	30	50	75
Pb	50	100	150
Zn	60	150	200
Mo	15	15	15
As	15	25	30
Co	30	50	60

Organic pollutant	**Concentration [mg kg^{-1}]**
Polycyclic Aromatic Hydrocarbons (PAHs) in sandy and skeletal soils (< 27% clay)	1
Polycyclic Aromatic Hydrocarbons (PAHs) in clay soils (> 27% clay)	2
Polychlorinated biphenyls – PCBs (PCB 28 + PCB 52 + PCB 101 + PCB 118 + PCB 138 + PCB 153 + PCB 180)	0.5
Chlorinated hydrocarbon insecticides: DDT + DDD + DDE	0.1
Chlorinated hydrocarbon insecticides: aldrin + dieldrin + endrin	0.1
Chlorinated hydrocarbon insecticides: hexachlorocyclohexane (alfa-HCH + beta-HCH + gama-HCH + delta-HCH)	0.1
Herbicides: Atrazin	0.01
Total hydrocarbons in sandy and skeletal soils (< 27% clay)	1,000
Total hydrocarbons in clay soils (> 27% clay)	2000

environmental health is allowing for the expansion of existing legislative documents to non-agricultural soils. Existing legislation in Croatia [108] is therefore considering only one possible pathway of human exposure to pollutants, i.e., by ingestion, and not taking into consideration other, for human health also possibly significant routes of exposure such as inhalation of soil particles or dermal exposure.

According to the Report on the state of the environment published by the Government of the Republic of Croatia [45], from so far recognized pollutants in soils of Croatia, particular pollutants are identified in percentages as follows (from a total of 100%): Polycyclic Aromatic Hydrocarbons (29%), metals (23%), followed by Chlorinated hydrocarbons (12%), mineral oils (12%) and Aromatic Hydrocarbons (BTEX)402 (12%), as well as asbestos (6%) and phosphogypsum (6%).

Furthermore, the use of pesticides in Croatia is regulated by the Law on sustainable use of plant protection products [109], and they also must be approved by the Croatian Ministry of Agriculture (e.g., from July 2001, HCH Lindane has been banned from use). In Croatia, the emission of persistent organic pollutants was reported for the period from 2013–2016. Mainly, Polycyclic Aromatic Hydrocarbons (PAHs), dioxins and furans (Polychlorinated dibenzo-p-dioxins – PCDD), polychlorinated dibenzofurans – PCDF) and pesticides (Hexachlorocyclohexane – HCH, Hexachlorbenzene – HCB, and Polychlorinated biphenyls – PCBs; [45]).

Limit values for organic pollutants in agricultural soils are presented in NN 71/19 [108] (Table 2) issued by the Croatian Ministry of Agriculture. In Croatia, organic fertilizers and various soil amendments can be mixed with agricultural soil if they do not contain more than five times of total concentration of pollutants presented in Table 2, except for total cadmium concentration, which should not exceed two times of total concentration presented in Table 2 [108]. However, if no mixing with soil occurs (e.g., growing of crops in various substrates), then limit values presented in Table 2 should be considered the allowable pollutant concentration [108].

Agriculture is identified as the main source of ammonia (NH_3) emission (i.e., from the total of 83.8%, 60% is originating from the organic fertilizer management sector), which is deposited in the surrounding environment and ammonium nitrate is generated (e.g., most of the ammonia emitted in Europe is also deposited in Europe; Government of the Republic of Croatia [45]). Additionally, agriculture greatly contributes to groundwater contamination with nitrates (NO_3^-), which are not a pollutant but a nutrient while still in the soil. However, after precipitation, inadequate irrigation, or if not applied in appropriate amounts, leaching of non-adsorbing to soil particles nitrates through the soil profile may occur and cause significant pollution of groundwater resources. For example, Ondrašek et al. [110] published a report in which a clear trend of increased use of nitrogen fertilizers is evident throughout the counties of Croatia for the years 2000, 2012, and 2017, suggesting that this issue could gain its relevance in the future.

6 Soil Contamination with Microplastics

Microplastics pollution of Croatian soils still lacks more extensive research and additional data to assess this issue adequately. However, there are available studies of plastic pollution in the Adriatic Sea, thus may be connected to soils in the Croatian Mediterranean area. Gomiero et al. [111] noted that the inland deposition of microplastics is not investigated thoroughly in Croatia. However, they presented the possible sources of pollution for Croatian coastal areas (and broader). Offshore oil and gas production activities, transport infrastructure network, inadequate waste management in coastal regions with high and growing population density, sewage treatment plants and runoff from urban, agricultural, tourist, and industrial areas are likely contributing to the number of microplastics. In the context of agricultural soils, it should be mentioned that various plastic mulches are used for certain agricultural productions in Croatia, and mainly depending on how demanding is their removal from the soil after the harvest, plastic particles may be left in topsoil after the removal of mulch, or even incorporated into deeper soil horizons after tillage [112]. Still, Croatian pedosphere contamination with microplastics is somewhat pioneer research and is rarely related to agricultural soils. Thus, microplastic type of soil degradation in Croatia will hopefully be presented in further editions of this handbook.

7 Conclusion

Land degradation processes affect Croatian agricultural soils, but they are still in a favourable position. The types of degradation are mostly distributed according to environmental, geomorphological, pedological, and farming systems characteristics. The most widespread types of degradation were soil erosion by water and soil compaction, which affects more than half of agricultural land. It can be expected that those processes will be exacerbated in the future under modified climatic conditions because of reoccurring climatic extremes that accelerate and spread soil degradation processes to larger areas. Croatian farmers are generally without higher education, so adopting conservation, preventing, alleviating, and reclamation measures will very likely be accepted with certain resistance and will require a relatively long period. Overgrazing is not recognized in Croatia as an important degradation process. However, the intensification of this process in the future will highly correspond with a national policy and subventions rates aiming to reverse the negative trends of cattle number. Secondary soil salinity is still locally distributed, and the experimental reclamation measures positively impact soil quality. The contamination of agricultural soils with chemicals, including metals and other inorganic and organic chemicals, is below the legislative limitation levels, with identified polluted areas mainly originating from point sources. Future agricultural practices should use precision farming technology to avoid excessive agrochemicals use and mitigate environmental pollution. Data about microplastics pollution of Croatian pedosphere are still missing, suggesting that this issue should be more researched and monitored in the future because of the expected development of the country (i.e., industrial, transportation, urbanization, agricultural, and tourism growth are expected to continue). The lack of national monitoring programmes for the most important soil degradation processes in Croatia indicates the need to adopt the national soil degradation mitigation policy. New policies and regulation acts related to agricultural land should include monitoring the severity and the extension of the particular degradation processes, the implementation of mitigation strategies, including the risk assessment or containment strategies, and remediation measures.

Acknowledgment This chapter was supported by the Croatian Science Foundation through the "Soil erosion and degradation in Croatia" project (UIP-2017-05-7834) (SEDCRO).

References

1. Bašić F (2013) The soils of Croatia. Springer, Dordrecht, p 179
2. Blum WE (2005) Functions of soil for society and the environment. Rev Environ Sci Biotechnol 4(3):75–79
3. Kisić I (2016) Anthropogenic soil erosion. University of Zagreb Faculty of Agriculture, Zagreb, p 275. (in Croatian)

4. Bogunović I, Kisić I, Mesić M, Zgorelec Ž, Šestak I, Perčin A, Bilandžija D (2018) Sustainable soil management measures in organic agriculture for climate conditions of Mediterranean Croatia. University of Zagreb Faculty of Agriculture, Zagreb (in Croatian), p 147
5. Birkás M, Szemők A, Antos G, Neményi M (2008) Environmentally-sound adaptable tillage. Akadémiai Kiadó, Budapest, p 353
6. Bogunović I (2015) Changes of physical characteristics of Pseudogley under different tillage systems on slopes. Doctoral thesis, University of Zagreb Faculty of Agriculture, Zagreb (in Croatian), p 179
7. Filipovic D, Husnjak S, Kosutic S, Gospodaric Z (2006) Effects of tillage systems on compaction and crop yield of Albic Luvisol in Croatia. J Terramech 43(2):177–189
8. Turšić I, Husnjak S, Žalac S (2008) Soil compaction as one of the causes of lower tobacco yields in the republic of Croatia. Cereal Res Commun 36:687–690
9. Bogunovic I, Pereira P, Kisic I, Sajko K, Sraka M (2018a) Tillage management impacts on soil compaction, erosion and crop yield in Stagnosols (Croatia). Catena 160:376–384
10. Bogunovic I, Andabaka Z, Stupic D, Pereira P, Galic M, Novak K, Telak LJ (2019a) Continuous grass coverage as a management practice in humid environment vineyards increases compaction and CO2 emissions but does not modify must quality. Land Degrad Dev 30(18):2347–2359
11. Bogunovic I, Pereira P, Galic M, Bilandzija D, Kisic I (2020a) Tillage system and farmyard manure impact on soil physical properties, CO2 emissions, and crop yield in an organic farm located in a Mediterranean environment (Croatia). Environ Earth Sci 79(3):1–11
12. Birkás M, Kisić I, Bottlik L, Jolánkai M, Mesić M, Kalmár T (2009) Subsoil compaction as a climate damage indicator. Agric Conspec Sci 74(2):91–97
13. Bogunovic I, Fernández MP, Kisic I, Marimón MB (2019b) Agriculture and grazing environments. In: Pereira P (ed) Advances in chemical pollution. Environmental management and protection, vol 4, pp 23–70
14. Keller T, Sandin M, Colombi T, Horn R, Or D (2019) Historical increase in agricultural machinery weights enhanced soil stress levels and adversely affected soil functioning. Soil Tillage Res 194:104293
15. Husnjak S, Filipović D, Košutić S (2002a) Influence of different tillage systems on soil physical properties and crop yield. Rostlina Vyroba 48:249–254
16. Turšić I, Mesić M, Kisić I, Husnjak S (2010) The influence of secondary tillage on soil compaction and the yield of flue-cured tobacco. Tutun 60(1–6):22–26
17. Vukadinović V, Jug D, Đurđević B, Jug I, Vukadinović V, Stipešević B, Lović I, Kraljičak Ž (2013) Agricultural compaction of some soil types in eastern Croatia. In: Vukadinović V, Đurđević B (eds) Soil and plant management: adaptation and mitigation of climate changes, Osijek, pp. 38–45
18. Bogunovic I, Telak LJ, Pereira P (2020b) Experimental comparison of runoff generation and initial soil erosion between vineyards and croplands of eastern Croatia: a case study. Air Soil Water Res 13:1178622120928323
19. Bogunovic I, Bilandzija D, Andabaka Z, Stupic D, Comino JR, Cacic M, Brezinscak L, Maletic E, Pereira P (2017a) Soil compaction under different management practices in a Croatian vineyard. Arab J Geosci 10(15):1–9
20. Telak LJ, Dugan I, Bogunovic I (2021a) Soil management and slope impacts on soil properties, hydrological response, and erosion in hazelnut orchard. Soil Syst 5(1):5
21. Telak LJ, Pereira P, Bogunovic I (2021b) Management and seasonal impacts on vineyard soil properties and the hydrological response in continental Croatia. Catena 202:105267
22. Taylor HM, Gardner HR (1963) Penetration of cotton seedlingn taproots as influenced by bulk density, moisture content, and strength of soil. Soil Sci 96(3):153–156
23. Jug D, Đurđević B, Birkás M, Brozović B, Lipiec J, Vukadinović V, Jug I (2019) Effect of conservation tillage on crop productivity and nitrogen use efficiency. Soil Tillage Res 194: 104327

24. Stipesevic B, Brozovic B, Jug D, Jug I, Vukadinovic V, Djurdevic B (2019) Effects of tillage method and fertilizer type on the yield of Sudan grass (Sorghum bicolor L.). Turk J Agric For 43(3):378–387
25. Håkansson I (2005) Machinery-induced compaction of arable soils. Swedish University of Agricultural Sciences, Uppsala, p 153
26. FAO (2021) Food and Agriculture Organisation, Faostat. Available online: http://www.fao.org/faostat/en/#data/QC. Accessed 18 Mar 2021
27. Hengl T, Heuvelink GB, Rossiter DG (2007) About regression-kriging: from equations to case studies. Comput Geosci 33(10):1301–1315
28. Bogunovic I, Trevisani S, Seput M, Juzbasic D, Durdevic B (2017b) Short-range and regional spatial variability of soil chemical properties in an agro-ecosystem in eastern Croatia. Catena 154:50–62
29. Bogunovic I, Kisic I, Mesic M, Percin A, Zgorelec Z, Bilandžija D, Jonjic A, Pereira P (2017c) Reducing sampling intensity in order to investigate spatial variability of soil pH, organic matter and available phosphorus using co-kriging techniques. A case study of acid soils in Eastern Croatia. Arch Agron Soil Sci 63(13):1852–1863
30. Durdevic B, Jug I, Jug D, Bogunovic I, Vukadinovic V, Stipesevic B, Brozovic B (2019) Spatial variability of soil organic matter content in Eastern Croatia assessed using different interpolation methods. Int Agrophys 33(1):31–39
31. Martinović J (2003) Management of forest soils in Croatia. Croatian Forest Research Institute, Croatian Forests Ltd, Zagreb, p 525
32. Dekemati I, Simon B, Bogunovic I, Vinogradov S, Modiba Maimela M, Csaba G, Birkás M (2021) Three-year investigation of tillage management on the soil physical environment, earthworm populations and crop yields in Croatia. Agronomy 11(5):825
33. Kisic I, Basic F, Nestroy O, Mesic M, Butorac A (2002) Chemical properties of eroded soil material. J Agron Crop Sci 188(5):323–334
34. Kisic I, Bogunovic I, Zgorelec Z, Bilandzija D (2018a) Effects of soil erosion by water under different tillage treatments on distribution of soil chemical parameters. Soil Water Res 13(1): 36–43
35. Basic F, Kisic I, Mesic M, Nestroy O, Butorac A (2004) Tillage and crop management effects on soil erosion in Central Croatia. Soil Tillage Res 78(2):197–206
36. Husnjak S (2000) Water erosion risk assessment by mapping method in Croatia. Doctoral thesis, University of Zagreb Faculty of Agriculture, Zagreb (in Croatian), p 138
37. Husnjak S, Bogunović M (2002b) Risk of soil erosion by water on agricultural land in agricultural regions of Croatia. Agron Glas 64:5–6
38. Kisic I, Bogunovic I, Birkás M, Jurisic A, Spalevic V (2017) The role of tillage and crops on a soil loss of an arable Stagnic Luvisol. Arch Agron Soil Sci 63(3):403–413
39. Ministry of Agriculture (2015) Studija određivanja područja pod utjecajem prirodnih ili drugih specifičnih ograničenje u poljoprivredi s kalkulacijama. Available online at: https://ruralnirazvoj.hr/files/documents/Studija-podru%C4%8Dja-sa-posebnim-ograni%C4%8Denjima-iz-2015.pdf
40. Telak LJ, Bogunovic I (2020) Tillage-induced impacts on the soil properties, soil water erosion, and loss of nutrients in the vineyard (Central Croatia). J Cent Eur Agric 21(3): 589–601
41. Croatian Bureau of Statistics (2017) Structure of agricultural farms – final data situation as on 1 June, 2016 (in Croatian). Available online at: https://www.dzs.hr/Hrv_Eng/publication/2017/01-01-29_01_2017.htm. Accessed 1 Jan 2021
42. Ministry of Agriculture (2016) Research on the structure of agricultural holdings 2016 (in Croatian). Available online at: https://poljoprivreda.gov.hr/UserDocsImages/dokumenti/poljoprivredna_politika/poljoprivreda_u_brojkama/Istra%C5%BEivanje%20o%20strukturi%20poljoprivrednih%20gospodarstava%202016.pdf. Accessed 10 Jan 2021

43. Ljubičić I, Britvec M, Jelaska SD, Husnjak S (2014) Plant diversity and chemical soil composition of rocky pastures in relation to the sheep grazing intensity on the northern Adriatic islands (Croatia). Acta Bot Croat 73(2):419–435
44. Bašić F (1994) Classification of soil damages in Croatia. Agron Glas 56(3–4):291–310
45. Government of the Republic of Croatia (2019) Report on the state of the environment in the Republic of Croatia for the period 2013–2016, p 532. Available online at: https://ec.europa.eu/environment/eir/pdf/report_hr_en.pdf. Accessed 18 Feb 2021
46. Šoštarić J, Dadić M, Bukvić G, Josipović M, Petošić D, Turšić I (2006) Soil preparation for grassland and pasture. Krmiva 48(4):221–225
47. Bogunovic I, Kisic I, Jurisic A (2015) Influence of wildfire and fire suppression by seawater on soil properties. Appl Ecol Environ Res 13:1157–1169
48. Kisić I, Bogunović I (2016) Wildfire induced changes in forest soils in southern Croatia. Radovi Šumarskog Fakulteta Univerziteta u Sarajevu 21(1):91–97
49. Bogunovic I, Vukadinovic V, Kisic I, Chiavalon S, Vucic H, Durdevic B (2018b) Tillage and soil amendments effect on soil physical properties and yield of oats (Avena sativa L.) in organic farm in Mediterranean Croatia. Agric Conspec Sci 83(1):17–23
50. Goreta S, Bučević-Popović V, Pavela-Vrančić M, Perica S (2007) Salinity-induced changes in growth, superoxide dismutase activity, and ion content of two olive cultivars. J Plant Nutr Soil Sci 170(3):398–403
51. Perica S, Goreta S, Selak GV (2008) Growth, biomass allocation and leaf ion concentration of seven olive (Olea europaea L.) cultivars under increased salinity. Sci Hortic 117(2):123–129
52. Pavlović I, Mlinarić S, Tarkowská D, Oklestkova J, Novák O, Lepeduš H, Vujčić Bok V, Radić Brkanac S, Strnad M, Salopek-Sondi B (2019) Early brassica crops responses to salinity stress: a comparative analysis between Chinese cabbage, white cabbage, and kale. Front Plant Sci 10:450
53. Romic D, Ondrasek G, Romic M, Josip B, Vranjes M, Petosic D (2008a) Salinity and irrigation method affect crop yield and soil quality in watermelon (Citrullus lanatus L.) growing. Irrig Drain 57(4):463–469
54. Ondrasek G, Romic D, Romic M, Duralija B, Mustač I (2006) Strawberry growth and fruit yield in a saline environment. Agric Conspec Sci 4:155–158
55. Biško A, Ćosić T, Jelaska S (2010) Reaction of three strawberry cultivars to the salinity: vegetative parameters. Agric Conspec Sci 75(2):83–90
56. Biško A, Leko M, Ćosić T (2013) Reaction of three strawberry cultivars to the salinity of growing substrate: generative parameters. Int J Agric Innov Res 2(3):378–383
57. Ondrasek G, Rengel Z, Veres S (2011) Soil salinisation and salt stress in crop production. In: Shanker A, Venkateswarlu B (eds) Abiotic stress in plants – mechanisms and adaptations. InTech Open, Rijeka, pp 171–190
58. Shahid SA, Zaman M, Heng L (2018) Introduction to soil salinity, sodicity and diagnostics techniques. In: Zaman M, Shahid SA, Heng L (eds) Guideline for salinity assessment, mitigation and adaptation using nuclear and related techniques. Springer, Cham, pp 1–42
59. Bogunovic M, Vidacek Z, Husnjak S, Sraka M (1998) Inventory of soils in Croatia. Agric Conspec Sci 63(3):105–112
60. Tomić F, Krička T, Matić S (2008) Available agricultural areas and the use of forests for biofuel production in Croatia. Šumarski List 132(7–8):323–330
61. Adam M (1981) Salt affected and salinized-alkalized soils of Slavonia and Baranya. Doctoral thesis, Faculty of Agriculture in Osijek, University of Osijek (in Croatian), p 239
62. Škorić A, Anić J, Beštak T, Bišof R, Bogunović M, Cestar D, Čižek J, Dekanić I, Kovačević J, Licul R, Malez M, Martinović J, Mihalić V, Miljković I, Pavlek P, Pelcer Z, Racz Z, Srebrenović D, Šilješ I, Taksić A, Vidaček Ž (1977) Soils of Slavonia and Baranja. Projektni savjet pedološke karte SR Hrvatske, Zagreb (in Croatian), p 256
63. Bogunović M, Bašić F, Adam M (1983) Pedological map of section Osijek-2. Posebno izdanje Projektnog savjeta za izradu OPK Hrvatske, Tisak-VGI, Beograd. (in Croatian)

64. Racz Z (1981) Meliorativna pedologija, II. dio. Sveučilišna naklada Liber, Zagreb (in Croatian), p 189
65. Bogunovic I, Pereira P, Brevik EC (2017d) Spatial distribution of soil chemical properties in an organic farm in Croatia. Sci Total Environ 584:535–545
66. NN 56/2013 (2013) Water law. Narodne novine d.d., Zagreb (in Croatian). Available online at: https://narodne-novine.nn.hr/clanci/sluzbeni/2013_05_56_1139.html. Accessed 6 Apr 2021
67. Romic D, Zovko M, Romic M, Ondrasek G, Salopek Z (2008b) Quality aspects of the surface water used for irrigation in the Neretva Delta (Croatia). J Water Land Dev 12:59–70
68. Matijević L, Romić D, Maurović N, Romić M (2012) Saline irrigation water affects element uptake by bean plant (Vicia faba L.). Eur Chem Bull 1(12):498–502
69. Romic D, Vranjes M, Ondrasek G (2003) Intrusion of seawater and its effect on quality of surface water in the Neretva River valley. In: Gereš D (ed) Croatian waters in 21st century. Croatian Waters, Osijek, pp 443–449. (in Croatian)
70. Zovko M (2015) Assessment of salinity risk and metal mobility in coastal river plain agricultural soils. Doctoral dissertation, University of Zagreb Faculty of Agriculture, Zagreb (in Croatian), p 135
71. Romic D, Romic M, Zovko M, Bakic H, Ondrasek G (2012) Trace metals in the coastal soils developed from estuarine floodplain sediments in the Croatian Mediterranean region. Environ Geochem Health 34(4):399–416
72. Romic D, Romic M, Zovko M, Bubalo M, Ondrasek G, Husnjak S, Stricevic I, Maurovic N, Bakic H, Matijevic L, Vranjes M (2014) Monitoring of soil and water salinity in the Neretva River valley – five-year (2009–2013) report (in Croatian)
73. Srzić V, Lovrinović I, Racetin I, Pletikosić F (2020) Hydrogeological characterization of coastal aquifer on the basis of observed sea level and groundwater level fluctuations: Neretva valley Aquifer, Croatia. Water 12(2):348
74. Bogunović I, Galić M, Vukadinović V (2019) Application of sulfur, gypsum and farmyard manure on ameliorated saline-sodic soils in River Raša Valley. Annual report, 2018. Department of General Agronomy of the Faculty of Agriculture, University of Zagreb
75. Ondrasek G, Rengel Z (2021) Environmental salinization processes: detection, implications & solutions. Sci Total Environ 754:142432
76. Racetin I, Krtalic A, Srzic V, Zovko M (2020) Characterization of short-term salinity fluctuations in the Neretva River Delta situated in the southern Adriatic Croatia using Landsat-5 TM. Ecol Indic 110:105924
77. Matosic S, Birkás M, Vukadinovic V, Kisic I, Bogunovic I (2018) Tillage, manure and gypsum use in reclamation of saline-sodic soils. Agric Conspec Sci 83(2):131–138
78. Galović V (1998) Effects of the drainage and agroreclamation at alkali soil of Slavonia. Doctoral dissertation, University of Zagreb Faculty of Agriculture, Zagreb (in Croatian), p 54
79. Romic M (2012) Bioavailability of trace metals in terrestrial environment: methodological issues. Eur Chem Bull 1(11):489–493
80. Kisić I (2015) Effects of soil contamination on the selection of remediation method. In: Gaurina-Medjimurec N (ed) Handbook of research on advancements in environmental engineering. IGI Global – Disseminator of Knowledge, pp 200–227
81. Kisić I, Zgorelec Ž, Perčin A (2018b) Soil treatment engineering. In: Tomašić V, Zelić B (eds) Environmental engineering – basic principles. De Gruyter, pp 277–316
82. Croatian Bureau of Statistics (2021) Statistical databases, PC-axis databases, agriculture, hunting, forestry and fishing, agricultural consumption of mineral fertilisers, mineral fertiliser consumption, in tonnes of active substances, Republic of Croatia. Available online at: https://www.dzs.hr/default_e.htm
83. Croatian Bureau of Statistics (2021) Statistical databases, pc-axis databases, agriculture, hunting, forestry and fishing, pesticide sales, pesticide sales by chemical classes, in kilograms, Republic of Croatia. Available online at: https://www.dzs.hr/default_e.htm

84. Bracher C, Frossard E, Bigalke M, Imseng M, Mayer J, Wiggenhauser M (2021) Tracing the fate of phosphorus fertilizer derived cadmium in soil-fertilizer-wheat systems using enriched stable isotope labeling. Environ Pollut 287:117314
85. Komar D, Dolenec M, Lambaša Belak Ž, Matešić SS, Lojen S, Kniewald G, Vrhovnik P, Dolenec T, Rogan Šmuc N (2015) Geochemical characterization and environmental status of Makirina Bay sediments (northern Dalmatia, Republic of Croatia). Geol Croat 68(1):79–92
86. Kisić I, Mesić S, Bašić F, Brkić V, Mesić M, Durn G, Zgorelec Ž, Bertović L (2009) The effect of drilling fluids and crude oil on some chemical characteristics of soil and crops. Geoderma 149(3–4):209–216
87. HAOP – Croatian Agency for Environment and Nature (2020) Environment on the palm of the hand I-2020. In: Zovko M, Mesić H (eds) Report. Croatian Ministry of Environmental Protection and Energy, Zagreb (in Croatian), p 35. Available online at: http://www.haop.hr/sites/default/files/uploads/publications/2020-06/2020%20Okolis%20na%20dlanu_Final_Draft%20f.pdf. Accessed 18 Jan 2021
88. Cuculić V, Cukrov N, Kwokal Ž, Mlakar M (2009) Natural and anthropogenic sources of Hg, Cd, Pb, Cu and Zn in seawater and sediment of Mljet National Park, Croatia. Estuar Coast Shelf Sci 81:311–320
89. Hrenovic J, Durn G, Seruga Music M, Dekic S, Troskot-Corbic T, Skoric D (2017) Extensively and multi drug-resistant Acinetobacter baumannii recovered from technosol at a dump site in Croatia. Sci Total Environ 607–608:1049–1055
90. Ribic I (2008) Sustainable redevelopment of hazardous waste landfills-the hazardous waste landfill of Sovjak (Rijeka, Croatia) as case study. Nat Croat 17:375–384
91. Lozo K (2015) Influence of extraction hydrocarbons on soil pollution by cadmium, mercury and arsenic. Master thesis, Faculty of Agriculture University of Zagreb, Zagreb (in Croatian), p 39
92. Kisić I, Nemčić-Jurec J, Bašić F, Zgorelec Ž (2018b) The origin of arsenic in soils and groundwater of the Pannonian part of Croatia. Holistic Approach Environ 8(1):23–36
93. Durn G (2003) Terra rossa in the mediterranean region: parent materials, composition and origin. Geol Croat 56(1):83–100
94. HAOP – Croatian Agency for Environment and Nature (2018) Report on data from the registry of facilities where dangerous substances are present – evidence of reported major accidents for 2017, Zagreb, Croatia (in Croatian). Available online at: http://www.haop.hr/sites/default/files/uploads/dokumenti/022_reg_oneciscivaca/Izvjesca/Publikacije_RPOT_OPVN_2017.pdf. Accessed 18 Jan 2021
95. City of Zagreb (2020) Statistical yearbook of the City of Zagreb 2020. City of Zagreb - City Office for the Strategic Planning and Development of the City, Office for Strategic Information and Research, Department for Statistical and Analytical Affairs, Zagreb, pp 1–412
96. Bašić F, Kisić I (1995) Soil damage in Primorsko-Goranska County. "Ecological and economic evaluation of the Primorsko-Goranska County for agricultural development" expert project report. University of Zagreb Faculty of Agriculture, Zagreb (in Croatian)
97. Tisma A (2021) Changes in soil quality under the influence of the bora in the lagoon environment in the area of ex-the Jadral alumina factory. Doctoral dissertation, University of Zagreb, Zagreb (in Croatian) p 83
98. Bakić H (2014) Organic carbon stabilization and metal retention by organo-mineral complexation in the soils of the karst fields ("polje"). Doctoral dissertation, University of Zagreb Faculty of Agriculture, Zagreb (in Croatian), p 250
99. Filipović L, Romić M, Romić D, Filipović V, Ondrašek G (2018) Organic matter and salinity modify cadmium soil (phyto) availability. Ecotoxicol Environ Saf 147:824–831
100. Halamić J, Miko S (2009) Geochemical atlas of the Republic of Croatia. Croatian Geological Survey, Zagreb, p 87
101. Vidosavljević D, Puntarić D, Gvozdić V, Jergović M, Miškulin M, Puntarić I, Puntarić E, Šijanović S (2013) Soil contamination as a possible long-term consequence of war in Croatia. Acta Agric Scand B Soil Plant Sci 63(4):322–329

102. Lončarić Z, Karalić K, Popović B, Rastija D, Vukobratovic M (2008) Total and plant available micronutrients in acidic and calcareous soils in Croatia. Cereal Res Commun 36:331–334
103. Ivezić V, Singh BR, Almås ÅR, Lončarić Z (2011) Water extractable concentrations of Fe, Mn, Ni, Co, Mo, Pb and Cd under different land uses of Danube basin in Croatia. Acta Agric Scand B Soil Plant Sci 61(8):747–759
104. Sollitto D, Romić M, Castrignanò A, Romić D, Bakić H (2010) Assessing heavy metal contamination in soils of the Zagreb region (Northwest Croatia) using multivariate geostatistics. Catena 80:182–194
105. NN 80/13 (2013) Law on environmental protection. Narodne novine d.d., Zagreb (in Croatian). Available online at: https://narodne-novine.nn.hr/clanci/sluzbeni/2007_10_110_3226.html. Accessed 15 Apr 2021
106. NN 20/18 (2018) Law on agricultural land. Narodne novine d.d., Zagreb (in Croatian). Available online at: https://narodne-novine.nn.hr/clanci/sluzbeni/2018_03_20_402.html. Accessed 14 Apr 2021
107. Husnjak S, Romić M, Poljak M, Pernar N (2011) Recommendations for soil management in Croatia. Agric Conspec Sci 76(1):1–8
108. NN 71/19 (2019) Regulation on the protection of agricultural land from pollution. Narodne novine d.d., Zagreb (in Croatian). Available online at: https://narodne-novine.nn.hr/clanci/sluzbeni/2014_01_9_167.html. Accessed 16 Apr 2021
109. NN 14/14 (2014) Law on sustainable use of plant protection products. Narodne novine d.d., Zagreb (in Croatian). Available online at: https://narodne-novine.nn.hr/clanci/sluzbeni/2014_02_14_269.html. Accessed 16 Apr 2021
110. Ondrašek G, Romić D, Bakić Begić H, Bubalo Kovačić M, Husnjak S, Mesić M, Šestak I, Salajpal K, Barić K, Bažok R, Pintar A, Romić M, Krevh V, Konjačić M, Vnučec I, Zovko M, Brkić Ž, Žiža I, Kušan V (2019) Determination of groundwater monitoring priority zones within intensive agricultural area (SAGRA 2). University of Zagreb Faculty of Agriculture, Zagreb (in Croatian), p 335
111. Gomiero A, Strafella P, Fabi G (2018) From macroplastic to microplastic litter: occurrence, composition, source identification and interaction with aquatic organisms. Experiences from the Adriatic Sea. In: Gomiero A (ed) Plastics in the environment. InTech Open, Rijeka, pp 1–20
112. Filipović V, Romić D, Romić M, Borošić J, Filipović L, Mallmann FJK, Robinson DA (2016) Plastic mulch and nitrogen fertigation in growing vegetables modify soil temperature, water and nitrate dynamics: experimental results and a modeling study. Agric Water Manag 176: 100–110

Agricultural Land Degradation in the Czech Republic

David Zumr

Contents

1 Introduction 36
2 Soil Erosion 38
2.1 Soil Erosion by Water 39
2.2 Soil Erosion by Wind 42
2.3 Soil Erosion Measures and Policy 42
3 Changes in Soil Properties, Soil Compaction 43
3.1 Soil Structure Loss 43
3.2 Soil Compaction 45
4 Soil Contamination 46
5 Microplastics in Soils 48
6 Effects of Fires 49
7 Conclusions 51
References 52

Abstract Soil degradation has been identified as a major threat to the productivity of agricultural land. In the Czech Republic, soils are threatened primarily by water and wind erosion, but compaction, loss of organic matter, loss of soil structure stability, pollution and over-fertilization, loss of biodiversity, and soil sealing are also major concerns. Poor soil health results in many off-site effects such as surface water siltation, groundwater pollution, loss of biodiversity in the countryside, and decreasing crop yields. The Czech agricultural landscape is characterized by large fields with a very small number of interrupting elements such as furrows, paths, or balks and the crop

The original version of this chapter was revised: this chapter was previously published non-open access. The correction to this chapter is available at https://doi.org/10.1007/978-3-031-32052-1_1016

D. Zumr (✉)
Department of Landscape Water Conservation, Faculty of Civil Engineering, Czech Technical University in Prague, Prague, Czech Republic
e-mail: david.zumr@cvut.cz

Paulo Pereira, Miriam Muñoz-Rojas, Igor Bogunovic, and Wenwu Zhao (eds.),
Impact of Agriculture on Soil Degradation II: A European Perspective,
Hdb Env Chem (2023) 121: 35–58, DOI 10.1007/698_2022_928,

Published online: 14 December 2022

structure is rather uniform. The state has a history of land collectivization which first took place during the twentieth century. The ongoing intensive and unsustainable industrial farming, which is often focused more on high yields of certain economically valuable crops rather than the environment, speeds up soil degradation. These problems are fortunately recognized by the stakeholders, legal authorities, and the public. There has been significant debate on sustainable landscape management and agricultural practices, and many positive examples already exist in the Czech Republic.

Keywords Land collectivization, Pesticides, Soil compaction, Soil erosion, Soil sealing, Wild fires, Wind erosion

1 Introduction

The Czech Republic is a landlocked country with an area of 78,887 km^2. Its relief is moderately hilly, with most of the area (78.6%) lying at an elevation between 200 and 600 m a. s. l. The climate is classified as warm temperate humid continental Cfb [1], with predominantly western circulation and mean annual precipitation around 600 to 800 mm. The most intensive rainstorms and peak surface runoff events usually occur in the late spring and early summer months. In general, the climate is rather dynamic, and the weather may change quickly. Moreover, several exceptionally wet and dry years have been recorded in the last two decades, resulting in local floods and droughts [2].

In comparison with other EU countries, the Czech Republic has a high percentage of arable land. Agricultural lands cover 42,002 km^2, which is approximately 53% of the total land area (arable land 42.2%), and forests cover 26,773 km^2, which is 34% of the total land area. Most of the agricultural lands are arable and grasslands (Fig. 1). Farmlands are typically found in less favorable soils and climatic conditions from the production perspective, and the conditions are classified as a submontane type of agriculture [3]. The soil profiles are usually deep, i.e. 40% are deeper than 60 cm, and 54% are between 30 and 60 cm, with soil profile depth considered as a soil layer with stoniness below 50% [4]. In some parts of the country, specifically in Southern

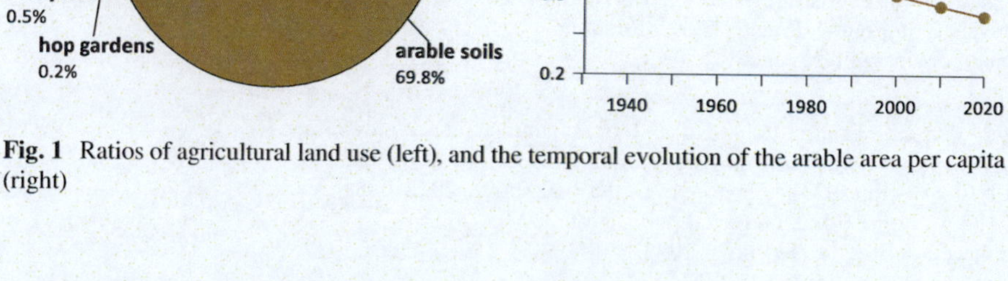

Fig. 1 Ratios of agricultural land use (left), and the temporal evolution of the arable area per capita (right)

Moravia and in Northwestern Bohemia, high-quality (and endangered) Chernosols can be found. The most abundant soil types in the Czech farmlands are Cambisols, followed by Luvisols, Chernosols, Stagnosols, and Fluvisols. The fields are predominantly rainfed, and only 0.6% of the agricultural area is irrigated. Agricultural lands were extensively evaluated during the extensive national mapping campaign of agricultural soils in Czechoslovakia (1961–1971), and in the 1970s and 1980s the lands were classified according to their productive capacity and economic value into Evaluated Soil-Ecological Units (ESEU, in Czech so-called BPEJ) [5]. The Soil-Ecological Units and the official soil prices have been gradually updated [6]. The Soil-Ecological Unit is assessed on the basis of the climatic region in which the parcel is located, the main soil unit, slope, exposure, soil stoniness, and soil profile depth. The State Land Office distributes the soil value map (1:5,000), which is spatially the most detailed resource, e.g., for USLE-based soil erosion modelling [7].

The arable land area decreased significantly between the 1950s and the 1970s, when agricultural fields were transformed to civil infrastructure and mines. In recent years, the area of arable land has decreased slowly and steadily at a rate of approximately 100 km^2 per year, mainly due to the conversion of arable lands to permanent grasslands and to urban uses, especially residential and industrial zones [8]. The farmland area ratio has been declining significantly faster than in the neighboring countries, although the market prices of farmland have been rising [9]. The forests, grasslands, and pasture areas keep increasing slightly [10]. A typical feature of farming in the Czech Republic is that a relatively small number of farms account for a substantial majority of the agricultural area. The average agricultural holding size of a farm is 152 ha [10], which is by far the largest among the EU countries [11]. Farming in the Czech Republic can therefore be characterized, to a considerable extent, as industrial farming.

In the Czech Republic, as in other countries in Central Europe, forestry and agriculture utilize a major portion of the land area, with both positive and negative impacts on the soils and on the environment. Because of the intensive management, the landscape maintains its land use diversity in terms of economical and cultural functions. However, overexploitation causes severe soil degradation resulting in lower soil fertility, soil and water contamination, loss of biodiversity, and changes in land use and in the landscape matrix. Most of the soil-related degradation processes are in line with the general trends in Central Europe. However, there are a few specific issues in the Czech Republic which are related mainly to the abrupt changes in agricultural management in the second half of the twentieth century [12]. Former privately-owned small arable parcels were merged into large soil blocks (through the process of collectivization), balks were ploughed, small landscape features were removed, subsurface tile drainage and systems of surface ditches were introduced into the landscape to extend the arable parcels in occasionally flooded areas (this process had already started at the beginning of the twentieth century), and fertilization and crop production were intensified [13]. As a result, fields have very large size (often over 30 ha), soils are currently threatened mainly by water- and wind-induced soil erosion, the soil organic matter quality has been deteriorating, soils have low hydraulic conductivity, and the subsoil is often compacted [14]. These conditions further lead to unfavorable changes in the physical, chemical, and biological properties of the soil [15].

The effects of past mismanagement of soils have not been eliminated until now, and the long-term trend of unsustainable agricultural land management is still clearly visible on arable soils. The average size of land parcels has not decreased dramatically (although it has been recognized that the extreme size is one of the main reasons of soil degradation and low biodiversity [16]), crops selection and crop rotation are driven by economic rather than by environmental concerns, soil productivity is maintained with the use of mineral fertilizers instead of organic amendments such as manuring, and mostly large agricultural enterprises farm on leased soils, often in an unsustainable manner (only 16% of farmland is owned by farmers who indeed manage the land). In addition, there is an increased pressure on new building plots around the agglomerations. These factors are causing sealing of the most fertile soils in the lowlands [13]. Kozák et al. [17] have identified soil sealing as the main current problem. They have also pointed out the loss of agricultural land due to open-cast mining, though some of this land has already been recultivated.

Acid deposition in the twentieth century and unfavorable forest management based on spruce monocultures have led to forest degradation in some areas in the Czech Republic. There had also been some concerns about the acidification of arable soil, as lower pH could mobilize hazardous trace elements, especially cadmium. Fortunately, however, no extensive acidification has been measured [18]. Salinization of soils is not an extensive problem in the Czech Republic and may occur only locally around roads treated with salt during winter, or in irrigated greenhouses [17]. There is serious concern about soil quality among land managers, academia, and also in the general public. Quantitative assessments are already available such as the study by Šarapatka and Bednář [19] that constructed a PCA (principal component analysis)-based model to estimate the vulnerability of Czech soils to degradation. The results were published in the form of a soil degradation threat map.

2 Soil Erosion

In comparison with the rest of the world, the Czech Republic has relatively low soil erosion potential, as it lacks high mountains and heavy convective rainstorms, and a considerable part of the area is covered by forests [17]. Nevertheless, soil erosion (especially by water) is the most frequently highlighted soil degradation process. A total of 43% of the arable land is on slopes ranging from 3° to 7°, and 10% of the land is on slopes exceeding 7°. Wind erosion occurs mainly, but not exclusively, in the south-eastern part of the Czech Republic, where the climate is warmer and the soils are generally lighter. The estimated annual soil loss from agricultural land for the whole country is estimated to be 21 million tons, which represents an economic loss of up to CZK 4.3 billion [20]. Soil translocation due to tillage operations has also been studied [21, 22]. Up to 16% of arable land is negatively affected by the tillage operations, especially in the most fertile regions of south and northeast

Moravia [23]. Soil erosion has been a prominent topic of national research in the last 20 years. According to the Central Register of Research & Development projects (https://www.isvavai.cz/cep, as of March 2021), 47 out of 116 grant-funded projects related to soils have dealt with the problematics of soil erosion.

2.1 Soil Erosion by Water

More than 55% of agricultural land in the Czech Republic is potentially threatened by surface runoff and subsequent soil erosion by water and out of this percentage, over 17% is extremely threatened [24, 25]. Similar or slightly lower values have been reported during the last 20 years [26], which indicates that the situation has not improved. The permissible soil loss limit on moderately deep to deep soils in the Czech Republic is 4 t ha^{-1} $year^{-1}$, but some argue that the real soil formation processes in the Central European region are even slower (estimated to be 1 cm of soil in approximately 100 years), and that the permissible soil loss limits should be reduced by at least a half. Single erosion events often lead to irreversible topsoil losses of several centimeters locally [14].

The Czech Republic contains densely populated areas where soil erosion is usually accompanied by the off-site negative impacts of considerable economic damage to water structures (e.g., ditches, rivers, and reservoirs affected by siltation and consequently by eutrophication) or to the civil infrastructure (local flash floods, mud flows into villages and gardens) [27–29]. Over 40% of reported erosion cases end up with damage to roads, over 30% with damage to the civil infrastructure, and more than 17% with damage to water bodies [24]. Eroded soils in the Czech Republic produce lower crop yields per hectare. The reported yields are lower by 15–20% on moderately eroded soils and by as much as 75% on heavily eroded soils [24]. Agriculture is the main non-point pollution source of surface waters and groundwater. Surface runoff containing detached soil particles introduces considerable loads of adsorbed pollutants, mainly phosphorus [28, 30] and nitrogen [30]. Soil erosion also has a significant influence on the price of affected parcels [31, 32].

The unsatisfactory present-day situation is connected with mass production in agriculture and with extreme changes in land use (the so-called collectivization process in the 1950s), when large units of arable land were consolidated and the soil degradation process accelerated dramatically in some areas (Fig. 2) [16]. Dostál et al. [33] identified the following dominant factors which have contributed to a dramatic increase in soil erosion:

- Establishment of very large fields (on an average 20 ha, but there are even parcels of 200 ha [34])
- Removal of the dense network of linear features and spot elements in the landscape (such as dirt roads, paths, ridges, grass belts, groves, etc.) which could potentially prevent or terminate surface runoff

Fig. 2 An example of the change in the landscape pattern between 1954 (left) and 2020 (right) due to the collectivization of agriculture (vicinity of the Nučice experimental catchment [35]) (source of the historical orthophotograph CENIA 2010, current situation ČÚZK, available from: http://geoportal.gov.cz)

- Extensive soil amelioration with introduction of dense networks of tile drains, straightening and deepening of streams
- Transformation of grasslands and pastures into arable areas in morphologically unfavorable landscape areas (foothills, slope areas).
- Drainage of inundation areas, leading to more arable land but to lower water retention capacity of the landscape.
- Utilization of heavy machinery, which has resulted in soil compaction and reduced soil infiltration capacity
- Planting of wide row crops (e.g., corn, potatoes)
- A drastic reduction in organic matter inputs, due to reduced livestock production since the 1990s.
- Increased application of mineral fertilizers since the 1970s
- Insufficient use of modern technologies, lack of knowhow, and a lack of political support for soil protective cultivation of the land

Since as early as the 1960s, severe erosion events have been reported, especially in the highlands [36, 37]. The reporting strategy has improved since then. The Research Institute for Soil and Water Conservation and the State Land Office have introduced an online tool for soil erosion event monitoring (accessible from https://me.vumop.cz/app/), which visualizes the spatial and temporal distribution of severe water soil erosion events on agricultural land in the Czech Republic [38]. The increased rates of erosion events, which are observed every year in May and June, are related to a combination of the rainfall pattern and the undeveloped vegetation, which provides insufficient cover for the soil surface at that time of the year (Fig. 3). Erosion events have recently also been recorded in the autumn and increase in the areas where oilseed rape is grown. Most of the erosion events occur on Cambisols (over 40% of the events), which are predominantly located in the hilly regions. Chernozems, which are in general susceptible to water erosion, do not exhibit an extreme number of events in the Czech Republic, as they are located in flatter parts of the country and cover a comparatively small area (7% of the events) [38].

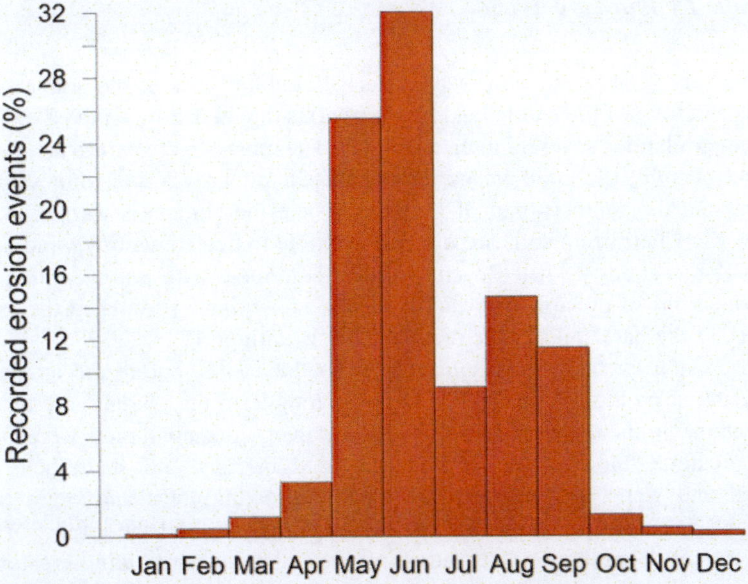

Fig. 3 Ratio of recorded water erosion events in the Czech Republic during a year (based on a monitoring from 2012 to 2015). This figure is based on data retrieved from the soil erosion event monitoring of the Research Institute for Soil and Water Conservation and the State Land Office [24]

Fig. 4 Sheet and rill erosion development near Přestavlky (left), to formation of gullies at Vlkov (right) (photographed in June 2020, courtesy of Josef Krása)

The most widespread crops in the Czech Republic are wheat, barley, and oilseed rape (the sowing areas vary annually between 40% and 65%), followed by fodder crops and maize. The diversity of agricultural crops has decreased in the last 25 years [39]. In the long term, maize has been the most problematic crop from the soil erosion perspective, with approximately one half of significant erosion events occurring on maize fields, followed by winter oilseed rape, winter cereals, and potatoes. By contrast with most other crops, soil erosion events on maize fields have been observed even when the canopy is already closed [38] (Fig. 4).

2.2 Soil Erosion by Wind

Approximately 10% of the area of the Czech Republic (23% of the arable land in Bohemia and 40% of the arable land in Moravia) is threatened by wind erosion, and 5% of the farmland is severely threatened [33]. The most-affected areas are the Elbe plain in Bohemia, the spoil heaps, and the open pit brown coal mines in NW Bohemia, and the southern part of Moravia, where the climate is warmer and the generally dry Chernozem soils are more susceptible to detachment by wind than in the rest of the country [40]. Recent weather extremes, with warm winters with reduced amounts of precipitation and almost no snow, have stimulated the erosion processes, as the bare soil surface dries quickly in spring [41].

Wind erosion events have occurred even in the 1950s, before the arable land collectivization process started [42], but in the long term the disruptive effects did not exceed the limits of natural erosion [43]. However, after small plots were merged into large arable blocks, the semi-natural wind barriers and shelterbelts were destroyed and were later replaced by quickly growing trees and often poorly maintained. The crop diversity was dramatically reduced, and the impact of wind erosion on soil degradation increased. In general, the danger of wind erosion was underestimated. In Southern Moravia, the mean topsoil loss due to wind erosion is currently estimated to be 0.4 mm of the plough layer per year, and locally up to 4–5 mm per year. In the case of severe dust storms, even 20 mm of topsoil loss have been reported [33].

2.3 Soil Erosion Measures and Policy

Effective measures against soil erosion are well known and they have been increasingly applied on a number of fields in the Czech countryside. Unfortunately, effective measures have usually been carried out on a local scale, and either on fields that have already been repeatedly affected by erosion or by a limited number of progressive farmers and land owners. Nevertheless, the ongoing process of land consolidation (aimed at reducing the high ownership fragmentation caused by the restitution of confiscated land in the 1990s) provides a great opportunity to reshape the rural landscape matrix, to introduce stabilizing elements, e.g., pathways, grass strips, alleys, ditches, green vegetation, among others [44], and to improve the resistance of the soil to degradation [45, 46].

The main policy tools implemented by the Ministry of Agriculture are the standards of Good Agricultural and Environmental Condition (GAEC), which support agricultural management in compliance with environmental protection [47]. Farmers and smallholders are motivated to take care of the soil. The Ministry of Agriculture, together with the Research Institute for Soil and Water Conservation, established, among others, a website (http://eagri.cz/public/web/mze/puda) with up-to-date information, guidelines, and interactive tools (an erosion calculator, soil maps, contaminated sites, etc.). In most cases, the goal is to apply a set of erosion measures to protect the soil. The set of anti-erosion measures includes organizational changes, such as proper crop distribution, strip tillage, and complex landscape consolidation. Agrotechnical measures include contour tillage and conservation tillage. Leveling, field balks, and terracing are some of the technical measures. Further measures are aimed at reducing erosion and transport in streams and rivers [48]. The amount of measures required by Cross Compliance is carefully balanced with the available funding (which comes in the form of subsidies to farmers) and the required measures and tolerable limits are therefore in general underestimated, and effective soil conservation would require more effort. However, best management and soil conservation practices, conservation tillage, mulching, direct seeding, and other practices are being implemented on larger and larger areas. Further positive development can be expected with the anticipated ban on glyphosates, which will bring fundamental changes in farming practices [49].

3 Changes in Soil Properties, Soil Compaction

Soils on farmlands are losing their structure, mostly due to intensive tillage and soil compaction. Soil compaction, caused by overuse of heavy machinery, intensive cropping, and inappropriate soil management, has been recognized as one of the major problems of modern agriculture [50]. Compaction results not only in soil aggregates deformation, but also in changes in the conductivity and connectivity of the pores. One can commonly observe the formation of dense soil layers with very low macroporosity and hydraulic conductivity in the shallow soil profile [51]. Consequently, water infiltration is reduced, and in addition this situation causes reduced gas and heat fluxes within the soil profile. In a global perspective, this can influence the global carbon and nitrogen cycles [52, 53].

3.1 Soil Structure Loss

The soil structure of arable land has a significant impact on water and soil air availability, nutrient uptake, and leaching [54]. Thereby, the soil structure indirectly affects the ground and surface water supply and water quality. The aggregation of soil particles and interconnected large pores increase the bypass flow in the soil. Healthy structured soils exhibit increased infiltration rates, reduced surface runoff, water percolating deeper into the soil profile, and usually, but not necessarily, also higher yields.

Agricultural management practices (the tillage system, crop rotation, fertilizers, etc.) can significantly impact the stability of the topsoil aggregates and soil hydraulic properties [55]. Growing crops, tillage, and subsequent reconsolidation due to natural wetting and drying cycles cause changes in the soil bulk density and porosity, the ratio of macropores, the soil hydraulic properties, the surface roughness, or the depression storage of rainwater. The stability of soil aggregates is maintained mainly by the organic matter content, the clay content, iron oxides, and biological activity [56].

A decline in the soil organic matter (SOM) and the microbial biomass in the topsoil has been considered a major agronomic and environmental problem, mainly due to its negative impact on soil properties [17, 57]. Several studies based on long-term monitoring of SOM on various soil types in the Czech Republic indicate a lower current SOM content with worse qualitative parameters than decades ago [58–61]. The SOM decline is attributed mainly to tillage, to the intensification of farming, and to reduced application of manure due to the reduced numbers of livestock (as shown in Fig. 6 in the next section). Bednář and Šarapatka (2018) showed high SOM losses on drained fields and on parcels affected by water erosion [62]. Soil type and the farmer's attitude are also significant factors for loss of SOM, as shown by [63]. Farmers often do not treat soil in a sustainable manner, because they usually do not own the parcels nor have a long tenure contract. As was noted earlier, farming of leased farmland is widespread in the Czech Republic, and a lack of a sense of responsibility for the soil is therefore often a problem.

There have been several research activities related to the soil physical properties of farmed soils in the Czech Republic. The effects of different soil and agricultural management on soil structure and soil hydraulic properties were analyzed by means of long-term monitoring and numerical modelling on Luvisol at the Hněvčeves experimental station (maintained by the Crop Research Institute in Prague) [55]. These studies showed that land use significantly influences the soil hydraulic properties in the upper part of the soil profile (A and Bt1 horizons, down to a depth of approximately 60 cm). Soil water retention and near-saturation hydraulic conductivity were higher in a soil profile with grassland compared to a soil with periodic tillage. Seasonal variability of soil bulk density, saturated water content, and unsaturated hydraulic conductivity were analyzed by Zumr et al. [64] on the Nučice experimental site. The soil water holding capacity generally decreased during the vegetation season as a result of the rainfall kinetic energy, poor soil structure stability, and a compacted shallow plough pan which caused frequent topsoil saturation.

A study by Podrázský et al. [65] proves a positive effect of the minimum tillage system on the soil aggregate stability, especially in the case of maize. A change of land use, such as afforestation or conversion of farmlands to grasslands, is a way to improve the soil water regime of degraded soil. However, the previous land use continues to affect the soil properties for many years, and the imprint of arable soils can prevail even 30 years after a change [44].

A poor soil structure accelerates other soil degradation processes. Agricultural uplands are very susceptible to soil water erosion when they are tilled repeatedly and are left without a protective cover. Erosion tends to preferentially remove low density or light particles, including both clay and soil organic carbon, which are two of the primary bonding agents in the aggregation process [66].

3.2 Soil Compaction

Approximately 38–45% of the Czech farmlands are negatively affected by topsoil and subsoil compaction [67]. This makes compaction, together with soil erosion, loss of organic matter and soil sealing, one of the most prominent soil degradation processes. The consequences of excessive soil compaction are very serious, as the most-affected soils are very fertile [68].

Less than 30% of the threatened soils are vulnerable to the so-called pedogenetic compaction, and more than 70% are vulnerable to the so-called technogenic compaction (VÚMOP 2021). Pedogenetic compaction arises during the formation of whitish illuvial or gley layers and is therefore typical for soil profiles with a comparatively high clay content. Technogenic soil compaction, resulting from mechanical pressure caused by field trafficking by agricultural machines, is dangerous mainly due to the possibility that it can occur in soils of any textural composition [69]. Current over-compaction has been caused mainly by excessively intensive farming in recent decades, mainly disproportionate doses and an incorrect assortment of mineral fertilizers, an insufficient supply of organic matter, and the use of heavy machinery. Conservation and minimum tillage technologies are the main practices in the Czech Republic, while direct seeding is marginal. Livestock trampling causes problems only locally on pastures. Kroulík et al. [70] showed that the ground area percentage that is trafficked at least once in a year is almost 90% for conventional tillage and 72% for conservation tillage [70]. Direct seeding technology requires approximately one half of the trafficked area. Controlled traffic farming with a fixed track system, which has been introduced on many farms, reduces the trafficked area to nearly 30% [71].

In general, soil compaction tends to be a more serious problem for soils with a high clay content [62]. At present, the situation is more complicated, as the long-term degradation has resulted in compaction in subsoil horizons which is very persistent and cannot be removed easily. So far, only minimal attention has been paid to finding an effective solution to this serious issue in agricultural enterprises. Compacted soils exhibit low infiltration capacity and water retention, reduced biological activity due to worse aeration and thermal regimes, higher bulk density, limited effective depth of the soil profile, fast soil drying, fast runoff, and often waterlogging. The direct consequences are increased power consumption during tillage, impaired nutrient utilization by plants, lower quality and a lower amount of yields [68].

A comprehensive study of the compacted subsoil layer and its spatial homogeneity was carried out by Jeřábek et al. [51] at the Nučice experimental catchment. The research was based on a combination of direct soil sampling, mechanical penetration resistance monitoring, geophysical methods (shallow electrical resistivity tomography), and remote sensing (delineation of wheel tracks). The measurements showed that the plough pan was homogeneous in a large part of the catchment, and its mean depth was between 11 and 14 cm (Fig. 5). Zumr et al. [72] showed by means of rainfall runoff monitoring and numerical modelling that the shallow plough

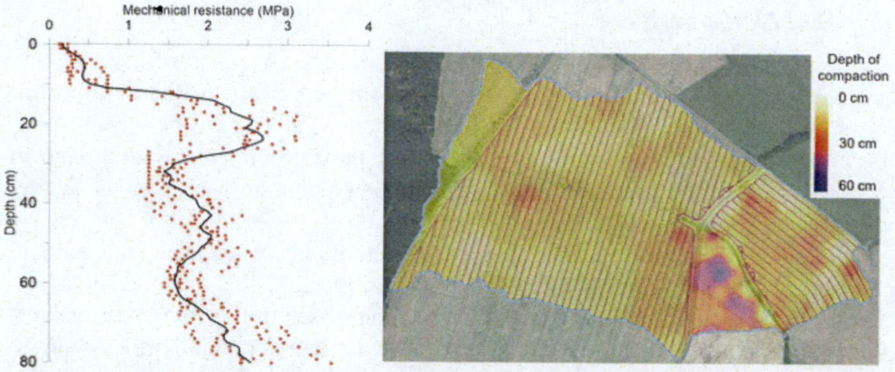

Fig. 5 Mechanical resistance against penetration at one of the observation points in the Nučice catchment. The red dots represent single measured values, and the black line shows the average resistance depth profile (left). The map of the plough pan position was reconstructed on the basis of over 100 measured soil profiles. The lines represent the wheel tracks which are mostly followed during trafficking (right), based on [51]

pan can explain immediate response to intense rainfall and the low soil water retention capacity of the Nučice catchment [73].

4 Soil Contamination

Pollution by various contaminants may cause disturbances in the functioning of the soil ecosystem and presents a risk to humans and to the environment. The main contaminants in the Czech soils are potentially toxic elements (PTASs) and man-made organic chemicals (xenobiotics), such as synthetic pesticides, dissolving agents, hydrocarbons, and drugs. The chemical elements of greatest concern are arsenic, cadmium, nickel, lead, and chromium. In most cases, soil contamination does not cause diffuse pollution, and the contaminated sites are usually small and disconnected. The most-affected areas are those with heavy industry and mining activities (West Bohemia, North Moravia) [17] and areas with high transportation (Prague and its surroundings) [74, 75]. Soils are locally also contaminated in the alluvial plains, due to occasional inundations containing wastewater [76]. Sewage sludge has only rarely been deposited or added to arable soils, as there are strict limits on its chemical composition. Contamination of the surrounding soils from modern landfills has also not been a serious problem [77].

The maximum tolerable values of risk elements and persistent organic pollutants are set in the Czech legislation [17, 78]. Since 1992, arable soils have been regularly tested for agrochemicals and hazardous substances within the Basal Soil Monitoring System organized by the Central Institute for Supervising and Testing in Agriculture (UKZUZ). The results of the monitoring, which has taken place within the area of the whole Czech Republic, show that the limiting values are only rarely (in approximately 1% of the samples) exceeded for cadmium and for arsenic [79], while the remaining tested elements exceed the limits even less often. The limits for the tested organic pollutants (mono and polyaromatic hydrocarbons, PCB, HCB, DDT, styrene, PCDD, and PCDF) were also exceeded only in exceptional cases [18, 78]. Long-term monitoring has not proven any significant temporal trends in soil contamination [66, 80, 81]. Nevertheless, a small number of hotspots remain where soils are strongly contaminated, mainly due to mining activities, industry, or historical landfills [82].

A present-day problem of Czech arable soils is insufficient manuring and unsustainable overdosing of agrochemicals, namely pesticides, herbicides, and mineral fertilizers. The main reason for the manure deficit is a dramatic decrease in livestock production, especially of cattle (Fig. 6). Central Institute for Supervising and Testing in Agriculture of the Czech Republic estimates that due to the lack of organic matter in arable soils, which is being recompensed with fertilizers, the soils will require at least 30 years to recover their function in the ecosystem [83]. In recent years, the average annual fertilizers consumption has been around 130–140 kg ha^{-1} of mineral fertilizers (approximately 75% of which is nitrogen, 15% is phosphorus, and 10% is potassium) and 2 kg ha^{-1} of pesticides (Czech Statistical Office). The statistical data show that the consumption of plant protection products has been declining in recent years, mainly due to lower application of glyphosates [84]. Nevertheless, pesticide residues in arable soils continue to pose an environmental threat,

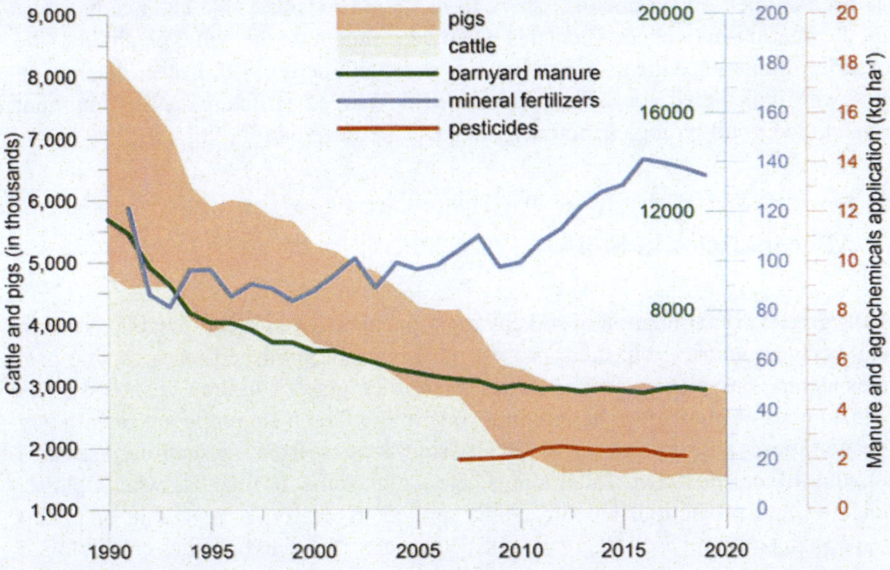

Fig. 6 A decline in numbers of livestock and available barnyard manure has resulted in increased application rates of mineral fertilizers (mainly N, P, K). Data for pesticides is not available for the period before 2006. The consumption of pesticides has decreased slightly in recent years (source of data, Czech Statistical Office, UKZUZ)

especially in the case of triazine and conazole fungicides [85]. In addition, residues of organochlorinated pesticides, which have not been applied since the 1990s, still persisted in the topsoil layers [86]. Pesticides can be transported by infiltrating water into deeper horizons, especially in soils with a heterogeneous structure with the presence of preferential flow [87].

A comprehensive analysis of the current situation and spatial and temporal trends of pesticides in the topsoil has been presented by Kodeš [88]. The study is based on the monitoring period of 2014–2017. Samples from 34 different localities with various soil types were analyzed for 64 currently-used fungicides and insecticides, including conazole and triazines, and recently-banned pesticides (e.g., atrazine, acetochlor, and linuron). The intensive use of protecting agents has resulted in their frequent and widespread occurrence in soils, both for currently-used products and for recently-banned products (and their transformation products). The highest numbers of pesticides have been observed on fields where rapeseed and wheat were cultivated (these are the most widely-grown crops in the Czech Republic). Kodeš [88] pointed out that glyphosate, which is of environmental concern and is applied in large doses, has not been evaluated within this study. In 2014–2019, monitoring performed by the Czech Hydrometeorological Institute detected above-limit concentrations, mostly of metabolites of herbicides, in approximately 30% of groundwater samples. Most of the affected samples were collected in the vicinity of fields with rapeseed planted for biofuels [89–94].

The problematics of veterinary and human pharmaceuticals in soils, their sorption characteristics, degradation rates, leakage to groundwater and root uptake by crops has been studied on Cambisols, Chernozems, and Luvisols [91]. The results show that the pharmaceutical persistence is mostly dependent on the soil type and that the sorption of pharmaceuticals generally decreases with depth [89]. Lower dissipation rates were calculated for soils with a well-developed structure, a high nutrient content, and good biological activity, such as Chernozems [95, 96].

5 Microplastics in Soils

The degree of contamination of soils by microplastics (MPs) is comparable to that in neighboring countries, but there has yet not been any published research that provides a quantitative analysis of the amount of microplastics in the Czech soils. The main sources of plastics on agricultural soils in the Czech Republic are most likely transportation (abrasion of tires) and plastic wastes from agriculture activities (remains of plastic bags, foils, straps, etc.). On some farmlands, where plastic mulching and protection nets are being applied, typically in vegetable or potato production, or in orchards, degraded plastic particles and macroplastics can be found in higher quantities, but this has been evaluated only visually so far.

Microplastics have been found in sludge from wastewater treatment plants and even in treated water. Pivokonsky et al. [97] analyzed water samples of untreated and treated water and showed that the concentration of MPs was approximately 83%

lower in the treated water compared to the raw wastewater [97]. Interestingly, even treated potable water had a rather high MPs concentration of 338 to 628 particles per liter. It should be pointed out that MPs from 1 μm in size were counted, which explains why such a large number of particles were observed. Further relevant research on MPs in the aquatic environment has been carried out by a team from the Institute of Hydrodynamics of the Czech Academy of Sciences. The research has been focused on drinking water and has produced promising results which indicate that the current treatment technology is capable of removing most of the MPs [98, 99]. It can therefore be concluded that a considerable proportion of the trapped MPs (difference between the microplastics amount in the raw and the treated water), which were mostly PET, PP, and PE fibers and fragments below 10 μm in size, most likely ends up in the sludge. Fertilization with sludge is currently not widespread, due to current legislation which sets strict limits on the composition of sludge. However, there is a prospect that sludge will soon be utilized more often as a way to enrich soils with organic matter. Sludge will then become another potential source of microplastics in soils.

Currently, a new research project is starting within the framework of the EU ITN Marie Sklodowska-Curie program entitled "Macro and microplastics in agricultural soil systems." The Czech Technical University in Prague will be evaluating MP fluxes from arable fields into surface waters.

6 Effects of Fires

Although wildfires are a natural phenomenon, their main cause in populated landscapes is human activity [100–102]. In Europe, 97% of forest fires of known origin within the period from 2006–2015 were directly or indirectly caused by humans. Similarly, the majority of recent wildfires in the Czech Republic were caused by humans [103]. The distribution of wildfires in the landscape is influenced by ignition triggers and also by environmental factors of both anthropogenic and natural origin. The presence of humans in the landscape usually serves as an ignition trigger, while environmental factors influence the probability that a wildfire will occur (Table 1). The environmental factors can be biotic, such as the vegetation cover influencing the fuel type, load and flammability, or abiotic, such as climate, topography, or soil types, which influence the moisture of the fuel and the spread of the fire. Anthropogenic factors influencing fire occurrence can be socio-economic, such as population density or the rate of unemployment, as well as socio-environmental, such as land use [104]. However, the effect of each of these factors on fire incidence varies among habitat types and depends on the temporal and spatial scale, as, for example, climatic variables usually operate on a scale that is broader than regional [105].

In the Czech Republic, the issue of wildfires and their influence on the soil, on human health, and on ecosystems is likely to grow in importance due to the climate change. The Czech Republic has a fragmented terrain and a dense network of forest paths, making it an area where forest fires seldom cause catastrophic damage.

Table 1 Statistics on forest fires in the Czech Republic in the last 10 years, which shows a generally positive correlation between the occurrence of wildfires and the mean temperature [107]

Year	Mean annual temperature (°C)	No. of forest fires	Burned area (ha)	Casualties	Injuries
2010	7.2	732	205	1	21
2011	8.5	1337	337	1	27
2012	8.3	1549	634	2	30
2013	7.9	666	92	0	7
2014	9.4	865	563	2	10
2015	9.4	1748	344	1	33
2016	8.7	892	141	0	6
2017	8.6	966	170	2	9
2018	9.6	2033	492	0	35
2019	9.5	1963	NaN	0	32

Because of this, the role of forest fires in local ecosystems was marginal in Central Europe in the past [106]. However, it has been proved that wildfires have affected the long-term development of forests even in the area of the Czech Republic [100]. The causes of forest fires in the Czech Republic have been analyzed by a small number of authors. The number of forest fires varied between 444 and 1,398 per year in the period from 2006–2018, with an average number of 725 per year. The burned area is usually not large, usually around 0.35 ha, and about 70% of all forest fires affect an area smaller than 0.05 ha. The incidence of forest fires in the Czech Republic is not uniform. In some municipalities no single forest fire was reported within a one-year period, while other municipalities reported more than ten forest fires [107].

Although forest fires are usually emphasized in the scientific literature and in the media, fires on agricultural fields have had a higher economic impact in recent years in the Czech Republic. The fires mostly occur during the hot and dry summer months during the harvest, and the fire is usually caused by contact between the harvester and stones on the soil surface. The fires damage the crops and reduce the yields significantly. In addition, expensive machinery is often damaged beyond repair in the fire. According to the Fire Rescue Service Statistical yearbook, there were over 600 fires in agricultural areas and nearly 2,000 fires in forest areas in 2019. Approximately 60 major fire events were recorded in 2019, which caused damage exceeding 10 million Czech crowns (EUR 400,000). One of the fires broke out on grassland, 11 in forests, and five were related to agricultural activities. 140 cases of self-ignition of agricultural crops occurred in 2019 [108].

Until now, little has been known about the consequences of fires on tilled soils and on the water regime, including the transport characteristics of various chemical components (such as PTEs, pesticides, fertilizers) in Czech conditions. The increased fire risk highlights the need to understand the processes occurring after a fire in terms of the changes to vegetation dynamics, soils, and water. In the Czech Republic, there is no data on post-fire contaminants and little attention has been paid to the long-term effects of fire, especially the effects on solutes and associated pollutants. The effectiveness of post-fire rehabilitation treatments is highly relevant in this context. Research projects funded by the Ministry of Youth and Education and by EU COST Action have recently started to study the effects of wildfires on agriculture soils and grasslands (Fire effects on soils, https://starfos.tacr.cz/cs/project/LTC20001, viewed December 2022).

The risks related to fires in forests, grasslands, shrublands, and croplands are expected to increase due to key factors: (i) the expansion of forests as a result of land abandonment and afforestation; (ii) the increase in fuel load and fuel continuity due to reduced land management; (iii) greater ignition potential due to population growth at the urban/rural interface; (iv) climate change, which will induce higher temperatures, increased wind speeds, an increased probability of prolonged dry periods affecting vegetation flammability [108]; (v) the growth of urban areas located closer to the land; and (vi) intensification of agriculture and harvesting during very dry conditions.

7 Conclusions

Overall, soil degradation is usually a result of more than one degradation process. It is often not possible to assess what degradation factor is the primary cause and what is the consequence, e.g., in the case of soil erosion and low organic matter content. The impacts of unhealthy soils are vast, ranging from biodiversity issues to groundwater quantity and quality. Soil sealing is another important problem for which inappropriate urban and landscape planning is mainly responsible. Soil degradation must therefore be regarded as a complex problem with multiple variables that involve many disciplines (including economics and ecology).

Farmers in the Czech Republic are aware that degraded soils pose a long-term threat to their livelihood. Many farmers have already encountered problems such as reduced yields due to lower amounts of precipitation, and decreased soil water retention capacity, difficult soil tillage and manuring, higher wear of machinery during dry soil cultivation, pests, water and wind erosion, difficulties in estimating the dosage of fertilizers and pesticides in uncertain weather conditions, etc.

The literature review presented here and also many research project reports, guidelines, methodologies, tools, NGO initiatives and politically-oriented activities, which are not fully listed in this chapter, suggest the following specific measures (in line with best management practices), which have a strong potential to slow down soil degradation and result in a more sustainable way for farming in the Czech Republic:

– To increase the soil organic matter (SOM) content in soils. SOM improves the soil structure and the stability of soil aggregates, and therefore reduces the vulnerability of the soil to erosion and improves the soil water regime. The optimal solution is high-quality barnyard manure. Compost, crop residues, and

green manuring are also very beneficial. First and foremost, however, the existing SOM must not be depleted.
- Providing support for small farms with sustainable livestock breeding.
- To change the landscape pattern toward higher spatial fragmentation, with smaller fields and with more small-scale landscape features to increase the ecological stability and the biodiversity of the landscape. A functional landscape will provide ecosystem services such as protection against wind and water erosion, better soil health, rainwater retention and purification, groundwater recharge, and many others.
- Seed catch crops and apply mulching.
- Using soil conservation tillage with a reduced number of passes, controlled trafficking and working soil at the optimal soil moisture. Apply occasional deep soil ripping to at least 40 cm to remove subsoil compaction. Reduce the use of pesticides.
- Providing support for organic farming and agroforestry, which usually results in higher biodiversity and more sustainable agriculture providing beneficial ecological services.
- To ensure sustainable urban growth

Although the current soil status may not look optimistic, there have been many examples of positive practices and trends. The public is aware of the situation and is concerned not only about food quality, but also about the impacts of agriculture on the environment. Organic food and products of local sustainable agriculture are becoming more and more popular. The state legislation follows the European agriculture framework, and stricter limits are expected on pesticide application and on large monoculture fields. We may therefore hope that the trend in the soil quality indicators will follow the air and surface water markers, which have been improving in recent decades.

Acknowledgment Valuable comments from my dear colleagues Tomáš Dostál, Josef Krása, Jakub Jeřábek, Magda Vaverková, and Daniel Žížala are greatly appreciated. This chapter has aimed to summarize current knowledge of the status of the soils and related problems in the Czech Republic, and it could not have been prepared without the use of publications prepared by many experts in the field of soil science. I sincerely apologize for any information or publication which I may have missed. I would like to thank the teams that worked on Tudi project no. 101000224, EU ITN SOPLAS project no. 955334, and on project no. LTC20001 "Fire effects on soils," which contributed to some of the results reported in this chapter.

References

1. Skalák P, Farda A, Zahradníček P, Trnka M, Hlásny T, Štěpánek P (2018) Projected shift of Köppen-Geiger zones in the central Europe: a first insight into the implications for ecosystems and the society. Int J Climatol 38:3595–3606. https://doi.org/10.1002/joc.5520
2. Trnka M, Semerádová D, Novotný I, Dumbrovský M, Drbal K, Pavlík F, Vopravil J, Štěpánková P, Vizina A, Balek J, Hlavinka P, Bartošová L, Žalud Z (2016) Assessing the

combined hazards of drought, soil erosion and local flooding on agricultural land: a Czech case study. Climate Res 70:231–249. https://doi.org/10.3354/CR01421
3. Hauptman I, Kukal Z, Posmourny K (2009) Soils in the czech Republic (in Czech).1st edn. Consult, Prague
4. VÚMOP (2021) Soils in numbers. In: https://statistiky.vumop.cz/
5. Zádorová T, Skála J, Žížala D, Vaněk A, Penížek V (2021) Harmonization of a large-scale national soil database with the World Reference Base for Soil Resources 2014. Geoderma 384: 114819. https://doi.org/10.1016/j.geoderma.2020.114819
6. Voltr V (2012) Concept of soil fertility and soil productivity: evaluation of agricultural sites in the Czech Republic. Arch Agron Soil Sci 58. https://doi.org/10.1080/03650340.2012.700511
7. Vopravil J, Janeček M, Tippl M (2007) Revised soil erodibility K-factor for soils in the Czech Republic. Soil Water Res 2:1–9. https://doi.org/10.17221/2100-swr
8. Janků J, Jakšík O, Kozák J, Marhoul AM (2016) Estimation of land loss in the Czech Republic in the near future. Soil Water Res 11:155–162. https://doi.org/10.17221/40/2016-SWR
9. Seeman T, Šrédl K, Prášilová M, Svoboda R (2020) The price of farmland as a factor in the sustainable development of Czech agriculture (a case study). Sustainability 12:5622. https://doi.org/10.3390/su12145622
10. Czech Office for Surveying M and C (2021) Cadastre of the Czech Republic Summarization Documents.1st edn. ČÚZK, Prague, p 8
11. Eurostat (2007) Farm structure survey. Structure of agricultural holdings 2007. European Communities, Luxembourg
12. Bičík I, Kupková L, Jeleček L, Kabrda J, Štych P, Janoušek Z, Winklerová J (2015) Land use changes in the Czech Republic 1845-2010: socio-economic driving forces. 1–215. https://doi.org/10.1007/978-3-319-17671-0
13. Szturc J, Karásek P, Podhrázská J (2017) Historical changes in the land use connected with appropriation of agricultural land - case study of Cadastral Areas Dolní Věstonice and Modřice (Czech Republic). Eur Countrys 9:658–678
14. Šarapatka B, Bednář M, Novák P (2010) Analysis of soil degradation in the Czech Republic: GIS approach. Soil Water Res 5:108–112. https://doi.org/10.17221/487-SWR
15. Bílá P, Šarapatka B, Horňák O, Novotná J, Brtnický M (2020) Which quality indicators reflect the most sensitive changes in the soil properties of the surface horizons affected by the erosion processes? Soil Water Res 15:116–124. https://doi.org/10.17221/71/2019-SWR
16. Devátý J, Dostál T, Hösl R, Krása J, Strauss P (2019) Effects of historical land use and land pattern changes on soil erosion – case studies from Lower Austria and Central Bohemia. Land Use Policy 82:674–685. https://doi.org/10.1016/j.landusepol.2018.11.058
17. Kozák J, Borůvka L, Kodešová R, Jacko K, Hladík J (2010) Soil atlas of the Czech Republic. CULS Prague, Prague
18. Podlešáková E, Nemecek J (2000) Contamination and degradation of soils in the Czech Republic - contemporary and future state. In: Soil quality, sustainable agriculture and environmental security in Central and Eastern Europe. Springer Netherlands, pp 79–86
19. Šarapatka B, Bednář M (2015) Assessment of potential soil degradation on agricultural land in the Czech Republic. J Environ Qual 44:154–161. https://doi.org/10.2134/jeq2014.05.0233
20. Podhrázská J, Kučera J, Karásek P, Konečná J (2016) Land degradation by erosion and its economic consequences for the region of South Moravia (Czech Republic). Soil Water Res 10: 105–113. https://doi.org/10.17221/143/2014-swr
21. Novák P, Hůla J, Kumhálová J (2006) Translocation of soil particles ad different speed of tillers. In: Proceedings of 6th international conference on trends in agricultural engineering, pp 433–437
22. Hrabalíková M, Huisová P, Holubík O, Žížala D, Ureš J, Kumhalová J (2018) Assessment of changes in topsoil depth redistribution by different tillage technologies. In: Zlatic M, Kostadinov S (eds) Soil and water resources protection in the changing environment, pp 200–210

23. Žížala D, Juřicová A, Kapička J, Novotný I (2021) The potential risk of combined effects of water and tillage erosion on the agricultural landscape in Czechia. https://doi.org/10.1080/17445647.2021.1942251
24. Kapička J, Žížala D, Novotný I (2020) Monitoring eroze zemědělské půdy -Závěrečná zpráva. Research Institute for Soil and Water Conservation, Praha
25. Žížala D, Juřicová A, Zádorová T, Zelenková K, Minařík R (2019) Mapping soil degradation using remote sensing data and ancillary data: South-East Moravia, Czech Republic. Eur J Remote Sens 52:108–122. https://doi.org/10.1080/22797254.2018.1482524
26. Janeček M, Bohuslávek J, Dumbrovský M, GerGergel J, Hrádek F, Kovář P, Kubátová E, Pasák V, Pivcová J, Tippl M, Toman F, Tomanová O, Váška J (2002) Ochrana zemědělské půdy před erozí.1st edn. ISV - Institut sociálních věcí, Praha
27. Bauer M, Dostal T, Krasa J, Jachymova B, David V, Devaty J, Strouhal L, Rosendorf P (2019) Risk to residents, infrastructure, and water bodies from flash floods and sediment transport. Environ Monit Assess 191. https://doi.org/10.1007/s10661-019-7216-7
28. Krasa J, Dostal T, Jachymova B, Bauer M, Devaty J (2019) Soil erosion as a source of sediment and phosphorus in rivers and reservoirs – watershed analyses using WaTEM/SEDEM. Environ Res 171:470–483. https://doi.org/10.1016/j.envres.2019.01.044
29. Jáchymová B, Krása J, Dostál T, Bauer M (2020) Can lumped characteristics of a contributing area provide risk definition of sediment flux? Water (Switzerland) 12. https://doi.org/10.3390/w12061787
30. Rosendorf P, Vyskoč P, Prchalová H, Fiala D (2016) Estimated contribution of selected non-point pollution sources to the phosphorus and nitrogen loads in water bodies of the Vltava river basin. Soil Water Res 11:196–204. https://doi.org/10.17221/15/2015-SWR
31. Podhrazska J, Szturc J, Karasek P, Kucera J, Konecna J (2019) Economic impacts of farmland degradation in the Czech republic – case study. Agric Econ (Czech Republic) 65:529–538. https://doi.org/10.17221/89/2019-AGRICECON
32. Sklenicka P, Molnarova K, Pixova KC, Salek ME (2013) Factors affecting farmland prices in the Czech Republic. Land Use Policy 30:130–136. https://doi.org/10.1016/j.landusepol.2012.03.005
33. Dostál T, Janecek M, Kliment Z, Krása J, Langhammer J, Váška J, Vrana K (2006) Czech Republic. In: Boardman J, Poesen J (eds) Soil erosion in Europe. Wiley, Chichester, pp 107–116
34. Podhrázská J, Karásek P (2014) Systém analýzy území a návrhu opatření k ochraně půdy a vody v krajině.1st edn. VUMOP, Brno
35. Li T, Jeřábek J, Noreika N, Dostál T, Zumr D (2021) An overview of hydrometeorological datasets from a small agricultural catchment (Nučice) in the Czech Republic. Hydrol Process 35:e14042. https://doi.org/10.1002/hyp.14042
36. Zachar D (1970) Erozia pody. SAV, Bratislava
37. Bučko Š, Holý M, Stehlík O (1967) Soil erosion in Czechoslovakia. Journal of the Czechoslovak Geographical Society 37–46
38. Žížala D, Kapička J, Novotný I (2015) Monitoring soil erosion of agricultural land in Czech Republic and data assessment of erosion events from spatial database. In: Proceedings from international conference soil – the non-renewable environmental resource, pp 354–370
39. Gebeltová Z, Malec K, Maitah M, Smutka L, Appiah-Kubi SNK, Maitah K, Sahatqija J, Sirohi J (2020) The impact of crop mix on decreasing soil price and soil degradation: a case study of selected regions in Czechia (2002–2019). Sustainability 12:444. https://doi.org/10.3390/su12020444
40. Podhrázská J, Kučera J, Chuchma F, Středa T, Středová H (2013) Effect of changes in some climatic factors on wind erosion risks - the case study of South Moravia. Acta Universitatis Agriculturae et Silviculturae Mendelianae Brunensis 61:1829–1837. https://doi.org/10.11118/actaun201361061829
41. Pechanec V, Prokopová M, Salvati L, Cudlín O, Procházka J, Samec P, Včeláková R, Cudlín P (2021) Moving toward the north: a country-level classification of land sensitivity to

degradation in Czech Republic. CATENA 206:105567. https://doi.org/10.1016/J.CATENA.2021.105567
42. Švehlík R (2002) Větrná eroze na jihovýchodní Moravě v obrazech. Přírodovědný klub v Uherském Hradišti, Uherské Hradiště
43. Podhrázská J, Kučera J, Středová H (2015) The methods of locating areas exposed to wind erosion in the South Moravia Region. Acta Universitatis Agriculturae et Silviculturae Mendelianae Brunensis 63:113–121. https://doi.org/10.11118/actaun201563010113
44. Jakšík O, Kodešová R, Kubiš A, Stehlíková I, Drábek O, Kapička A (2015) Soil aggregate stability within morphologically diverse areas. Catena 127:287–299. https://doi.org/10.1016/j.catena.2015.01.010
45. Homoláč L, Tomsik K (2016) Historical development of land ownership in the Czech Republic since the foundation of the Czechoslovakia until present. Agric Econ (Czech Republic) 62:528–536. https://doi.org/10.17221/250/2015-AGRICECON
46. Moravcová J, Koupilová M, Pavlíček T, Zemek F, Kvítek T, Pečenka J (2017) Analysis of land consolidation projects and their impact on land use change, landscape structure, and agricultural land resource protection: case studies of Pilsen-South and Pilsen-North (Czech Republic). Landsc Ecol Eng 13:1–13. https://doi.org/10.1007/s11355-015-0286-y
47. Novotný I, Žížala D, Kapička J, Beitlerová H, Mistr M, Kristenová H, Papaj V (2016) Adjusting the CPmax factor in the Universal Soil Loss Equation (USLE): areas in need of soil erosion protection in the Czech Republic. J Maps 12:58–62. https://doi.org/10.1080/17445647.2016.1157834
48. Zumr D, Dostál T, Devátý J, Valenta P, Rosendorf P, Eder A, Strauss P (2017) Experimental determination of the flood wave transformation and the sediment resuspension in a small regulated stream in an agricultural catchment. Hydrol Earth Syst Sci 21. https://doi.org/10.5194/hess-21-5681-2017
49. Kudsk P, Mathiassen SK (2020) Pesticide regulation in the European Union and the glyphosate controversy. Weed Sci 68:214–222. https://doi.org/10.1017/WSC.2019.59
50. Hamza MA, Anderson WK (2005) Soil compaction in cropping systems: a review of the nature, causes and possible solutions. Soil Tillage Res 82:121–145
51. Jeřábek J, Zumr D, Dostál T (2017) Identifying the plough pan position on cultivated soils by measurements of electrical resistivity and penetration resistance. Soil Tillage Res 174. https://doi.org/10.1016/j.still.2017.07.008
52. Beare MH, Gregorich EG, St-Georges P (2009) Compaction effects on CO2 and N2O production during drying and rewetting of soil. Soil Biol Biochem 41:611–621. https://doi.org/10.1016/J.SOILBIO.2008.12.024
53. Novara A, Armstrong A, Gristina L, Semple KT, Quinton JN (2012) Effects of soil compaction, rain exposure and their interaction on soil carbon dioxide emission. Earth Surf Process Landf 37:994–999. https://doi.org/10.1002/ESP.3224
54. Sněhota M, Sobotková M, Císlerová M (2008) Impact of the entrapped air on water flow and solute transport in heterogeneous soil: experimental set-up. J Hydrol Hydrodynam 56:247–256
55. Kodesova R, Jirku V, Kodes V, Muhlhanselova M, Nikodem A, Žigová A (2011) Soil structure and soil hydraulic properties of Haplic Luvisol used as arable land and grassland. Soil Tillage Res 111:154–161. https://doi.org/10.1016/j.still.2010.09.007
56. Kodešová R, Rohošková M, Žigová A (2009) Comparison of aggregate stability within six soil profiles under conventional tillage using various laboratory tests. Biologia 64:550–554. https://doi.org/10.2478/s11756-009-0095-6
57. Hofman J, Dušek L, Klánová J, Bezchlebová J, Holoubek I (2004) Monitoring microbial biomass and respiration in different soils from the Czech Republic - a summary of results. Environ Int 30:19–30. https://doi.org/10.1016/S0160-4120(03)00142-9
58. Menšík L, Hlisnikovský L, Kunzová E (2019) The state of the soil organic matter and nutrients in the long-term field experiments with application of organic and mineral fertilizers in different soil-climate conditions in the view of expecting climate change. In: Organic fertilizers - history, production and applications. IntechOpen

59. Usowicz B, Lipiec J (2020) The effect of exogenous organic matter on the thermal properties of tilled soils in Poland and the Czech Republic. J Soil Sediment 20:365–379. https://doi.org/10.1007/s11368-019-02388-2
60. Horáček J, Novák P, Liebhard P, Strosser E, Babulicová M (2017) The long-term changes in soil organic matter contents and quality in chernozems. Plant Soil Environ 63:8–13. https://doi.org/10.17221/274/2016-PSE
61. Horáček J, Kolář L, Čechová V, Hřebečková J (2008) Phosphorus and carbon fraction concentrations in a cambisol soil as affected by tillage. Commun Soil Sci Plant Anal 39: 2032–2045. https://doi.org/10.1080/00103620802134867
62. Bednář M, Šarapatka B (2018) Relationships between physical–geographical factors and soil degradation on agricultural land. Environ Res 164:660–668. https://doi.org/10.1016/j.envres.2018.03.042
63. Walmsley A, Azadi H, Tomeckova K, Sklenicka P (2020) Contrasting effects of land tenure on degradation of Cambisols and Luvisols: the case of Central Bohemia Region in the Czech Republic. Land Use Policy 99:104956. https://doi.org/10.1016/j.landusepol.2020.104956
64. Zumr D, Jeřábek J, Klípa V, Dohnal M, Sněhota M (2019) Estimates of tillage and rainfall effects on unsaturated hydraulic conductivity in a small central European agricultural catchment. Water (Switzerland) 11. https://doi.org/10.3390/w11040740
65. Podrázský V, Holubík O, Vopravil J, Khel T, Moser WK, Prknová H (2015) Effects of afforestation on soil structure formation in two climatic regions of the Czech Republic. J For Sci 61:225–234. https://doi.org/10.17221/6/2015-JFS
66. Weissmannová HD, Mihočová S, Chovanec P, Pavlovský J (2019) Potential ecological risk and human health risk assessment of heavy metal pollution in industrial affected soils by coal mining and metallurgy in Ostrava, Czech Republic. Int J Environ Res Public Health 16. https://doi.org/10.3390/ijerph16224495
67. Lhotský J (2000) Soil compaction and measures against it. Ústav zemědělských a potravinářských informací, Prague
68. Javůrek M, Vach M (2008) The negative impacts of soil overcompaction and set of measures for their remedy. Výzkumný ústav rostlinné výroby, v.v.i., Praha
69. Pražan R, Kubín K, Gerndtová I (2014) Key factors of technogenic soil compaction. Mechanizace zemědělství 64:11–13
70. Kroulík M, Kumhála F, Hůla J, Honzík I (2009) The evaluation of agricultural machines field trafficking intensity for different soil tillage technologies. Soil Tillage Res 105:171–175. https://doi.org/10.1016/j.still.2009.07.004
71. Kroulík M, Kvíz Z, Kumhála F, Hůla J, Loch T (2011) Procedures of soil farming allowing reduction of compaction. Precis Agric 12:317–333. https://doi.org/10.1007/s11119-010-9206-1
72. Zumr D, Dostál T, Devátý J (2015) Identification of prevailing storm runoff generation mechanisms in an intensively cultivated catchment. J Hydrol Hydromech 63. https://doi.org/10.1515/johh-2015-0022
73. Noreika N, Li T, Zumr D, Krasa J, Dostal T, Srinivasan R (2020) Farm-scale biofuel crop adoption and its effects on in-basin water balance. Sustainability 12:10596. https://doi.org/10.3390/SU122410596
74. Vácha R, Sánka M, Skála J, Čechmánková J, Horváthová V (2016) Soil contamination health risks in Czech proposal of soil protection legislation. In: Environmental health risk - hazardous factors to living species. InTech
75. Skála J, Vácha R, Hofman J, Horváthová V, Sáňka M, Čechmánková J (2017) Spatial differentiation of ecosystem risks of soil pollution in floodplain areas of the Czech Republic. Soil Water Res 12:1–9. https://doi.org/10.17221/53/2016-SWR
76. Vaverková MD, Zloch J, Radziemska M, Adamcová D (2017) Environmental impact of landfill on soils – the example of the Czech Republic. Polish J Soil Sci 50:93–105. https://doi.org/10.17951/pjss.2017.50.1.93

77. Ministry of the Environment of the Czech Republic (2019) Decree no. 271/2019 on the procedures for ensuring the protection of agricultural land. Collection of Laws
78. Poláková Š, Kubík L, Malý S, Němec P (2010) Monitoring of agricultural soils in the Czech Republic, 1992-2007.1st edn. Central Institute for Supervising and Testing in Agriculture, Brno
79. Dušek J, Vogel T, Dohnal M, Lichner L, Čipáková A (2006) Simulated cadmium transport in macroporous soil during heavy rainstorm using dual-permeability approach. Biologia (Poland) 61:S251–S254. https://doi.org/10.2478/s11756-006-0167-9
80. Holubová K, Zumr D, Císlerová M (2011) Numerical simulation of water dynamics and contaminant transport at Quarry Dump Hájek. Waste Forum 2:82–91
81. Kozák J, Janku J, Jehlicka J (1995) The problems of heavily polluted soils in the Czech Republic: a case study. In: Heavy metals. Springer, Berlin Heidelberg, pp 287–300
82. Sáňka M (2018) Je hospodaření s půdou udržitelné? Veronica 1:2–5
83. UKZUZ (2020) Data on the consumption of active substances contained in Plant Protection Products. Brno
84. Vašíčková J, Hvězdová M, Kosubová P, Hofman J (2019) Ecological risk assessment of pesticide residues in arable soils of the Czech Republic. Chemosphere 216:479–487. https://doi.org/10.1016/j.chemosphere.2018.10.158
85. Shegunova P, Klánová J, Holoubek I (2007) Residues of organochlorinated pesticides in soils from the Czech Republic. Environ Pollut 146:257–261. https://doi.org/10.1016/j.envpol.2006.03.057
86. Kodešová R, Kočárek M, Kodeš V, Šimůnek J, Kozák J (2008) Impact of soil micromorphological features on water flow and herbicide transport in soils. Vadose Zone J 7:798–809. https://doi.org/10.2136/vzj2007.0079
87. Kosubová P, Škulcová L, Poláková, Hofman J, Bielská L (2020) Spatial and temporal distribution of the currently-used and recently-banned pesticides in arable soils of the Czech Republic. Chemosphere 254:126902. https://doi.org/10.1016/j.chemosphere.2020.126902
88. Kodeš V (2020) Směsi cizorodých látek v podzemních vodách. TZB-info
89. Kodešová R, Kočárek M, Klement A, Golovko O, Koba O, Fér M, Nikodem A, Vondráčková L, Jakšík O, Grabic R (2016) An analysis of the dissipation of pharmaceuticals under thirteen different soil conditions. Sci Total Environ 544:369–381. https://doi.org/10.1016/j.scitotenv.2015.11.085
90. Kodešová R, Grabic R, Kočárek M, Klement A, Golovko O, Fér M, Nikodem A, Jakšík O (2015) Pharmaceuticals' sorptions relative to properties of thirteen different soils. Sci Total Environ 511:435–443. https://doi.org/10.1016/j.scitotenv.2014.12.088
91. Kočárek M, Kodešová R, Vondráčková L, Golovko O, Fér M, Klement A, Nikodem A, Jakšík O, Grabic R (2016) Simultaneous sorption of four ionizable pharmaceuticals in different horizons of three soil types. Environ Pollut 218:563–573. https://doi.org/10.1016/j.envpol.2016.07.039
92. Klement A, Kodešová R, Bauerová M, Golovko O, Kočárek M, Fér M, Koba O, Nikodem A, Grabic R (2018) Sorption of citalopram, irbesartan and fexofenadine in soils: estimation of sorption coefficients from soil properties. Chemosphere 195:615–623. https://doi.org/10.1016/j.chemosphere.2017.12.098
93. Fer M, Kodešová R, Golovko O, Schmidtová Z, Klement A, Nikodem A, Kočárek M, Grabic R (2018) Sorption of atenolol, sulfamethoxazole and carbamazepine onto soil aggregates from the illuvial horizon of the haplic luvisol on loess. Soil Water Res 13:177–183. https://doi.org/10.17221/82/2018-SWR
94. Kodešová R, Klement A, Golovko O, Fér M, Kočárek M, Nikodem A, Grabic R (2019) Soil influences on uptake and transfer of pharmaceuticals from sewage sludge amended soils to spinach. J Environ Manage 250. https://doi.org/10.1016/j.jenvman.2019.109407
95. Hiller E, Šebesta M (2017) Effect of temperature and Soil pH on the sorption of ibuprofen in agricultural soil. Soil Water Res 12:78–85. https://doi.org/10.17221/6/2016-SWR

96. Paula S, Arianoutsou M, Kazanis D, Tavsanoglu Ç, Lloret F, Buhk C, Ojeda F, Luna B, Moreno JM, Rodrigo A, Espelta JM, Palacio S, Fernández-Santos B, Fernandes PM, Pausas JG (2009) Fire-related traits for plant species of the Mediterranean Basin. Ecology 90:1420–1420. https://doi.org/10.1890/08-1309.1
97. Pivokonsky M, Cermakova L, Novotna K, Peer P, Cajthaml T, Janda V (2018) Occurrence of microplastics in raw and treated drinking water. Sci Total Environ 643:1644–1651. https://doi.org/10.1016/j.scitotenv.2018.08.102
98. Pivokonský M, Pivokonská L, Novotná K, Čermáková L, Klimtová M (2020) Occurrence and fate of microplastics at two different drinking water treatment plants within a river catchment. Sci Total Environ 741:140236. https://doi.org/10.1016/j.scitotenv.2020.140236
99. Novotna K, Cermakova L, Pivokonska L, Cajthaml T, Pivokonsky M (2019) Microplastics in drinking water treatment – current knowledge and research needs. Sci Total Environ 667:730–740
100. Holuša J, Berčák R, Lukášová K, Hanuška Z, Agh P, Vaněk J, Kula E, Chromek I (2018) Forest fires in the Czech Republic - definition and classification: review. For Res Rep 63:102–111
101. Adámek M, Jankovská Z, Hadincová V, Kula E, Wild J (2018) Drivers of forest fire occurrence in the cultural landscape of Central Europe. Landsc Ecol 33:2031–2045. https://doi.org/10.1007/s10980-018-0712-2
102. Kula E, Jankovská Z (2013) Forest fires and their causes in the Czech Republic (1992-2004). J For Sci 59:41–53. https://doi.org/10.17221/36/2012-jfs
103. Niklasson M, Zin E, Zielonka T, Feijen M, Korczyk AF, Churski M, Samojlik T, Jędrzejewska B, Gutowski JM, Brzeziecki B (2010) A 350-year tree-ring fire record from Białowieża Primeval Forest, Poland: implications for Central European lowland fire history. J Ecol 98:1319–1329. https://doi.org/10.1111/j.1365-2745.2010.01710.x
104. Jazebi S, de Leon F, Nelson A (2020) Review of wildfire management techniques-Part I: causes, prevention, detection, suppression, and data analytics. IEEE Trans Power Deliv 35: 430–439. https://doi.org/10.1109/TPWRD.2019.2930055
105. Moritz MA, Parisien M-A, Batllori E, Krawchuk MA, van Dorn J, Ganz DJ, Hayhoe K (2012) Climate change and disruptions to global fire activity. Ecosphere 3:1–22. https://doi.org/10.1890/ES11-00345.1
106. Adámek M, Bobek P, Hadincová V, Wild J, Kopecký M (2015) Forest fires within a temperate landscape: a decadal and millennial perspective from a sandstone region in central Europe. For Ecol Manage 336:81–90. https://doi.org/10.1016/j.foreco.2014.10.014
107. Fire Rescue Service of the Czech Republic (2020) Statistical Yearbook 2019. Ministry of the Interior – General Directorate Fire Rescue Service of the Czech Republic
108. Žalud Z, Trnka M, Hlavinka P (2020) Agricultural drought in the Czech Republic (in Czech). Agrární komora České republiky, Prague

Agricultural Soil Degradation in Estonia, Latvia and Lithuania

Paulo Pereira, Miguel Inacio, Igor Bogunovic, Lyudmyla Symochko, Damia Barcelo, and Wenwu Zhao

Contents

1 Background 60
2 Study Area 61
3 Soil Compaction 61
4 Soil Erosion 67
5 Soil Pollution 73
5.1 Heavy Metals 73
5.2 Nitrogen and Phosphorus 75
5.3 Organic Pollutants 77
6 Acidification 78
7 Conclusion 79
References 79

Abstract Agricultural soil degradation is a global phenomenon that is expected to increase in the future due to agricultural intensification and climate change. In Baltic states (Estonia, Latvia and Lithuania), after the Soviet Union fall, the impacts of

P. Pereira (✉) and M. Inacio
Environmental Management Laboratory, Mykolas Romeris University, Vilnius, Lithuania

I. Bogunovic
Faculty of Agriculture, University of Zagreb, Zagreb, Croatia

L. Symochko
Uzhhorod National University, Uzhhorod, Ukraine

Coimbra University, Coimbra, Portugal

D. Barcelo
Catalan Institute for Water Research (ICRA-CERCA), Girona, Catalonia, Spain
e-mail: dbcqam@cid.csic.es

W. Zhao
Faculty of Geographical Science, Beijing Normal University, Beijing, China
e-mail: zhaoww@bnu.edu.cn

Paulo Pereira, Miriam Muñoz-Rojas, Igor Bogunovic, and Wenwu Zhao (eds.),
Impact of Agriculture on Soil Degradation II: A European Perspective,
Hdb Env Chem (2023) 121: 59–86, DOI 10.1007/698_2023_967,

Published online: 3 March 2023

agriculture on the ecosystems decreased. However, soil compaction, erosion, pollution and acidification are critical threats that hurt the environment. Unsustainable practices are responsible for increasing compaction, erosion, pollution and acidification. Often, they have dramatic impacts on water resources. The most important is the increased eutrophication in lakes and coastal areas. Nowadays, the Baltic sea is one of the most polluted seas in the world. Therefore, it is essential to halt the agriculture intensification in Baltic countries and establish more sustainable practices to reverse this trend. This book chapter aims to study the most important impacts of agriculture practices on soil degradation in Estonia, Latvia and Lithuania.

Keywords Agriculture, Baltic countries, Eutrophication, Intensification, Soil degradation

1 Background

Soils are the base of life and supply many ecosystem services (e.g., carbon sequestration, water purification, nutrient cycling, food, raw material, archaeological archive) essential for humans. Nevertheless, due to human activities' expansion and intensification, degradation has increased [1, 2]. According to FAO, "Soil degradation is defined as a change in the soil health status resulting in a diminished capacity of the ecosystem to provide goods and services for its beneficiaries [1]". For instance, it is estimated that 33% of the soils are already degraded, and 90% can be degraded by 2050. These numbers are especially dramatic since 95% of the food that we eat comes from soil. [2] In agricultural areas, soil degradation is a global problem produced by high food demand [3]. Soil degradation in agricultural areas occurs due to unsustainable practices (e.g., tractor trafficking, tillage, short rotation periods, application of pesticides and herbicides, pollution, and overgrazing) and climate change. In the long term, it is expected to threaten food security. Normally degradation is perceived when there is a decrease in soil biodiversity and soil fertility [4, 5]. There are a large number of works that highlight the perverse impacts of agriculture intensification in compaction (e.g., [6]), acidification (e.g., [7]), pollution (e.g., [8]), erosion and water losses (e.g., [9]). Overall, it is widely known that agricultural intensification has a dramatic impact on soil functions and ecosystem services, and sustainable practices are needed to reverse this trend [2]. Different initiatives at the global level were conducted to halt and reverse soil degradation such as United Nations Decade on Ecosystem Restoration, [3] Land Degradation Neutrality [4] or the

[1] https://www.fao.org/soils-portal/soil-degradation-restoration/en/.

[2] https://www.fao.org/about/meetings/soil-erosion-symposium/key-messages/en/.

[3] https://www.decadeonrestoration.org/.

[4] https://www.unccd.int/land-and-life/land-degradation-neutrality/overview.

Sustainable Development Goals (SDGs). [5] Soil conservation practices are key to achieving this. They are mostly related to SDG 6, 13 and 6 [10]. In Europe, soil degradation is a major problem, as highlighted in different works (e.g., [11, 12]). Major initiatives are being conducted to decrease human footprint and make agriculture more sustainable (e.g., Green Deal [6]) due to the negative impacts of agriculture intensification on soil degradation [13]. Soil-sustainable practices are key to achieving green deal targets [14]. Although soil degradation has not been addressed in the Baltic countries (e.g., Estonia, Latvia and Lithuania), the impacts on the economic losses are very high [15]. Therefore, this chapter aims to assess agriculture's impacts on soil degradation in Estonia, Latvia and Lithuania.

2 Study Area

Estonia, Latvia and Lithuania are known as the Baltic countries. They are located in the eastern part of the Baltic sea. Lithuania is the country with the largest area, and Estonia is the one that has the lowest. The region is relatively flat and does not have a high orography (Fig. 1). Estonia, Latvia and Lithuania are included in the boreal/hemiboreal bioclimatic zone. In 2018, forest and seminatural areas covered most of the areas, while wetlands covered less. Regarding Estonia and Latvia, forest and seminatural areas covered most of the areas as well. The areas that had the slightest land use were artificial surfaces. Lithuania showed a different dynamic compared to the previous countries. Agricultural areas covered most of the territory and wetlands were the less common land use (Table 1). Agricultural areas occupied a larger area in Lithuania compared with the other countries. In all Baltic countries, from 1990 to 2000, there was a decrease in the agricultural areas. An increase followed this decrease in 2006. In the following years (2012 and 2018), the agricultural area stabilised (Fig. 2). According to the Eurostat, [7] in January of 2021, there were 1,330,068 inhabitants in Estonia, while in Latvia and Lithuania were 1,893,233 and 2,795,680 inhabitants, respectively.

3 Soil Compaction

Soil compaction affects approximately 33 million hectares in Europe. 20 million hectares are located in Eastern Europe. Soil compaction is a highly variable problem, and agriculture depends essentially on soil susceptibility to compaction, tractor traffic intensity and tillage practices [16]. The implications of compaction are

[5] https://sdgs.un.org/goals.

[6] https://www.switchtogreen.eu/the-eu-green-deal-promoting-a-green-notable-circular-economy/.

[7] https://ec.europa.eu/eurostat/cache/digpub/regions/#total-population.

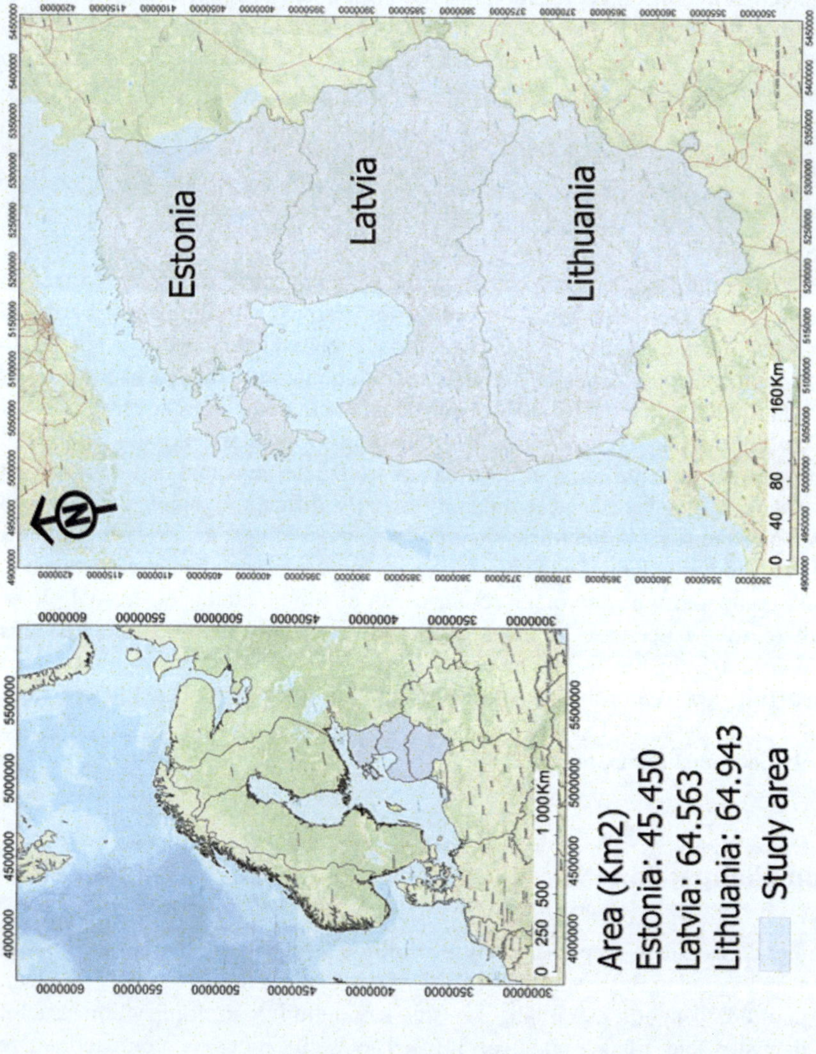

Fig. 1 Study area

Table 1 Land uses in Baltic countries according to Corine Land Cover (2018) in %. Source: https://land.copernicus.eu/pan-european/corine-land-cover/clc2018

Land use	All countries	Estonia	Latvia	Lithuania
Artificial surfaces	2.43	2.10	2.03	3.04
Agriculture areas	42.11	31.60	39.57	58.84
Forests and seminatural areas	50.48	56.65	53.73	34.91
Wetlands	2.30	4.61	2.53	0.86
Water bodies	2.67	5.04	2.11	1.99

reduced soil water infiltration, aeration, porosity [17], biota, root development, structure stability, cultivation, yield, plant germination and nutrient uptake [16, 18, 19], fertiliser efficiency and increased overland flow, erosion, nutrient losses, soil density, greenhouse gases emission and tractor fuel demand [19–21]. Previous works highlighted that soil compaction was a concern in the three Baltic states [12]. In this region, agricultural soils have a high susceptibility to compaction, especially in some areas in the south of Latvia. Lithuania's susceptibility is lower than the two other countries (Fig. 3). In agriculture areas, the soil compaction problem is very great in Lithuania and Latvia but less in Estonia. On average, the bulk density in agricultural lands in Estonia, Latvia and Lithuania is 1.28 (±0.158), 1.38 (±0.106) and 1.38 (±0.099), respectively (Fig. 4). In some areas of the Baltic countries (e.g., central and southwest of Lithuania), the soil has a reduced susceptibility to compaction (Fig. 3). However, the same areas have high compaction (Fig. 4). This may be due to the unsustainable practices related to agriculture intensification in these areas.

In Estonia, soil compaction in agricultural areas is a recognised problem [22]. Several works were developed to assess the impacts of soil compaction. Krebestain et al. [23] found that tractor trafficking increased topsoil (Cambisol) compaction in a Lucerne (*Medicago sativa L.*), decreasing soil air capacity, plant-available water, saturated hydraulic conductivity, and air conductivity. Khut et al. [24] found that penetration resistance increased significantly compared to the non-compacted soil in a strongly compacted Haplic Stagnosol. The high soil compaction decreases the conditions for plant germination. Reintam et al. [25] conducted a 5-year experiment in a *Stagnic Luvisol* planted with spring barley (*Hordeum vulgare L.*) monoculture. Two treatments were considered compacted (MTZ-82 tractor; total weight 4.84 mg) and non-compacted. During the experiment, tractor trafficking decreased grain yield (80%), barley shoots quantity, phytomass, and aboveground biomass. The most damaging impact was detected in the 2 initial years of the experiment. Also, in a *Stagnic Luvisol*, Kuht and Reintam [26] found that heavy compaction significantly decreased spring barley and wheat nitrogen, phosphorus, calcium and magnesium uptake. Trukmann et al. [27] identified that soil compaction decreased a spring barley monoculture's roots and shot weight in a *Stagnic Luvisol*. On the other hand, oilseed rape and narrow-leafed lupine shoot and root weight increased with soil compaction, showing that these species are more adapted to compacted soils. In Latvia, few international literature studies focused on the impacts of agriculture

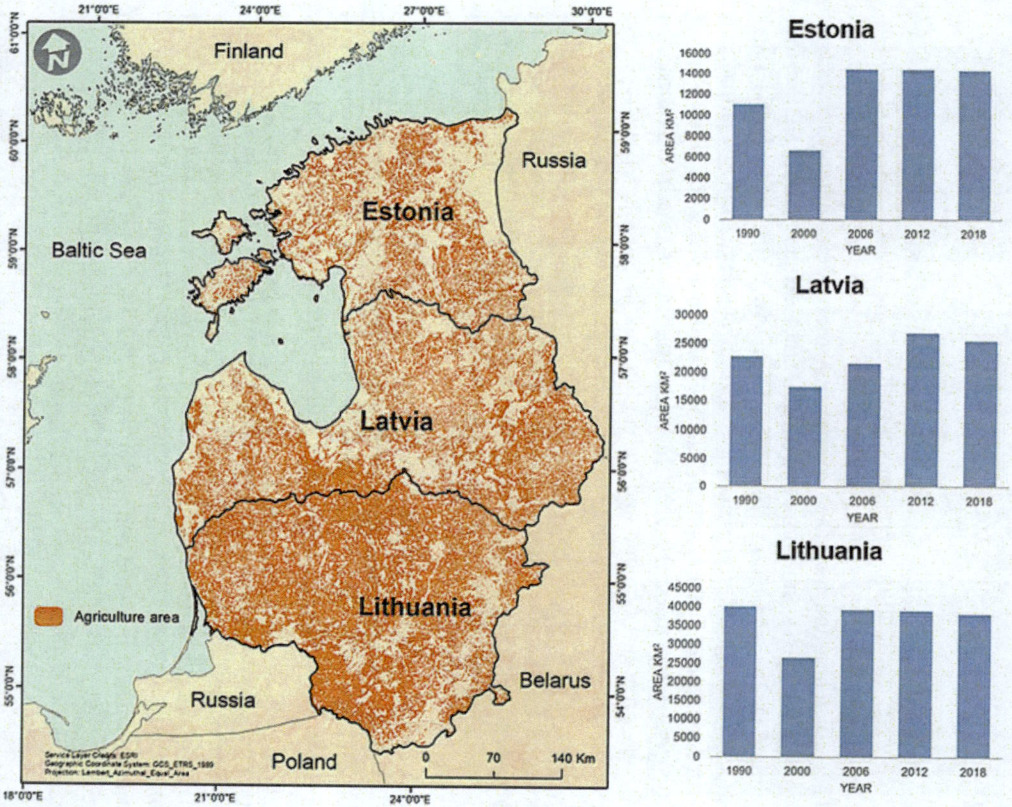

Fig. 2 Agricultural land use in 2018 (map) and evolution from 1990 to 2018 in Estonia, Latvia and Lithuania. Corine Land Cover data from https://land.copernicus.eu/

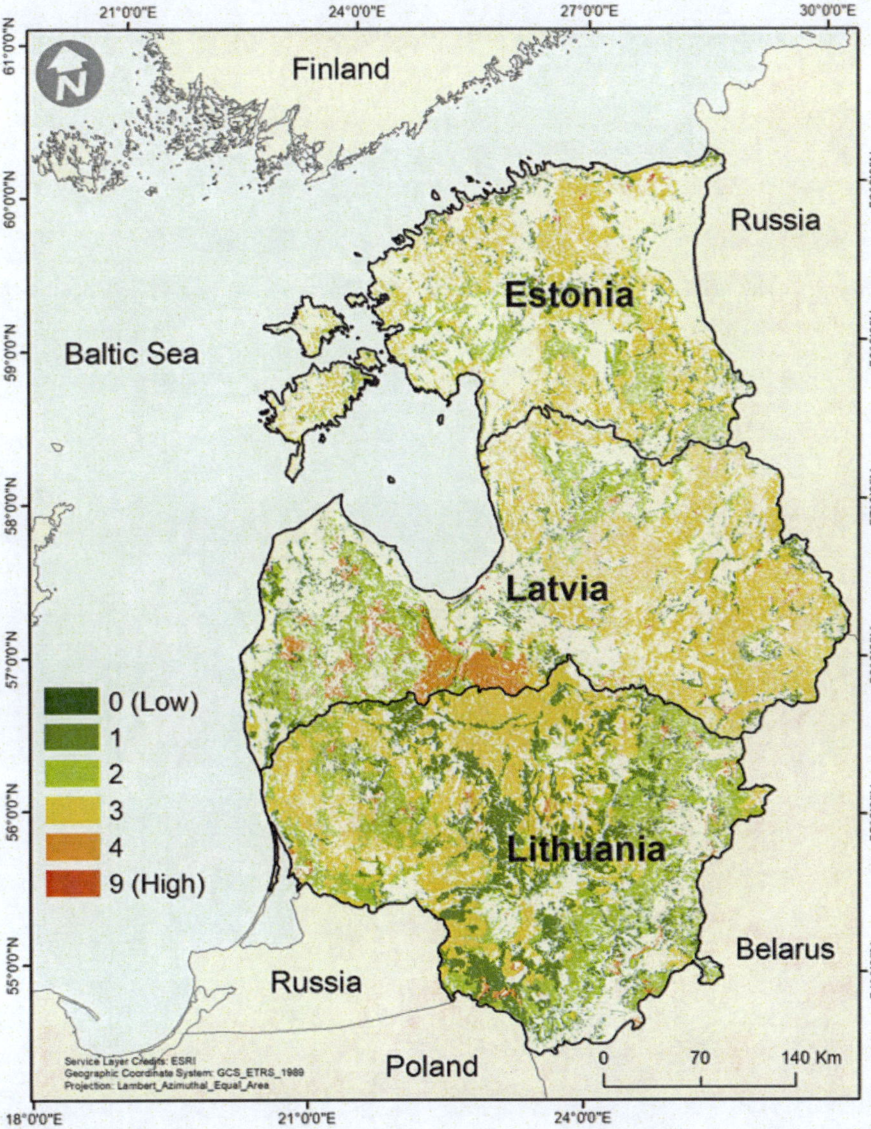

Fig. 3 Soil susceptibility to compaction in Estonia, Latvia and Lithuania. Data from: https://esdac.jrc.ec.europa.eu/themes/soil-susceptibility-compaction

intensification on soil compaction. However, a report delivered by the Ministry of Agriculture [28] highlighted the importance of soil compaction in the degradation of agrochemical properties. In Lithuania, several experiments were conducted to study the impact of agriculture on soil compaction. Previous works also observed that soil compaction induces enormous losses related to yield, greenhouse gases emission,

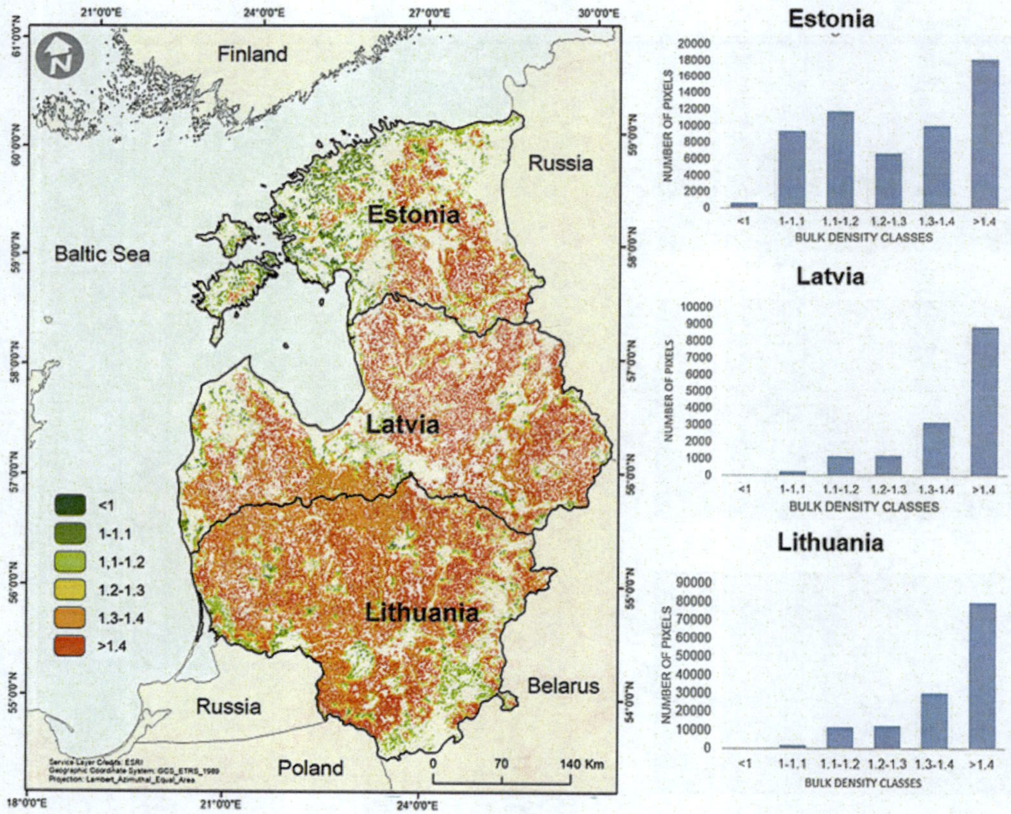

Fig. 4 Soil compaction (g/m^3) map in Estonia, Latvia and Lithuania (left) and histogram for each country (500 m resolution). Data from https://esdac.jrc.ec.europa.eu/themes/soil-compaction

erosion, flooding, fuel demand, nitrogen, phosphorous and potassium and water pollution; approximately 27 million Euros per year [29]. Šarauskis et al. [30] observed in experiments conducted in several parts of Lithuania that tillage practices reduced soil compaction. Non-tilled soils have the highest soil compaction. Nevertheless, although tillage decreases soil hardness, high soil tillage intensity reduces topsoil moisture and the conditions for plant germination. Also, Šarauskis et al. [31] after applying different tillage methods (deep cultivation, shallow cultivation, deep ploughing, shallow ploughing and no-tillage) found that autumn soil tillage by shallow cultivation effects on topsoil compaction was lower than the identified deep ploughing. Experiments carried out by Feiziene et al. [32] in an *Epihypogleyic Cambisol* found that conventional tillage practices had improved soil properties (low bulk density and high air permeability). Also, Romaneckas et al. [33] recorded that deep tillage decreased soil mesopores. Overall, the studies performed in the Baltic countries highlighted that soil compaction decreases substantially soil quality and food production.

4 Soil Erosion

In European Union, approximately 2.46 t/ha of soil is lost annually. Also, 12.7% of European arable lands have an extremely high soil loss (>5 t/ha) and need to be restored. Soil erosion has tremendous impacts on the environment, society and economy. A soil affected by erosion has difficulty storing water and nutrients, making them more vulnerable to droughts and reducing its fertility, biodiversity, functions substantially, and ES supplied (e.g., floods and erosion regulation) [2, 35, 36]. Therefore, crop growth and yield will negatively affect farmers' income [37]. Wind erosion also decreases soil fertility and increases dust storms that harm the environment and human health (e.g., respiratory diseases) [38, 39]. Compared to southern European mountainous areas, soil losses in the Baltic countries are reduced [34]. Soil water erosion in these countries is not severe (Fig. 5). On average, annually, Estonia soil losses are 0.48 t/ha (±0.50), in Latvia 0.54 t/ha (±0.68) and in Lithuania 0.75 t/ha (±0.73). All the countries have a reduced agricultural area affected by severe erosion [40]. This confirms the idea that water erosion in the Baltics is minimal compared to other European countries. According to Panagos et al. [40], between 2010 and 2016, soil water erosion in arable lands decreased in Estonia (−11%) and Lithuania (−0.6%), while in Latvia increased slightly (1.6%). The crop yields and economic losses in agricultural areas due to soil erosion are higher in Lithuania (0.079 million Euros) than in Latvia (0.019 million Euros) and Estonia (0.006 million Euros) [41]. This represents a land productivity loss of 0.006%, 0.009% and 0.0018% in Estonia, Latvia and Lithuania, respectively. Land productivity loss causes Estonia a GDP decrease of −0.049 million Euros, Latvia −0.095 million Euros and Lithuania −0.179 million Euros [41]. Regarding wind erosion, in Estonia on average the losses in agricultural areas are 0.25 t/ha (±0.46), in Latvia are 0.071 t/ha (±0.27) and in Lithuania 0.097 t/ha (±0.28) (Fig. 6).

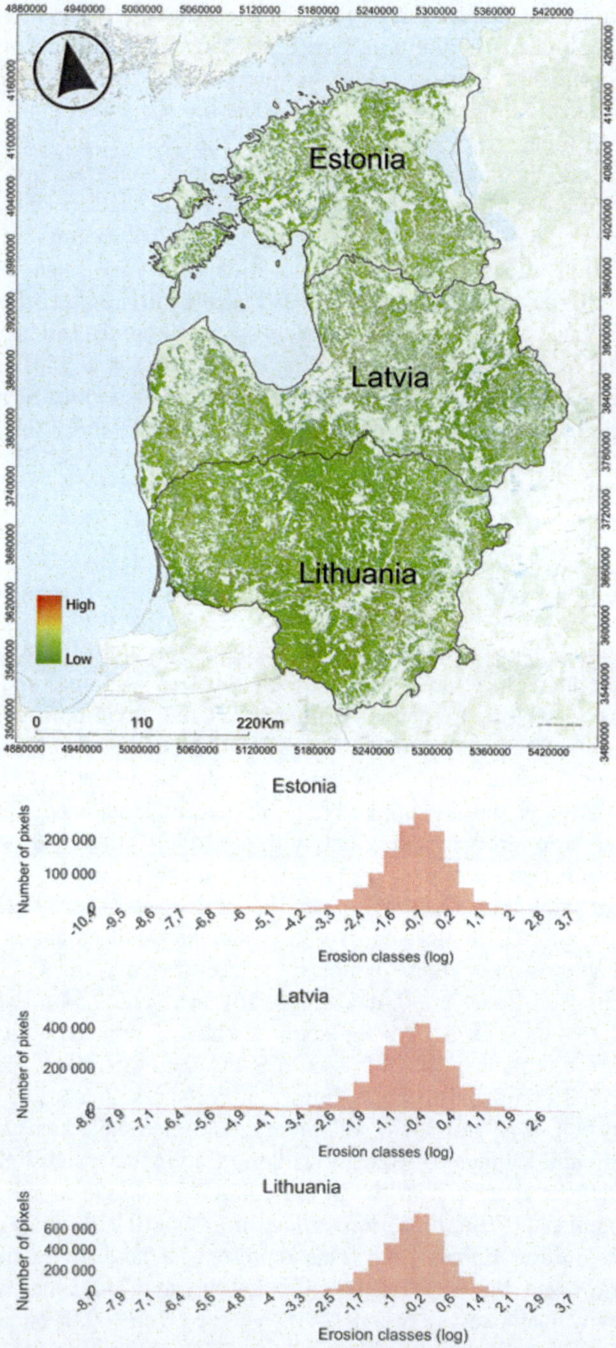

Fig. 5 Agricultural soil water erosion (t/ha) in Estonia, Latvia and Lithuania (100 m resolution). Below the soil erosion histograms for each country. Data from: https://esdac.jrc.ec.europa.eu/content/soil-erosion-water-rusle2015

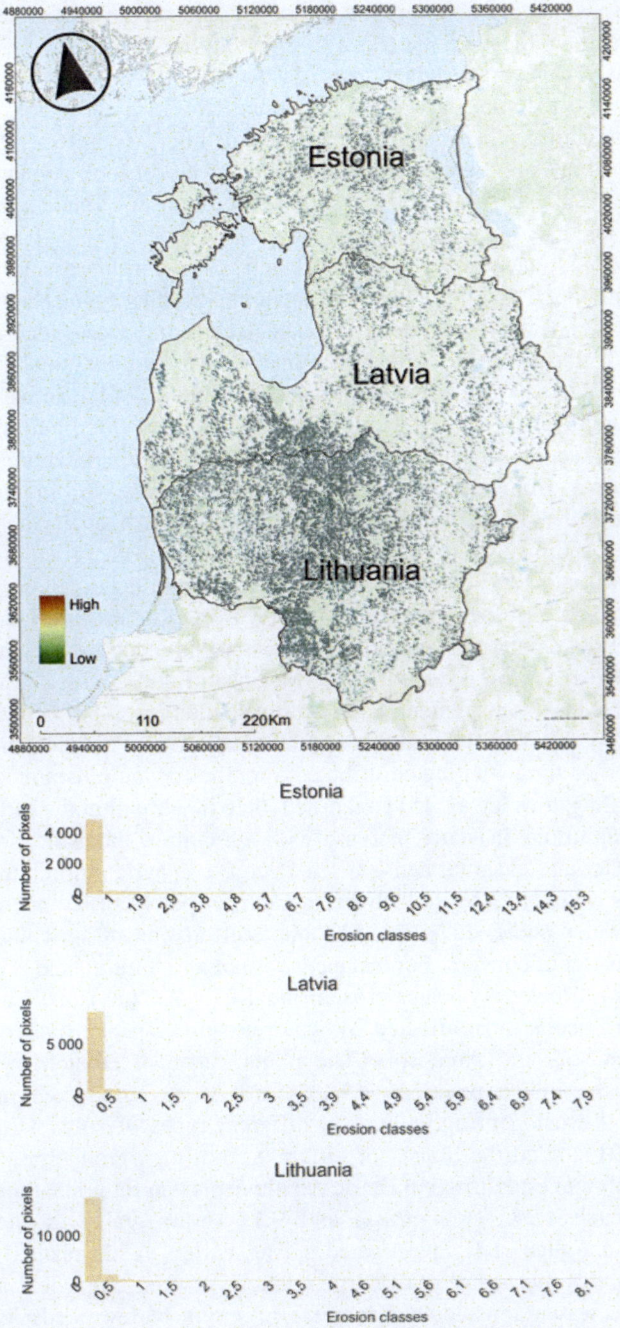

Fig. 6 Agricultural soil wind erosion (t/ha) in Estonia, Latvia and Lithuania. Below the soil erosion histograms for each country (1,000 m resolution). Data from: https://esdac.jrc.ec.europa.eu/themes/wind-erosion

Although these values are not high, the soil cover factor index is high in some areas of Lithuania and south of Latvia (Fig. 7), meaning the soil is unprotected against erosion agents. The average cover factor is 0.152 (±0.063), 0.144 (±0.074) and 0.184 (±0.069) in Estonia, Latvia and Lithuania. However, there are many pixels, mainly in Lithuania, where the cover factor is high. Therefore, the averages abovementioned are not representative of reality. A closer look confirms this idea. Bare soil index analysis conducted at a better resolution in Lithuania shows that in agriculture fields, bare soils are common (Fig. 7).

Since Estonia is flat, water and wind erosion is not considered one of the most relevant natural hazards [42]. Nevertheless, some studies highlight that 40% of agricultural areas are at risk of water erosion [43], and some works identified erosion associated with tillage practices [44]. Regional-focused works (southeast) showed that 6–37% are affected by erosion [45]. Another evidence of the active soil erosion from agricultural lands is the sediments and nutrients transported into water bodies. Previous works recognised that there are very few field experiments in Estonia to assess the transport from agricultural areas into water bodies [46]. In our search, we only found the work of Laas and Kull [47], which found in a small catchment that a high amount of sediments and nitrogen was transported from agricultural areas. In Latvia, soil erosion is considered the most critical form of degradation (Fig. 8). It is estimated that 20.8% of the agricultural area is affected by erosion [48]. The main causes for soil degradation in agricultural areas are drainage system management, not appropriated rent-tenure relationships, rare use of soil conservation measures, lack of crop residues management, short crop rotations and a high cover of annual crops (e.g., [49]). Soms [50] found that agricultural practices coupled with intensive rainfall and spring snow melting contributed to gullies development in southeastern Latvia. Also, Lagzdins et al. [51] recorded that in agricultural catchments, the nutrient concentrations in water increase with agriculture intensification increase. However, due to land abandonment and the decrease in agricultural intensification[8] in some areas of Latvia, several nutrients such as phosphorus and nitrogen are decreased in water bodies [52]. In Lithuania, soil erosion in agricultural areas is considered a matter of concern. For instance, 17% of agricultural land is eroded. The soil types most affected by water erosion are Regosols, Luvisols, Cambisol and Albeluvisol. The soils most affected by wind erosion (especially after tillage and dried) are Histosols and Arenosols. The major cause of erosion in Lithuanian agriculture is the tillage practices [53]. Several works conducted in Lithuania focused on soil erosion in agriculture in different types of soils. Jarasiunas and Kinderiene [53] studied the impact of different land (e.g., grain–grass and grass–grain crop rotations) established in Eutric Albeluvisols and found that under erosion rates could reach 1.38 (grain–grass) and 0.11 (grass–grain) $m^3\ ha^{-1}\ year^{-1}$. Jankauskas and Fullen [54] investigated spring barley yields planted in Dystric Albeluvisols and found a strong negative relation between crop yield and erosion severity. There was a considerable decrease in spring barley yields in slopes (–

[8] https://www.eea.europa.eu/soer/2010/countries/lv/land-use-state-and-impacts-latvia.

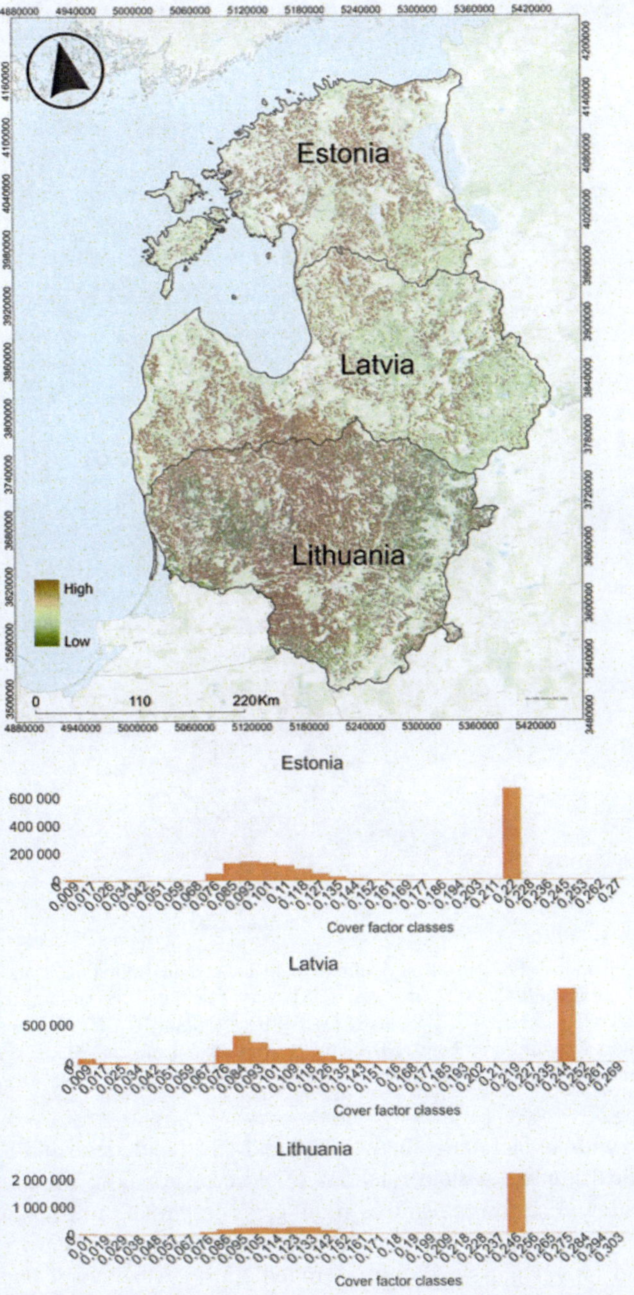

Fig. 7 Agricultural soil cover factor in Estonia, Latvia and Lithuania (100 m resolution). Below the soil erosion histograms for each country. Data from: https://esdac.jrc.ec.europa.eu/content/cover-management-factor-c-factor-eu

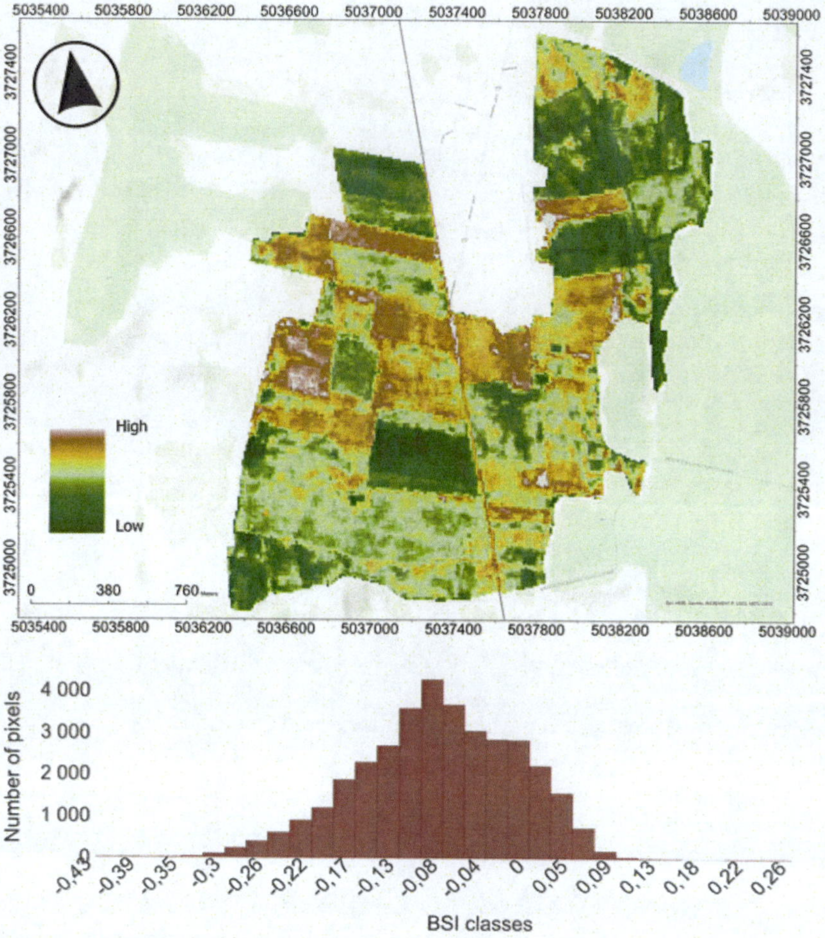

Fig. 8 Bare soil index (BSI) in an agricultural field located in Lithuania. Data from Sentinel 2 (10 m resolution) (2018–2022 median). BSI was calculated according to the formula BSI = ((RED+SWIR) – (NIR + BLUE))/((RED+SWIR) + (NIR + BLUE)). RED is the red band, SWIR is the short-wave infrared band, NIR is the near-infrared band and BLUE is the blue band. Data from Sentinel 2: https://scihub.copernicus.eu/

17.7–26.3%) with an inclination between 10 and 15°. Kinderiene and Karcauskiene [55] established numerous field rotations (Field crop rotation; Field crop rotation with black fallow; Erosion-preventive grain–grass crop rotation; Erosion-preventive grass–grain crop rotation; Not fertilised and not used (not mown) grassland; Fertilised and mown grassland) in an eroded Eutric Albeluvisol and found that erosion processes were accelerated under reduced vegetation cover and intense rainfall. The highest levels of soil erosion were identified in Field crop rotation with black fallow (13.21 t ha^{-1} $year^{-1}$). Also, in Eutric Albeluvisol, Jankauskas et al. [56] found high erosion rates in winter rye (3.2–8.6 m^3 ha^{-1} $year^{-1}$) and spring barley (9.0–27.1 m^3 ha^{-1} $year^{-1}$). They were especially very high under potatoes (24.2–87.1 m^3 ha^{-1} $year^{-1}$). Jarasiunas et al. [57] studied the long-term effects (1995–2012) of different field crops (Winter crops; Potatoes; Spring barley + under crop; Red clover + timothy perennial + Spring barley and Black fallow) planted in Luvisols. During the studied period, the erosion was especially high in Black fallow (89.34 ± 121.29 t ha^{-1} $year^{-1}$).

5 Soil Pollution

Soil pollution is a global problem that dramatically affects food security and human health since it reduces crop yields and makes food unsafe for consumption. The most significant contributors to soil pollution in agriculture are pesticides, amendments with solid waste, livestock wastes, irrigation with poor quality water, over fertilisation and irrigation with untreated wastewater. Agriculture is one of the main sources of soil and diffuse pollution. The diffuse pollution increase is usually connected to the increase of heavy metals, pesticides, persistent organic pollutants or other inorganic pollutants. Contrary to soil erosion or sealing, soil contamination is a degradation process that is difficult to assess [1, 58]. In the Baltic countries, soviet agricultural management was highly intensive, based on high agrochemical use leading to very high soil pollution. After the independence, Estonia, Latvia and Lithuania were facing several environmental problems, among them soil and water contamination from agricultural areas [59, 60]. After the regime transition and the decrease in agriculture intensification, the environmental impacts decrease importantly. However, they remain a problem [61, 62]. After the independence, several agricultural fields were developed in former military fields with high soil contamination [63, 64]. Although it is recognised that the situation is better than during soviet times, the agriculture sector is still an important source of pollutants in Baltic countries [65–67].

5.1 Heavy Metals

Agrochemicals [68, 69] and other products/compost used in soils, such as sewage sludge [70], manure [71], or fly ash [72], are responsible for the release of numerous heavy metals into the environment. Baltic countries are not an exception. Using as an example Cadmium (Fig. 9), on average, the concentration of this element in Estonian agriculture soils is 0.97 mg/kg (±0.02), while in Latvia and Lithuania was 0.97 mg/kg (±0.01), respectively. In some agricultural areas of the Baltic countries, there are also concerns regarding copper accumulation [73]. In Estonia, Reintam et al. [74] highlighted that heavy metal pollutant in agricultural areas is a serious problem. Lepane et al. [75] found that one of the main source's metal (lead, copper, zinc and mercury) pollution in lake Liinjarv were the agricultural areas. Roots and Roose [76]

Fig. 9 Agricultural soil cadmium (mg/kg) in Estonia, Latvia and Lithuania. Below the soil cadmium histograms for each country. Data from: https://esdac.jrc.ec.europa.eu/content/maps-heavy-metals-soils-eu-based-lucas-2009-hm-data-0

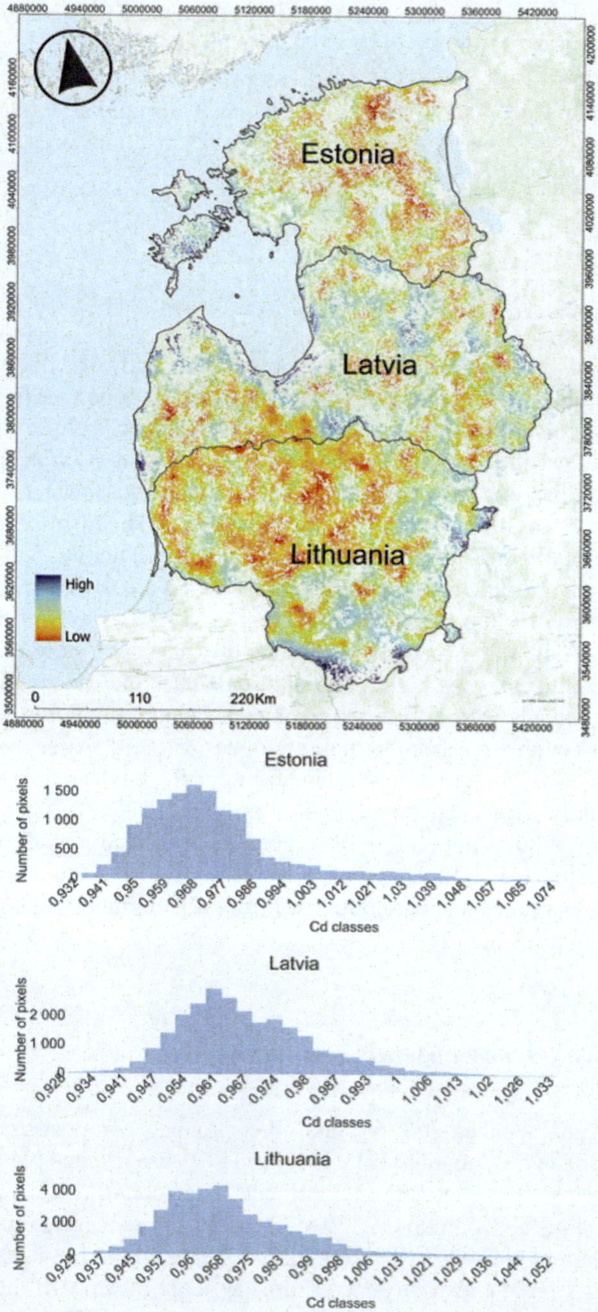

reported that there were high levels of nickel, cadmium and lead in areas with intensive agricultural practices near water bodies. In Latvia, Karklins [48] observed that in agricultural soils, the levels of chromium, nickel, cadmium, lead, zinc and copper were not different from those observed in the parent material. Therefore, it could be considered natural. Consistent with the previous work, Klavins et al. [77] found that rivers' heavy metal contents were similar to regional background levels. However, a report conducted by the World Bank [78] observed that soil pollution with toxic metals was high in major agricultural exploitations. Also, Vincevica-Gaile et al. [79] identified a high concentration of cadmium, copper, nickel and selenium in vegetables grown in farmlands. Overall, the results are not consistent and likely are site-specific. Marcinkonis et al. [80] found that long-term liming of acid soils increased manganese, strontium, boron and copper levels in Lithuania. Also, Petraitis et al. [81] found that in soils treated with liquid swine waste, the levels of copper, zinc, chromium and lead increased dramatically. Gregorauskiene and Kadunas [82] identified that agricultural practices (e.g., fertilisation) increased molybdenum, titanium and strontium in topsoil. The heavy metals in soils are likely to be transported to water bodies. Valskys et al. [83] found that the bottom sediments of the Sventoji River were rich in arsenic, likely due to pollution from agricultural areas. In the Curonian lagoon and Lithuanian coast, Remeikaitė-Nikienė et al. [84] found sediments with high concentrations of lead, copper, cadmium and zinc attributed to the application of agrochemicals in agriculture areas.

5.2 Nitrogen and Phosphorus

According to Eurostat[9] (Fig. 10), among the Baltic countries, Lithuania is the country that uses more inorganic nitrogen in agriculture, followed by Latvia and Estonia. In all the countries, between 2012 and 2020, there was an increase in the use of this fertiliser in agriculture. High levels of inorganic nitrogen application on soils have detrimental impacts on other nutrients plant (e.g., calcium, magnesium) uptake [85], water repellency [86], bacterial richness [87] and affect nitrogen cycling (e.g., denitrification) [88]. Also, excess inorganic nitrogen and phosphorous fertilisers are dramatically impacting surface and groundwater pollution. Aquatic ecosystems may promote (1) acidification, (2) eutrophication, and (3) increased toxic water levels. High inorganic nitrogen contents may also cause negative impacts on human health and the economy [89]. In the Baltic region, nitrogen and phosphorus applied in agricultural areas have a dramatic impact on the Baltic sea, one of the most polluted seas in the world.[10] Among the Baltic states, Lithuania is the one where nitrogen leaching is high, increasing Baltic Sea eutrophication [90]. In the Baltic states, few

[9] https://ec.europa.eu/eurostat/.

[10] https://www.euronews.com/green/2021/09/28/europe-starts-cleaning-up-its-act-to-save-the-baltic-sea.

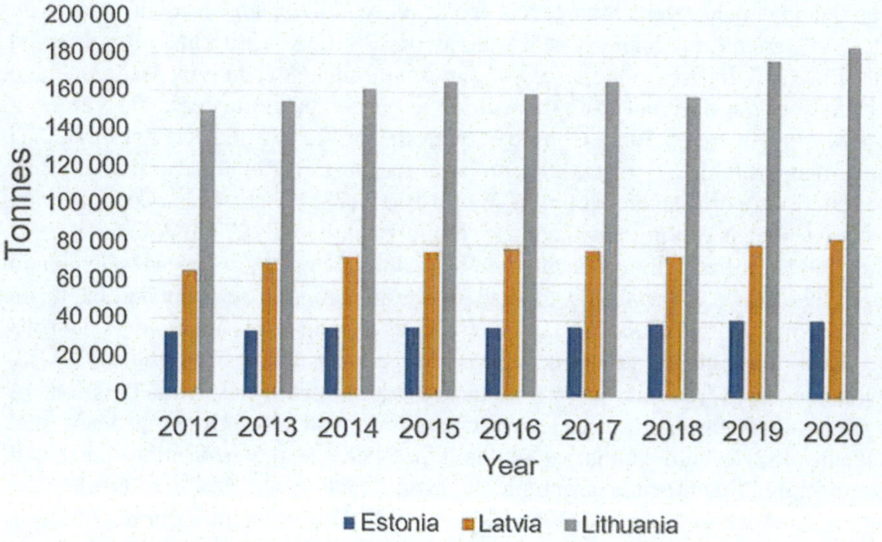

Fig. 10 Inorganic nitrogen used for agriculture (toons) in Estonia, Latvia and Lithuania. Source: https://ec.europa.eu/eurostat/

works focus on the impacts of inorganic nitrogen and phosphorus on soils. However, research was conducted on the impacts on water resources. Overall, there has been a reduction in phosphorus and nitrogen loading into the water resources since the end of the Soviet Union due to agriculture intensification decrease and land abandonment increase [91–93]. In Estonia, Mander et al. [94] found that agricultural intensification was responsible for groundwater pollution in the Selja River catchment. Several streams were nourished by groundwater, polluted with high levels of mineral nitrogen. Nõges et al. [91] observed that although phosphorous loads in two large lakes (Võrtsjärv and Peipsi) have decreased with time, climate change may reinforce eutrophication. Therefore, it is key to use fewer mineral fertilisers in agriculture. Also, in lake Peipsi, Kangur and Mols [95] found different phosphorous dynamics in the north and south of the lake. In the north, it was identified a reduction between 1970 and 2005. In the south, an increase was identified due to anthropogenic activities, including agriculture. A great part of the discharges of phosphorus and nitrogen into the lake are due to rivers. Buhvestova et al. [96] found that between 1992 and 2007 were discharged annually in lake Peipsi 5,600 tonnes of nitrogen and 179 tonnes of phosphorus. Also, Nõges et al. [97] found that in lake Võrtsjärv, nitrogen and phosphorous concentration depended on agriculture management and climate change (temperature and precipitation). In Latvia, Jansons et al. [98] observed that nitrogen and phosphorous dynamics in small catchments depended on agriculture intensity and climate and soil conditions. Although farmers applied several mitigation measures to reduce nitrogen and phosphorous losses, time is needed to improve water quality. Siksane and Lagzdins [99] investigated the trends of total and mineral nitrogen in small agricultural catchments. They found that between 1995 and 2018, there was an increase in the variables studied (subsurface drainage field) in 5 out 6 monitoring stations. Overall, there is a need to reduce fertiliser applications in the studied areas. Also, Jansons et al. [100] observed that in areas where high doses of pig manure were applied in the soil and where soviet style animal production farms, water pollution (phosphorus and nitrogen) is still a serious problem. Finally, poor application of fertilisers in the soil can induce losses to farmers. In Berze and Mellupite research sites, Siksane and Lagzdins [101] found that, on average, 13.2% and 15.4% of the nitrogen applied is lost through subsurface drainage. This implies to the farmers an average loss of 61.13 EUR ha^{-1} $year^{-1}$. Finally, in Lithuania, the dynamic is similar to that observed in the other Baltic countries. For instance, Povilaitis [102] identified a decrease in phosphorous concentration in rivers between 1991 and 2002. Also, Tumas [103] found a substantial decrease in nitrogen in rivers after 1991 (independence from the Soviet Union). In Susve River, Aksomaitienė and Berankienė [104] observed that between 1960 and 2000, that nitrogen and phosphorous used in agriculture were the most important polluters. Česonienė et al. [105] studied nitrogen and phosphorus in 10 ponds and 26 rivers across Lithuania. They found that 20% of these water bodies did not have a good or very good ecological status. The agricultural areas were the ones that contributed more to the pollution identified. Finally, Sileika et al. [106] found that the amount of phosphorous exports in the Nevezis River is related to the number of livestock in agricultural areas.

5.3 *Organic Pollutants*

Agriculture intensification is also responsible for the increase of other pollutants, such as persistent organic pollutants (e.g., pesticides, herbicides, fungicides [107], polycyclic aromatic hydrocarbons [108], pharmaceuticals, antibiotics [109] and microplastics [110]). The presence of these pollutants degrades soil status. For instance, persistent organic pollutants negatively affect soil microbiology [111] and biodiversity [112] and can be uptaken by plants [113]. Previous works observed that persistent organic pollutants were identified in high quantities in Baltic countries [114]. In Estonia, the use of pesticides contributes to environmental degradation.[11] Kumar et al. [115] found residues of organochlorine pesticides in agricultural areas due to intensive management. Also, Truu et al. [116] identified that applying pesticides in Calcaric Regosols, Calcaric Cambisols and Stagnic Luvisols negatively impacts soil microbial biomass and respiration. In an experiment conducted in a sandy loam Albeluvisol, Järvan et al. [117] found that dehydrogenase activity and microbe number were negatively affected by pesticides. Pesticides were also found to be present in beehive matrices [118] and honey [119] located near agricultural fields (oilseed rape). Some (tebuconazole) may affect bee development

[11] https://www.fao.org/3/ad238e/ad238e0d.htm.

[120]. Another important concern in agriculture is the use of sewage sludge and its use as an agricultural compost. For instance, Haiba et al. [121] found that ciprofloxacin, norfloxacin, ofloxacin, sulfadimethoxine and sulfamethoxazole were present in sewage sludge that could be used in agricultural management. These pharmaceuticals inhibit microbe development and reduce compost quality [122]. In Lithuania, a high amount of pesticides is used in agriculture [123]. For instance, Balezentiene [124] found that applying different chemical fertilisers and pesticides reduced plant biodiversity. Finally, Barčauskaitė et al. [125] identified that municipal composts had a high content of PAHs, which could be harmful to soils.

6 Acidification

Soil acidification is a natural process that can be accelerated by some plants or human management [126]. In this process, soluble organic and inorganic acids accumulate faster than they can be neutralised. When these acids are ionised, there is the creation of free H+ in the soil solution [126]. Under acid conditions, several elements, mainly aluminium, are more soluble and negatively impact plant roots [127]. In humid climates, the leaching of basic cations (e.g., calcium, magnesium, potassium or sodium) due to abundant rainfall is a natural cause of soil acidity [128]. Soils developed on acid parent material, such as granite, may also be acid [129, 130]. Other drivers of soil acidity are related to acid rain due to the release of pollutants into the atmosphere (e.g., sulphur dioxide and nitrogen oxides) [131], intensive crop exploitation or grazing [132], root nutrient uptake [133] and the application of mineral fertilisers [134]. Soil acidification is a critical problem due to agriculture intensification in Baltic countries. For instance, in Estonia, Sutri et al. [135] found that the application of some surface residues and nitrogenous fertilisers in a Stagnic Luvisol increased soil acidity. Ivask et al. [136] observed that soil acidification was a cause of the earthworm population decrease. In Latvia, approximately 63% of the agricultural soils are vulnerable to acidification [137]. Latvian agricultural soils tend to acidify [138] and must be limed [139]. Like the other Baltic countries, soil acidification is a severe problem in Lithuania. One of the major constraints to agricultural production is soil acidification. Approximately 16% of the soils are in accelerated acid conditions, and this number is increasing [140]. However, it is recognised that this is a natural process in the territory [141], although previous works identified that management practices increase soil acidification [142, 143].

7 Conclusion

Soil agricultural degradation is a pervasive phenomenon with negative impacts on food production and the environment. After the disintegration of the Soviet Union, the impacts of agriculture decreased in all the Baltic states. Nevertheless, the impacts of agriculture practices are still a concern, especially in soils with a high vulnerability to compaction and acidification. As observed in different works, unsustainable practices can accentuate these natural conditions. Soil water is not a relevant problem when compared to mountainous and semi-arid countries. However, intensive agricultural practices may trigger this problem. Pollution is a paramount concern, mainly because of the use and abuse of agrochemicals that harm soil and water. These pollutants are then transported and accumulated in the Baltic sea, one of the most polluted in the world. Overall, reducing agriculture intensification and establishing more sustainable practices to reduce soil degradation are important.

References

1. FAO and ITPS (2015) Status of the World's soil resources (SWSR) – main report. Food and Agriculture Organization of the United Nations and Intergovernmental Technical Panel on Soils, Rome, Italy
2. Pereira P, Bogunovic I, Munoz-Rojas M, Brevik E (2018) Soil ecosystem services, sustainability, valuation and management. Curr Opin Environ Sci Health 5:7–13
3. Viana C, Freire D, Abrantes P, Rocha J, Pereira P (2022) Agricultural land systems and related research fields supporting food security and the sustainable development goals: evidence from a systematic review. Sci Total Environ 806:150718
4. Maximillian J, Brusseau ML, Glenn EP, Matthias AD (2019) Pollution and environmental perturbations in the global system. In: Brusseau ML, Pepper IL, Gerba CP (eds) Environmental and pollution science, chapter 25. Elsevier, Amsterdam, pp 457–476
5. Kopttke PM, Menzies NW, Wang P, McKenna BA, Lombi E (2019) Soil and the intensification of agriculture for global food security. Environ Int 132:105078
6. Telak LJ, Pereira P, Ferreira CSS, Filipovic V, Filipovic L (2020) Short-term impact of tillage on soil and the hydrological response within a fig (Ficus Carica) orchard in Croatia. Water 12: 3295
7. Xu D, Carswell A, Zhu Q, Zhang F, de Wries W (2020) Modelling long-term impacts of fertilisation and liming on soil acidification at Rothamsted experimental station. Sci Total Environ 713:136249
8. Huang Y, Wang L, Wang W, Li T, He Z, Yang X (2019) Current status of agricultural soil pollution by heavy metals in China: a meta-analysis. Sci Total Environ 651:3034–3042
9. Vanwalleghem T, Gomez JA, Infante Amate J, González de Molina M, Vanderlinden K, Guzmán G, Laguna A, Giráldez JV (2017) Impact of historical land use and soil management change on soil erosion and agricultural sustainability during the anthropocene. Anthropocene 17:13–29
10. Yin C, Zhao W, Pereira P (2022) Soil conservation service underpins sustainable development goals. Glob Ecol Conserv 33:e01974
11. Borrelli P, Van Oost K, Meusburger K, Alewell C, Lugato E, Panagos P (2018) A step towards a holistic assessment of soil degradation in Europe: coupling on-site erosion with sediment transfer and carbon fluxes. Environ Res 161:291–298

12. Ronchi S, Salata S, Arcidiacono A, Montanarella L (2019) Policy instruments for soil protection among the EU member states: a comparative analysis. Land Use Policy 82:763–780
13. Sands RD, Suttles SA (2022) World agricultural baseline scenarios through 2050. Appl Econ Perspect Policy 44:2034–2048
14. Montanarella L, Panagos P (2021) The relevance of sustainable soil management within the European Green Deal. Land Use Policy 82:763–780
15. Von Braun J, Mirzabaev A (2016) Land use change and economics of land degradation in Baltic region. Baltic Region 8:33–44
16. Alaoui A, Diserens E (2018) Mapping soil compaction – a review. Curr Opin Environ Sci Health 5:60–66
17. Rayhan Shaheb M, Venkatesh R, Shearer SA (2021) A review on the effect of soil compaction and its management for sustainable crop production. J Biosyst Eng 46:417–439
18. Colombi T, Keller T (2019) Developing strategies to recover crop productivity after soil compaction – a plant eco-physiological perspective. Soil Tillage Res 191:156–161
19. Sonderegger T, Pfister S (2021) Global assessment of agricultural productivity losses from soil compaction and water erosion. Environ Sci Technol:12162–12171
20. Alskaf K, Mooney SJ, Sparkes P, Wilson P, Sjögersten S (2021) Short-term impacts of different tillage practices and plant residue retention on soil physical properties and greenhouse gas emission. Soil Tillage Res 206:104803
21. Moinfar A, Shahgholi G, Abbaspour Gilandeh Y, Kaveh M, Szymanek M (2022) Investigating the effect of the tractor driving system type on soil compaction using different methods of ANN, ANFIS and step wise regression. Soil Tillage Res 222:105444
22. Reintam E, Penu P, Köster T, Trükmann K, Krebstein K, Kuht J (2010) Soil compaction survey in Estonia. 19th world congress of soil science, soil solutions for a changing world
23. Krebstein K, von Janowsky K, Reintam E, Horn R, Leeduks J, Kuht J (2013) Soil compaction in a Cambisol under grassland in Estonia. Zemdirbyste-Agriculture 100:33–38
24. Kuht J, Reitam E, Edesi L, Nugis E (2012) Influence of subsoil compaction on soil physical properties and on growing conditions of barley. Agron Res 10:329–334
25. Reintam E, Trukmann K, Kuht J, Nugis E, Edesi L, Astover A, Noormets M, Kauer K, Krebstein K, Rannik K (2009) Soil compaction effects on soil bulk density and penetration resistance and growth of spring barley (Hordeum vulgare L.). Acta Agric Scand B Soil Plant Sci 59:265–272
26. Kuht J, Reintam E (2004) Soil compaction effect on soil physical properties and thecontent of nutrients in spring barley (Hordeum vulgare L.) and spring wheat (Triticum aestivum L.). Agron Res 2:187–194
27. Trukmann K, Reintam E, Kuht J, Nugis E, Edesi L (2008) Effect of soil compaction on growth of narrow-leafed lupine, oilseed rape and spring barley on sandy loam soil. Agron Res 6:101–108
28. Ministery of Agriculture of Latvia Republic (2014) Testing the biophysical criteria for Areas with Natural Handicaps. Summary report prepared by Ltd L.U. Consulting (2010) and updated in accordance with criteria definitions and thresholds as agreed on Regulation (EU) No 1305/2013
29. Zabrodskyi A, Šarauskis E, Kukharets S, Juostas A, Vasiliauskas G, Andriušis A (2021) Analysis of the impact of soil compaction on the environment and agricultural economic losses in Lithuania and Ukraine. Sustainability 13:7762
30. Šarauskis E, Romaneckas E, Buragienė S (2009) Impact of conventional and sustainable soil tillage and sowing technologies on physical-mechanical soil properties. Environ Res Eng Manag 3:36–43
31. Šarauskis E, Buragienė S, Romaneckas E, Masilionyte L (2014) Deep, shallow and no-tillage effects on soil compaction parameters. 13th international scientific conference "Engineering for Rural Development". Proceedings, vol 13, May 29–30, pp 31–36

32. Feiziene D, Feiza V, Lazauskas S, Kadziene G, Simanskaite D, Deveikyte I (2007) The influence of soil management on soil properties and yield crop rotation. Zemdirbyste Agric 94:129–145
33. Romaneckas K, Kimbirauskienė R, Sinkevičienė A (2022) Impact of tillage intensity on Planosol bulk density, pore size distribution, and water capacity in Faba Bean cultivation. Agronomy 12:2311
34. Panagos P, Borrelli P, Poesen J, Ballabio C, Lugato E, Meusberger K, Montanarella L, Alewell C (2015) The new assessment of soil loss by water erosion in Europe. Environ Sci Policy 54: 438–447
35. Borrelli P, Robinson DA, Panagos P, Ballabio C (2020) Land use and climate change impacts on global soil erosion by water (2015-2070). Proc Natl Acad Sci U S A 117:21994–22001
36. Nguyen KA, Liou YA (2019) Global mapping of eco-environmental vulnerability from human and nature disturbances. Sci Total Environ 664:995–1004
37. Barbier EB, Hochard JP (2018) Land degradation and poverty. Nat Sustain 1:623–631
38. Joshi JR (2021) Quantifying the impact of cropland wind erosion on air quality: a high-resolution modeling case study of an Arizona dust storm. Atmos Environ 263:118658
39. Broomandi P, Karaca F, Guney M, Fathian A, Geng X, Kim JR (2021) Destinations frequently impacted by dust storms originating from Southwest Iran. Atmos Res 248:105264
40. Panagos P, Ballabio C, Poesen J, Lugato E, Scarpa S, Montanarella L, Borrelli P (2020) A soil erosion indicator for supporting agricultural, environmental and climate policies in the European Union. Remote Sens (Basel) 12:1365
41. Panagos P, Standardi G, Borrelli P, Lugato E, Montanarella L, Bosello F (2018) Cost of agricultural productivity loss due to soil erosion in the European Union: from direct cost evaluation approaches to the use of macroeconomic models. Land Degrad Dev 29:471–484
42. Reintam L, Rooma I, Kull A (2001) Map of soil vulnerability and degradation in Estonia. In: Stott DE, Mohtar RH, Steinhardt GC (eds) Sustaining the global farm. Selected papers from the 10th international soil conservation organization meeting held May 24–29, 1999 at Purdue University and the USDA-ARS National Soil Erosion Research Laboratory, pp 1068–1074
43. Koster T, Penu P, Kikas T (2010) Estimation of soil erosion risk areas by GIS analysis and land use maps: an Estonian case study. Anadolu J Agric Sci 25:58–62
44. Reitam L., Rooma I, Kull A, Kolli R (2005) Soil information and its application in Estonia. European Soil Bureau – Research Report No. 9
45. Kotildelli R, Ellermäe O, Koster T (2010) Erosion-affected soils in the Estonian landscape: humus status, patterns and classification. Arch Agron Soil Sci 56:149–164
46. Lital A, Pachel K, Deelstra J (2008) Monitoring of diffuse pollution from agriculture to support implementation of the WFD and the nitrate directive in Estonia. Environ Sci Policy 11:185–193
47. Laas A, Kull A (2003) Application of GIS for soil erosion and nutrient loss modelling in a Small River catchment. WIT Trans Ecol Environ 67:10
48. Karklins A (1997) Soil degradation status and data availability in Latvia. In: implementation of a soil degradation and vulnerability database for central and Eastern Europe. In: Batjes NH, Bridges EM (eds) Proceedings of an international workshop (Wageningen, 1–3 October 1997), pp 51–53
49. Maximillian J, Brusseau ML, Glenn EP, Matthias AD (2019) Pollution and environmental perturbations in the global system. In: Brusseau ML, Pepper IL, Gerba CP (eds) Environmental and pollution science. Elsevier, Amsterdam, pp 457–476
50. Ministry of the Environment of the Republic of Latvia (2006) National report on the implementation of the United Nations Convention to combat desertification/land degradation (UNNCD)
51. Soms J (2006) Regularities of gully erosion network development and spatial distribution in southeastern Latvia. Baltica 19:72–79
52. Lagzdins A, Jansons V, Sudars R, Abramenko K (2012) Scale issues for assessment of nutrient leaching from agricultural land in Latvia. Hydrol Res 43:383–399

53. Stålnacke P, Grimvall A, Libiseller C, Kokorite I (2003) Trends in nutrient concentrations in Latvian rivers and the response to the dramatic change in agriculture. J Hydrol 283:184–205
54. Jankauskas B, Fullen M (2002) A pedological investigation of soil erosion severity on undulating land in Lithuania. Can J Soil Sci 82:311–321
55. Kinderiene I, Karcauskiene D (2012) Effects of different crop rotations on soil erosion and nutrient losses under natural rainfall conditions in Western Lithuania. Acta Agric Scand B Soil Plant Sci 62:199–205
56. Jankauskas B, Jankauskiene G, Fullen MA (2006) Soil erosion and changes in the physical properties of Lithuanian Eutric Albeluvisols under different land use systems. Acta Agric Scand B Soil Plant Sci 58:66–76
57. Jarasiunas G, Świtoniak M, Kinderiene I (2020) Dynamics of slope processes under changing land use conditions in young morainic landscapes, Western Lithuania. Int Agrophys 34:43–55
58. Rodríguez-Eugenio N, McLaughlin M, Pennock D (2018) Soil pollution: a hidden reality. FAO, Rome. 142 p
59. Eckberg K (1994) Environmental problems and policy options in the Baltic states: learning from the west? Environ Polit 3:445–478
60. Idzelis A (1979) Response of soviet Lithuania to environmental problems in the coastal zone. J Baltic Stud 10:299–308
61. Kern K (2011) Governance for sustainable development in the Baltic Sea region. Journal of Baltic Studies 42:21–35
62. Astover A, Roostalu H, Lauringson E, Lemetti I, Selge A, Talgre L, Vasilev N, Motildette M, Totilderra T, Penu P (2006) Changes in agricultural land use and in plant nutrient balances of arable soils in Estonia. Arch Agron Soil Sci 52:223–231
63. Salay J, Fenhann J, Jaanimiigi K (1993) Energy and environment in the Baltic States. Annu Rev Energy Environ 18:169–216
64. Raukas A (2004) Past pollution and its remediation in Estonia. Baltica 17:71–78
65. OECD (2019) Agriculture and water policies: main characteristics and evolution from 2009 to 2019. Estonia. OECD Publishing, Paris
66. OECD (2019) Agriculture and water policies: main characteristics and evolution from 2009 to 2019. Lithuania. OECD Publishing, Paris
67. OECD (2019) Agriculture and water policies: Main characteristics and evolution from 2009 to 2019. Latvia. OECD Publishing, Paris
68. Naccarato A, Tassone A, Cavaliere F, Elliani R, Pirrone N, Sprovieri F, Tagarelli A, Giglio A (2020) Agrochemical treatments as a source of heavy metals and rare earth elements in agricultural soils and bioaccumulation in ground beetles. Sci Total Environ 749:141438
69. Singh S, Kumar V, Kaur Sidhu G, Datta S, Singh Dhanjal D, Koul B, Singh Janeja H, Singh J (2019) Plant growth promoting rhizobacteria from heavy metal contaminated soil promote growth attributes of Pisum sativum L. Biocatal Agric Biotechnol 17:665–671
70. Agoro MA, Adeniji AO, Adefisoye MA, Okoh OO (2020) Heavy metals in wastewater and sewage sludge from selected municipal treatment plants in eastern Cape Province, South Africa. Water 12:2746
71. Liu WR, Zeng D, She L, Su WX, He DC, Wu GY, Ma XR, Jiang S, Jiang CH, Ying GG (2020) Comparisons of pollution characteristics, emission situations, and mass loads for heavy metals in the manures of different livestock and poultry in China. Sci Total Environ 734:139023
72. Jambhulkar HP, Montaha S, Shaikh S, Suresh Kumar M (2018) Fly ash toxicity, emerging issues and possible implications for its exploitation in agriculture; Indian scenario: a review. Chemosphere 213:333–344
73. De Vries W, Römkens PFAM, Kros J, Voogd JC, Schulte-Uebbing LF (2022) Impacts of nutrients and heavy metals in European agriculture. Current and critical inputs in relation to air, soil and water quality, ETC-DI, p 72
74. Reintam L, Karblane H, Petersell V (1997) Short report on SOVEUR status in Estonia. In: Batjes NH, Bridges EM (eds) Implementation of a soil degradation and vulnerability database

for central and Eastern Europe (SOVEUR project). Proceedings of an international workshop (Wageningen, 1–3 October 1997). FAO, pp 39–41
75. Lepane V, Varvas M, Viitak A, Alliksaar T, Heinsalu A (2007) Sedimentary record of heavy metals in Lake rouge Liinjarv, southern Estonia. Estonian J Earth Sci 56:221–232
76. Roots O, Roose A (2013) Hazardous substances in the aquatic environment of Estonia. Chemosphere 93:196–200
77. Klavins M, Briede A, Rodinov V, Kokorite I, Parele E, Klavina I (2000) Heavy metals in rivers of Latvia. Sci Total Environ 262:175–183
78. Laplante B, Smits K (1993) Estimating industrial pollution in Latvia. ECSSD Rural Development and Environment Sector Working Paper No. 4. p 51
79. Vincevica-Gaile Z, Klavins M, Rudovica V, Viksna A (2013) Research review trends of food analysis in Latvia: major and trace element content. Environ Geochem Health 35:693–703
80. Marcinkonis S (2008) Assessing trace element accumulation and depletion in agricultural soils in Lithuania. Acta Agric Scand B Soil Plant Sci 58:114–123
81. Petraitis E (2007) Research into heavy metal concentrations in agricultural soils. Ekologija 53: 64–69
82. Gregorauskiene V, Kadunas V (2006) Vertical distribution patterns of trace and major elements within soil profile in Lithuania. Geol Q 50:229–237
83. Valskys V, Motiejunas M, Ignatavicius G, Sinkevicius S (2016) The contamination of Sventoji River bottom sediments by heavy metals in Ukmerge, Lithuania. J Environ Sci Int 25:1–10
84. Remeikaitė-Nikienė N, Garnaga-Budrė G, Lujanienė G, Jokšas K, Stankevičius A, Malejevas V, Barisevičiūtė R (2018) Distribution of metals and extent of contamination in sediments from the southeastern Baltic Sea (Lithuanian zone). Oceanologia 60:193–206
85. Yanai J, Linehan DJ, Robinson D, Young IM, Hackett CA, Kyuma K, Kosaki T (1996) Effects of inorganic nitrogen application on the dynamics of the soil solution composition in the root zone of maise. Plant and Soil 180:1–9
86. Blanco-Canqui H, Schlegel AJ (2013) Implications of inorganic fertilization of irrigated corn on soil properties: lessons learned after 50 years. J Environ Qual 42:861–871
87. Kavamura VN, Hayat R, Clark IM, Rossmann M, Mendes R, Hirsch PR, Mauchline TH (2018) Inorganic nitrogen application affects both taxonomical and predicted functional structure of wheat rhizosphere bacterial communities. Front Microbiol 9:1074
88. Zhu WX, Hope D, Gries C, Grimm NB (2006) Soil characteristics and the accumulation of inorganic nitrogen in an arid urban ecosystem. Ecosystems 9:711–724
89. Camargo JA, Alonso A (2006) Ecological and toxicological effects of inorganic nitrogen pollution in aquatic ecosystems: a global assessment. Environ Int 32:831–849
90. Jansson T, Andersen HE, Gustafsson BG, Hasler B, Höglind L, Choi H (2019) Baltic Sea eutrophication status is not improved by the first pillar of the European Union Common Agricultural Policy. Reg Environ Chang 19:2465–2476
91. Nõges T, Janatian N, Laugaste R, Nõges P (2020) Post-soviet changes in nitrogen and phosphorus stoichiometry in two large non-stratified lakes and the impact on phytoplankton. Glob Ecol Conserv 24:e01369
92. Stålnacke P, Vandsemb SM, Vassiljev A, Grimvall A, Jolankai G (2004) Changes in nutrient levels in some eastern European rivers in response to large-scale changes in agriculture. Water Sci Technol Water Supply 79:29–36
93. Stålnacke P, Andreas Aakerøy P, Blicher-Mathiesen G, Iital A, Jansons V, Koskiaho J, Kyllmar K, Lagzdins A, Pengerud A, Povilaitis A (2014) Temporal trends in nitrogen concentrations and losses from agricultural catchments in the Nordic and Baltic countries. Agric Ecosyst Environ 198:94–103
94. Mander U, Oja T, Jarvet A (1994) Groundwater quality and eco-engineering measures in Selja river catchment, Estonia. Future groundwater resources at risk (proceedings of the Helsinki conference, June 1994). IAHSPubl. no. 222, pp 157–167
95. Kangur K, Mols T (2007) Changes in spatial distribution of phosphorus and nitrogen in the large north-temperate lowland Lake Peipsi (Estonia/Russia). Hydrobiologia 599:31–39

96. Buhvestova O, Kangur K, Haldna M, Mols T (2010) Nitrogen and phosphorus in Estonian riversdischarging into Lake Peipsi: estimation of loadsand seasonal and spatial distributionof concentrations. Estonian J Ecol 60:18–38
97. Nõges P, Kägu M, Nõges T (2007) Role of climate and agricultural practice in determining matter discharge into large, shallow Lake Võrtsjärv, Estonia. In: Qin B, Liu Z, Havens K (eds) Eutrophication of Shallow Lakes with special reference to Lake Taihu, China. Developments in hydrobiology, vol 194. Springer, Dordrecht, pp 125–134
98. Jansons V, Lagzdins A, Berzina L, Sudars R, Abramenko K (2011) Temporal and spatial variation of nutrient leaching from agricultural land in Latvia: long term Trendsin retention and nutrient loss in a drainage and small catchment scale. Environ Clim Technol 7:54–65
99. Siksnane I, Lagzdins A (2020) Temporal trends in nitrogen concentrations and losses from agricultural monitoring sites in Latvia. Environ Clim Technol 24:163–173
100. Jansons V, Vagstad N, Sudars R, Deelstra J, Dzalbe I, Kirsteina D (2011) Nutrient losses from point and diffuse agricultural sources in Latvia. Landbauforschung Völkenrode 1:9–17
101. Siksnane I, Lagzdins A (2017) Assessment of economic losses associated with nitrogen leaching in agriculture. Proceedings of the 8th international scientific conference rural development 2017
102. Povilaitis A (2004) Phosphorus trends in Lithuanian Rivers affected by agricultural non-point pollution. Environ Res Eng Manag 4:17–27
103. Tumas R (2000) Evaluation and prediction of nonpoint pollution in Lithuania. Ecol Eng 14: 443–451
104. Aksomaitienė R, Berankienė L (2003) Agricultural pollution in the river Susve basin in middle Lithuania. Water Manag Eng 23:14–22
105. Česonienė L, Šileikienė D, Marozas V, Čiteikė L (2021) Influence of anthropogenic loads on surface water status: a case study in Lithuania. Sustainability 13:4341
106. Sileika AS, Kutra S, Berankiene L (2002) Phosphate run-off in the Nevezis River (Lithuania). Environ Monit Assess 78:153–167
107. Nkontcheu Kenko DB, Tchamadeu Ngameni N, Nkontcheu Kamta P (2022) Environmental assessment of the influence of pesticides on non-target arthropods using PRIMET, a pesticide hazard model, in the Tiko municipality, Southwest Cameroon. Chemosphere 308:136578
108. Islam R, Kumar S, Karmoker J, Kamruzzaman M, Aminur Rahman M, Biswas N, Anh Tran TK, Mahmudur Rahman M (2018) Bioaccumulation and adverse effects of persistent organic pollutants (POPs) on ecosystems and human exposure: a review study on Bangladesh perspectives. Environ Technol Innov 12:115–131
109. Gros M, Mas-Pla J, Boy-Roura M, Geli I, Domingo F, Petrovic M (2019) Veterinary pharmaceuticals and antibiotics in manure and slurry and their fate in amended agricultural soils: findings from an experimental field site (Baix Empordà, NE Catalonia). Sci Total Environ 654:1337–1349
110. Ding L, Zhang S, Wang X, Yang X, Zhang C, Qi Y, Guo X (2020) The occurrence and distribution characteristics of microplastics in the agricultural soils of Shaanxi Province, in north-western China. Sci Total Environ 720:137525
111. Chen Y, Wang C, Wang Z (2005) Residues and source identification of persistent organic pollutants in farmland soils irrigated by effluents from biological treatment plants. Environ Int 31:778–783
113. Wang J, Liu X, Li Y, Powell T, Wang X, Wang G, Zhang P (2019) Microplastics as contaminants in the soil environment: a mini-review. Sci Total Environ 691:848–857
113. Gaur N, Narasimhulu K, Y, P. (2018) Recent advances in the bio-remediation of persistent organic pollutants and its effect on environment. J Clean Prod 189:1602–1631
114. Tuomisto J, Hagmar L (1999) Environmental health in the east Baltic region – pesticides and persistent organic compounds. Scand J Work Environ Health 3:65–71
115. Kumar KS, Priya M, Sajwan KS, Kolli R, Roots O (2009) Residues of persistent organic pollutants in Estonian soils (1964-2006). Estonian J Earth Sci 58:109–123

116. Truu M, Truu J, Ivask M (2008) Soil microbiological and biochemical properties for assessing the effect of agricultural management practices in Estonian cultivated soils. Eur J Soil Biol 44: 231–237
117. Järvan M, Edesi L, Adamson A, Võsa T (2014) Soil microbial communities and dehydrogenase activity depending on farming systems. Plant Soil Environ 60:459–463
118. Raimets R, Bontšutšnaja A, Bartkevics V, Pugajeva I, Kaart T, Puusepp L, Pihlik P, Keres I, Viinalass H, Mand M, Karise R (2020) Pesticide residues in beehive matrices are dependent on collection time and matrix type but independent of proportion of foraged oilseed rape and agricultural land in foraging territory. Chemosphere 238:124555
119. Karise R, Raimets R, Bartkevics V, Pugajeva I, Pihlik P, Keres I, Williams IH, Viinalass H, Mänd M (2017) Are pesticide residues in honey related to oilseed rape treatments? Chemosphere 188:389–396
120. Raimets R, Naudi S, Mänd M, Bartkevics V, Smagge G, Karise R (2022) Translocation of tebuconazole between bee matrices and its potential threat on honey bee (Apis mellifera Linnaeus) Queens. Insects 13:45
121. Haiba E, Nei L, Ivask M, Peda J, Järvis J, Lillenberg M, Kipper K, Herodes K (2016) Sewage sludge composting and fate of pharmaceutical residues – recent studies in Estonia. Agron Res 14:1583–1600
122. Nei L, Haiba E, Kutti S, Kipper K, Herodes K, Lillenberg M (2014) Sewage sludge compost, microbial activity and pharmaceuticals. Global J Adv Pure Appl Sci 3:30–37
123. Buivydaite VV (2005) Soil survey and available soil data in Lithuania. European Soil Bureau – Research Report No. 9, pp 211–223
124. Balezentiene L (2011) Alpha-diversity of differently managed agro-ecosystems assessed at a habitat scale. Pol J Environ Stud 20:1387–1394
125. Barčauskaitė K, Žydelis R, Mažeika R (2020) Screening of chemical composition and risk index of different origin composts produced in Lithuania. Environ Sci Pollut Res 27:24480–24494
126. Freedman B (1995) Acidification. In: Freedman (ed) Environmental ecology. The ecological effects of pollution, disturbance, and other stresses2nd edn. Elsevier, Amsterdam, pp 94–43
127. Prescott CE, Katzensteiner K, Weston C (2021) Soils and restoration of forested landscapes. In: Stanturf JA, Callaham Jr MA (eds) Soils and landscape restoration. Elsevier, Amsterdam, pp 299–331
128. Gaiser T, Stahr K (2013) Soil organic carbon, soil formation and soil fertility. In: Lal R, Lorenz K, Hüttl R, Schneider B, von Braun J (eds) Ecosystem services and carbon sequestration in the biosphere. Springer, Dordrecht, pp 407–418
129. Sun T, Deng L, Fei K, Zhang L, Fan X (2020) Characteristics of phosphorus adsorption and desorption in erosive weathered granite area and effects of soil properties. Environ Sci Pollut Res 27:28780–28793
130. Posselt Martins A, Andrade Costa SEVG, Anghinoni I, Robinson Kunrath T, Balerini F, Cecagno D, De Faccio Carvalho PC (2014) Soil acidification and basic cation use efficiency in an integrated no-till crop–livestock system under different grazing intensities. Agric Ecosyst Environ 195:18–28
131. Grennfelt P, Engleryd A, Forsius M, Hov O, Rodhe H, Cowling E (2020) Acid rain and air pollution: 50 years of progress in environmental science and policy. Ambio 49:849–864
132. Hao T, Zhu Q, Zeng M, Shen J, Shi X, Liu X, Zhang F, de Vries W (2019) Quantification of the contribution of nitrogen fertilisation and crop harvesting to soil acidification in a wheat-maise double cropping system. Plant and Soil 434:167–184
133. Goulding KWT (2016) Soil acidification and the importance of liming agricultural soils with particular reference to the United Kingdom. Soil Use Manage 32:390–399
134. Raza S, Miao N, Wang P, Ju X, Chen Z, Zhou J, Kuzyakov Y (2020) Dramatic loss of inorganic carbon by nitrogen-induced soil acidification in Chinese croplands. Glob Chang Biol 15:3738–3751

135. Sutri M, Shanskiy M, Ivask M, Reitam E (2022) The assessment of soil quality in contrasting land-use and tillage systems on farm fields with Stagnic Luvisol soil in Estonia. Agriculture 12:2149
136. Ivask M, Kuu A, Sizov E (2007) Abundance of earthworm species in Estonian arable soils. Eur J Soil Biol 43:S39–S42
137. Karklin A, Livmanis J, Nikodemus O (1999) Mapping of soil and terrain vulnerability in Latvia – project implementation. In: Batjes NH (ed) Soil degradation status and vulnerability assessment for central and Eastern Europe – preliminary results of the SOVEUR project. FAO, pp 57–60
138. Makutėnienė D, Perkumiene D, Makutėnas V (2022) Logarithmic mean Divisia index decomposition based on Kaya identity of GHG emissions from agricultural sector in Baltic States. Energies 15:1195
139. Kreišmane D, Naglis-Liepa K, Popluga D, Lēnerts A, Rivža P (2016) Liming effect on nitrogen use efficiency and nitrogen oxide emissions in crop farming. Res Rural Dev 1:30–36
140. Buivydaite VV (1999) Soil degradation, contamination and pollution status in Lithuania. In: Batjes NH (ed) Soil degradation status and vulnerability assessment for central and Eastern Europe – preliminary results of the SOVEUR project. FAO. pp 61–68
141. Eidukevičienė M, Ožeraitienė D, Tripolskaja L, Volungevičius J (2007) Change of soil pH in the territory of Lithuania: spatial and temporal analysis. ŽEMĖS ŪKIO MOKSLAI 3:1–8
142. Marcinkonis S, Booth CA, Fullen MA, Tripolskaja L (2011) Soil acidity indices in East Lithuania. Commun Soil Sci Plant Anal 42:1565–1580
143. Tripolskaja L, Kazlauskaite-Jadzevice A, Baksiene E, Razukas A (2022) Changes in organic carbon in mineral topsoil of a formerly cultivated Arenosol under different land uses in Lithuania. Agriculture 12:488

Agricultural Soil Degradation in Germany

Manuel Seeger

Contents

1 Introduction 88
2 Soil Compaction and Erosion (Water and Wind) 89
2.1 Soil Erosion by Water 89
2.2 Soil Erosion by Wind 93
3 Soil Compaction 95
4 Soil Pollution 96
5 Microplastics 97
6 Conclusions 98
References 99

Abstract Germany is a densely populated and highly developed country with multiple threats on soils still causing their degradation. Soil erosion by wind and water has been the most important process in Germany, which has been studied since the end of the nineteenth century. Soil erosion starts with the Neolithic Revolution. But erosion rates were in general low, even during Roman times. They increased during medieval times due to the strong expansion of agriculture and deforestation. Nowadays, at least 19% of Germany's agricultural land is affected by very high soil erosion, which reaches values higher than tolerable. Intensification of agriculture and the use of heavy machinery have led to this substantial increase in the most evident and widespread soil degradation process. Soil erosion by wind is mainly found in northern Germany and is the result of the interaction of flat topography, sandy to loamy soils, and large agricultural fields.

In modern times additional threats contribute to soil degradation. Pollution by pesticides or heavy metals is ubiquitous and endangers soil health and agricultural land use. Microplastics are reaching the soils by multiple pathways. The main problems here are the lack of knowledge on a methodology for quantification and

M. Seeger (✉)
Physical Geography, University of Trier, Trier, Germany
e-mail: seeger@uni-trier.de

Paulo Pereira, Miriam Muñoz-Rojas, Igor Bogunovic, and Wenwu Zhao (eds.),
Impact of Agriculture on Soil Degradation II: A European Perspective,
Hdb Env Chem (2023) 121: 87–104, DOI 10.1007/698_2022_948,

Published online: 29 January 2023

on the effect of microplastics in soils. Additionally, one of the major threats to German soils is the destruction by sealing of settlements and infrastructure. Despite having knowledge of soil, the different threats on soils, the pressure on them, and the dynamics of degradation are still high on Germany's agricultural soils.

Keywords Microplastics, Soil erosion, Soil sealing

1 Introduction

Despite being a highly industrialized land, more than half (52.3%) of Germany's surface is agricultural land, 30.2% under forest use, and 13.4% is covered by settlements and traffic infrastructure [1]. But there is an intense conflict of interest in these activities in how to use land and the functions of soil [2, 3]. As a consequence, there is a rapid trend to an increase in sealed soil surface (14% on 31.12.2020) in detriment of agricultural land (50.6%) and, on a minor way, forests (29.8%) [4]. There has been a substantial reduction in the daily land consumption between the year 2000 (129 ha day^{-1}) and 2019 (52 ha day^{-1}) [5], but the target has been missed to reduce the soil sealing to 30 ha a^{-1}. Data shows that especially the increase rate of industrial and commercial spaces has been lowered substantially [6], but the author emphasizes the fact that the data is only of weak quality, due to an ongoing change in the methodology of data collection. Nevertheless, it is interesting to see that in the last years land consumption has increased substantially (up to 12% of the overall land cover) due to the implementation of renewable energy production [6]. As these include biomass production, areas covered for photovoltaic sources, and wind farms the effect on soil quality and the degradation processes will be highly variable and different from the uses known until now and are an emerging subject of research. However, the dimension of the changes in soil use shows that sealing is one of the biggest threats on soils in Germany. Assessing the degradation intensity of soils needs a well-established and comprehensive system for evaluating the quality status of soil. The estimation of soil quality is regulated in Germany by the "Ackerzahl," which was established first in 1934 for tax purposes (revised last as [7]), based on the methodology proposed by [8]. This is used, e.g., for evaluation of the soil degradation by erosion (see below).

Anyhow, awareness and research on soil degradation status and processes have a long tradition in Europe and Germany [9, 10, see, e.g., 11]. As we will see in the following chapters, soil degradation is highly related to soil management practices. The intensification of agricultural practices has led to a substantial increase in soil erosion and degradation in Germany [12] is any more an issue in Germany [13]. However, the effects of the actual changes in soil management are still not known in the long term. We will address here most of the common agricultural soil degradation processes in Germany.

2 Soil Compaction and Erosion (Water and Wind)

2.1 Soil Erosion by Water

Physical degradation of soil by compaction and especially erosion was identified early as a major threat to soil resources, and thus, has been the focus of international and German research. Especially soil erosion has been in the focus of research, leading to early empirical and long-lasting soil erosion measurements [9, 14–16]. An exhaustive review of many of the plot measurements performed in Germany can be found in [17].

Regarding soil degradation by soil erosion by water, the maximum tolerable amount has to be established. For sustainable soil conservation, erosion rates should not exceed soil formation rates. These equilibrium rates have been estimated as ranging between 0.3 and 1.4 t ha^{-1} a^{-1} [18] for Europe, or 0.5 and 8.8 t ha^{-1} a^{-1} for Alpine grasslands [19], depending on soil age, type, and stability. However, limiting soil erosion to these values is unsuitable for agricultural land, as any kind of removal of vegetation or soil management leads generally to a loss of protective cover and reduction of aggregate stability, infiltration capacity, and other soil characteristics, consequently to soil erosion rates higher than the values mentioned above. Therefore, 20 proposed to relate tolerable soil erosion rates to the soil quality index, assuming a sustainability of the soils' agricultural productivity should not be substantially reduced within the upcoming 300–500 years. This would lead to a maximum tolerable erosion rate of around 10 t ha^{-1} a^{-1} on soils with very high soil quality identified in Germany. But to estimate locally the tolerable erosion rate, this is then inverse linear correlated to the German soil quality index [7, 20]. Therefore, the risk of soil degradation is not only depending on erosion intensity, but also on initial soil quality, especially the possible rooting depth.

As soil degradation is in general a slow process directly linked to human activity, a historical view on soil erosion is extraordinarily important, as it is the most relevant degradation process. There are several investigations on historical and even older soil erosion rates in Germany [12, 21–23]. A comprehensive reconstruction of soil erosion rates in Germany throughout the Holocene was compiled by analyzing lake and alluvial sediments as well as slope deposits [24]. It is clear that increased soil erosion became evident with the start of the Neolithic [25] but kept constant with very low rates <1 t ha^{-1} a^{-1}. Afterward, in the transition to the Bronze Age (and later to the Iron Age), soil erosion rates started to oscillate substantially between 0.4 and 7 t ha^{-1} a^{-1}, but with a clear trend to increase Contrasting with the intense land use during Roman times, the erosion intensities were only very low then

(0.4–0.7 t ha^{-1} a^{-1}). Extreme high erosion rates (10–19 t ha^{-1} a^{-1}) as reported by 26 are locally limited and related to scarce gully erosion events. During the transition to the Middle Ages, the amount of sheet erosion decreased even further. However, at the end of the Medieval Times, they increase both, sheet and gully erosion. The first one ranged between 3.3 and 7 t ha^{-1} a^{-1}, and the gully erosion up to more than 220 t ha^{-1} a^{-1}. The high level of erosion (2.5–5.5 t ha^{-1} a^{-1}) was still active since the beginning of modern times. This shows that soil degradation by water erosion has been an important and still ongoing issue and that human impact is the cause of most of it. Anyhow, the impact of climate patterns cannot be denied, as it is responsible for the large gullying events identified and reported during climate transition periods [26, 27]. From this very short analysis, we can see that soil erosion has been an important process in forming landscapes in Germany. However, soil erosion is still a present-day problem.

The Soil Atlas of Germany [28] focuses an extensive chapter on actual soil threats. The maps are also accessible online [29]. Here, potential susceptibility for soil erosion has been defined for Germany by applying the ABAG [30], which is the adapted version of USLE [20] for Germany. Soil erosion risk of agricultural land is classified into seven classes, ranging from extremely low (≤1 t ha^{-1} a^{-1}) to extremely high (≥55 t ha^{-1} a^{-1}). Approximately 9.5% of the total surface of Germany is classified as at least highly endangered (≥15 t ha^{-1} a^{-1}) by soil erosion, which corresponds to approximately 19% of the agricultural surface. On the other hand, low soil erosion risk (<10 t ha^{-1} a^{-1}) affects 24% of Germany's surface (respectively 48% of the agricultural surface). Older estimations of soil erosion reach 8 t ha^{-1} a^{-1} as a long-term average for Germany but emphasizing the large regional variability [31, 32]. The results showed that a closer look at these regional disparities and process intensities is needed.

The facts presented until now are the result of modeling or evaluation of sediment records. Direct measurement of soil erosion is difficult due to the high spatial variability of soil erosion, but also due to the erratic relation of magnitude and frequency of erosion events [17, 33]. Fiener et al. [17] evaluated a large amount of plot measurements in Germany and corrected the measurements to obtain standardized values of soil erosion. These reached an average of 15.2 t ha^{-1} a^{-1} on arable land, whereas it increases up to 88.6 t ha^{-1} a^{-1} for row crops. But the authors criticize the experimental setup of the soil erosion plots, as well as the standardization procedure of the measurements, indicating that both assume higher slope gradients than can be found in average throughout Germany. In consequence, the authors estimate an average soil loss of 2.7 t ha^{-1} a^{-1} when correcting the data to real slope gradients and the contribution of the different land use systems to overall erosion in Germany.

The map of Germany of soil erosion (by water) on agricultural land (Fig. 1) shows all over northern Germany low erosion rates. This is due to the low slopes predominant there. Moderate to high soil erosion rates are modeled by the ABAG [28]

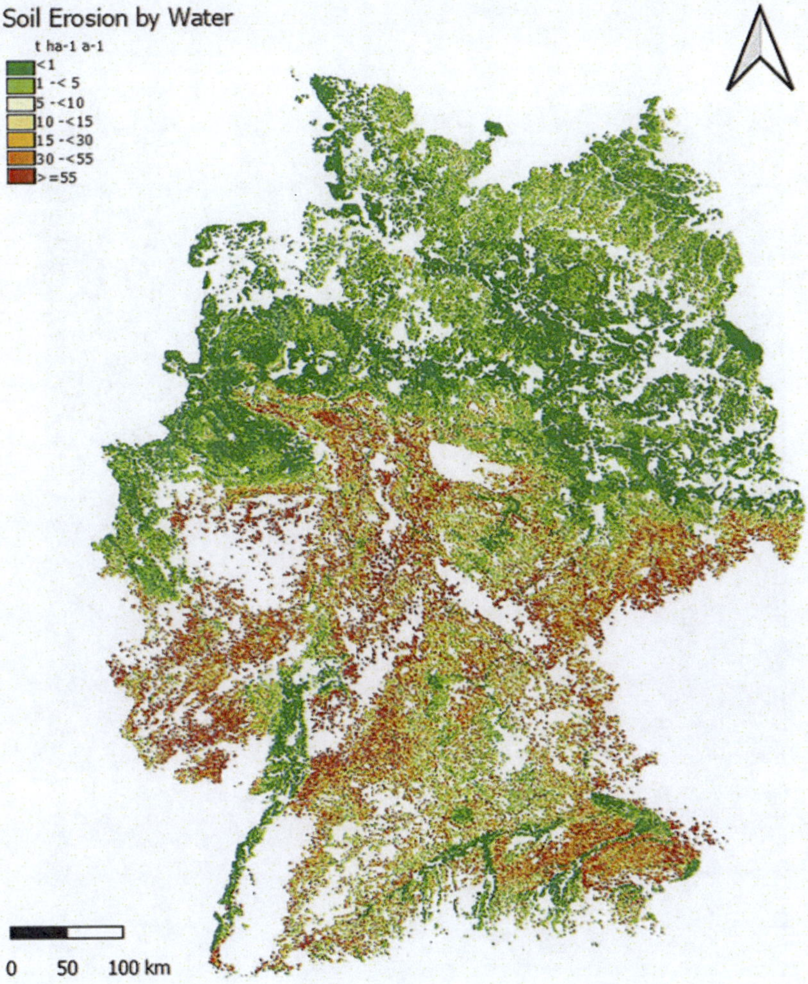

Fig. 1 Soil erosion by water as modeled by ABAG [29]. Map generated by original WGS data source. Source: BGR Bodenatlas Deutschland

throughout the mid-mountain ranges. The highest values are related to the intensive agricultural areas established on the rolling hills of loess deposition landscapes. These are also the areas with the soils of the highest quality, so the on-site damage is relevant for agriculture.

Besides this, the presented data, and new measurements and evaluation of pre-existing studies show that a more differentiated look is needed. Bug and Mosimann [34] showed that in 11 years of observations, rill erosion dominated the processes on agricultural soils of Lower Saxony, and thus, the corrections applied by [17] to the plot measurements are not valid for all the area of Germany's agricultural

Fig. 2 Soil erosion on a conventional vineyard in the Mosel Area after a single rainfall with ~90 mm in 8 h (1.5.2018). Rills develop on compacted wheel tracks with very low vegetation cover. The lanes with vegetation cover remained considerably more stable

soils. The long-lasting measurement period on the Mosel vineyards [14–16, 35–37] showed clearly that almost all overall soil erosion within 25 years was caused by very few events. Nevertheless, the extensive study leads to the recommendation and the nowadays widespread implementation of the greening of the driving lanes in steep sloping vineyards [16, 38], which is crucial for the prevention of erosion (Fig. 2) as well as for nutrient management. Conventional wine growers usually keep green cover on alternating lanes, while organic wine growers usually keep all lanes vegetated. Here we can see how the results of empirical studies on soil degradation processes may lead to changes in traditional soil management strategies.

Recent studies showed the effect of soil and vineyard management on soil erosion [39–44]. Organic vineyard management leads to a substantial reduction of runoff generation and soil erosion rates, even during high-intensity precipitation events. A combination of several positive effects could be identified as the cause of it: generally increased earthworm activity in the topsoil as well as in the subsoil, higher accumulation of organic matter and the consequent increase of both, infiltration capacity and aggregate stability. All these indicators show an increase in soil health status, and thus, soils less susceptible to degradation. However, it is also evident that land use changes have also a high impact on erosion intensity. Climate change and economic pressure are forcing different changes on agricultural lands, but especially on steep-sloping vineyards. Seeger et al. [39] showed with rainfall simulation experiments that the extraction of the grapevines, as it is needed for re-plantation, or required for abandonment, leads to the highest erosion rates. They observed also that the erodibility experiences a considerable reduction within 5 years. Therefore, the intensified dynamics of changes in cultivation, abandonment, and re-activation of vineyards demand the development of new management strategies. Diversification shows here promising results, even on steep vineyards with only little-developed soils [43, 44].

2.2 *Soil Erosion by Wind*

As soil degradation by wind erosion is related in general to semi-arid or even dryer landscapes, this process has not been the focus of soil erosion research in Germany. However, wind erosion is not considered to be the cause of severe soil degradation in Europe [45]. Also in Germany, wind erosion is not considered a big problem when compared with soil degradation by water [46]. Only when the off-site damage is obvious, like a large accident with numerous fatalities [47, 48], the concern grows. It affects only 15% of the entire agricultural surface endangered by soil erosion [49]. The soil atlas of Germany published a map of soil erosion risk by wind erosion (Fig. 3). Here we observe, contrasting with the previous map of soil erosion by water a considerably higher risk in northern Germany. The main causes are the relatively flat topography, a high proportion of sandy soils from Pleistocene glacial till and fluvioglacial sediments and, especially in eastern Germany, large agricultural fields. Consequently, measurements of wind erosion are still scarce [50]. They have shown the effectiveness of single wind erosion events, reaching soil losses >100 t ha^{-1}. Recently, monitoring sites for soil development are analyzed for understanding wind erosion in Germany, and for generating the needed database for correct model application [51].

Some studies point toward soil management as a large source of dust emissions in Germany [52, 53]. Quantification of dust transport, as well as its seasonality, indicate that dust emissions, this is Aeolian soil erosion, is likely to be up to 6.6 times higher than the one caused by erosive wind events [52]. Recent studies show even a mobilization higher by at least three orders of magnitude [53]. Therefore,

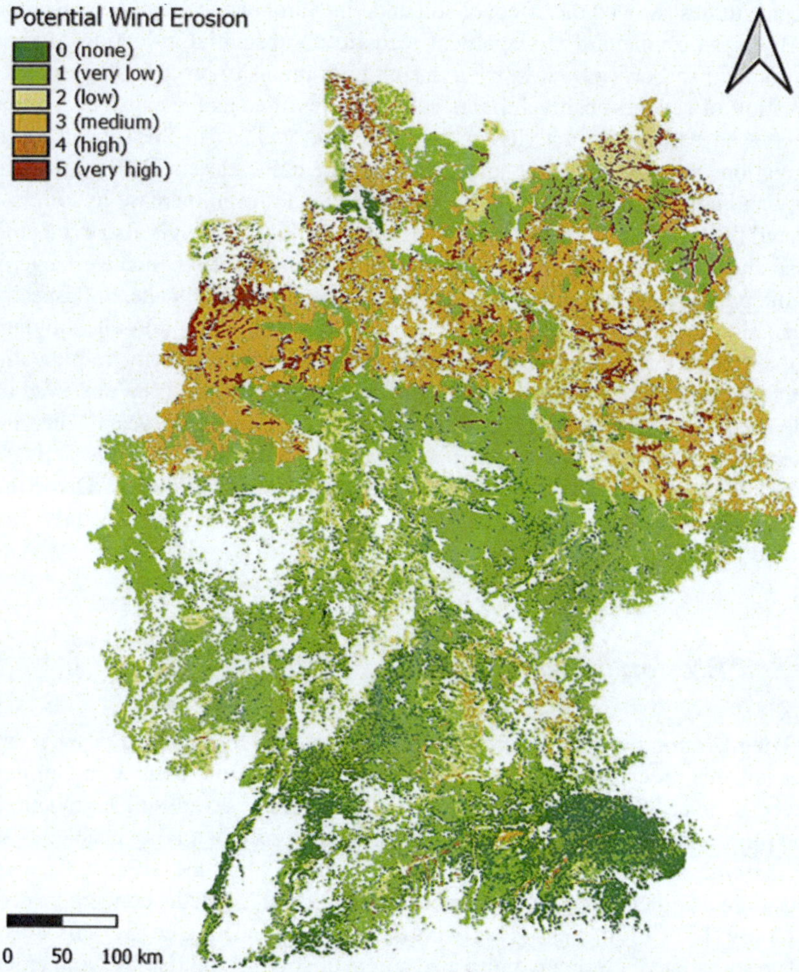

Fig. 3 Map of potential soil erosion by wind in Germany [29]. Map generated from original data source. Source: BGR Bodenatlas Deutschland

conservation agriculture is proposed as a methodology to apply for the reduction of soil degradation by wind (and water) erosion [49]. This agricultural practice summarizes all known methods of reduced tillage, like strip tillage, mulch, and direct seeding, etc.

Within the last decades, research on the combined effect of rainfall with wind has emerged [54–57]. They showed that for a proper prediction of soil erosion, and the resulting soil degradation, the combined effect of wind and rainfall needs to be taken into account. The effect is especially high on sandy soils, where short-range detachment and transport of soil particles is increased by up to one order of magnitude.

3 Soil Compaction

Physical degradation of soils is especially an issue of concern in Europe [58] due to the high level of mechanization of agriculture. The newest surveys showed a high awareness at the farmer's level on soil degradation by compaction and destruction of the soil's physical structure [59]. The soil atlas of Germany (Fig. 4) shows a predominance of moderate to high bulk densities (>1.6 g cm^{-3}) of soils on vast areas of the low mountain ranges. Also, large parts of the highly productive agricultural areas in the floodplains of the big streams (Rhine, Elbe), as well as the loess areas are characterized by these elevated bulk densities. Besides lithological reasons for elevated bulk densities, three mention soil compaction as one of the most relevant degradation types related to conventional soil management. Moreover, it has become an increasing threat of modern agriculture on soil due to the increasing size and weight of agricultural machinery, which has increased the overall economic efficiency on one hand but adds big loads on the soil. As with most of the degradation processes, its intensity and typology depend on the land use type: on agricultural land, compaction occurs mainly directly below the plowing depth. Sugar beet and maize are most vulnerable to compaction due to high frequency of driving on the fields, the increased weight of machinery, and the intense topsoil disturbance [60–63]. However, soil compaction is not only limited to agriculture or grassland. Modern forest management with heavy machinery has also been identified as the cause for soil compaction in forests [64–66], in general with different effects on soil functions and vegetation structure [67]. Recent works have also shown the impact of soil compaction on other soil degradation processes, mainly soil erosion [68]. The authors reveal the fundamental impact of highly compacted tramline structures on agricultural fields on soil erosion intensity.

The susceptibility of soils to compaction depends not only on the wheel load of the machinery and the tire pressure [69], but is highly dependent on the actual soil conditions, like moisture content at the time of overpassing, soil texture, and structural stability. This means that a precise planning of the work intensity may help to reduce soil compaction in agricultural fields. Therefore, different tools have been developed to help farmers to identify times with a lower risk of compaction [70], depending on their own needs and applied machinery. In addition, accurate knowledge on the spatial distribution and frequency of machinery passes on individual fields [71] adds valuable information to reduce the impact of heavy-load machinery on soils.

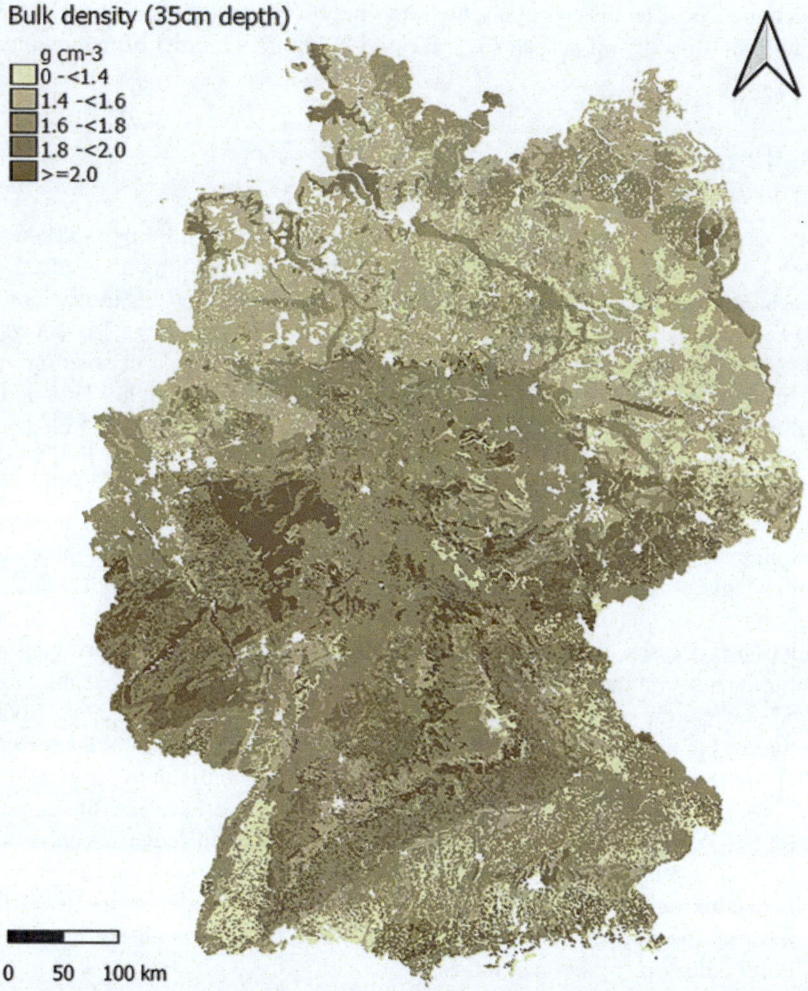

Fig. 4 Bulk density at 35 cm depth of soils in Germany [29]. Map generated from original data source. Source: BGR Bodenatlas Deutschland

4 Soil Pollution

As Germany is a country with high industrial development, high population density, and intense land use, soil contamination by heavy metals is also a process of soil degradation. The main pathways into the soils are dry and wet deposition from air-transported heavy metals [1, 72]. With this, soil contamination reaches very high levels in areas far away from industrial emitters, such as the low and mid-mountain ranges, where high precipitation rates lead to high concentrations. This is the case, e.g., for Pb, where its concentrations are up to the limits established by the EU (up to 300 ppm). The origin of the contaminants is highly variable. Due to the geo-tectonical history of Central Europe, in many areas, there can be found high background values of heavy metal concentrations. Especially the varistic morphotectonic unit, comprising the Rhenish slate mountain range, and the crystalline Saxothuringicum and Moldanubicum show, e.g., Pb-concentrations far beyond 60 ppm. Industrial emissions are, of course, the source of most of the dry and wet deposited materials, and the nuclear accident of Tchernobyl added radionuclides. In addition, the application of sewage sludge or irrigation with sewage effluents has also caused an increase in heavy and precious metals in soils [73]. Fluvial sediments are often highly contaminated with heavy metals, especially in the low mountain ranges, as the result of former mining and industrial activities [72]. This may constitute a severe threat to soils in the alluvial plains when devastating, and highly erosive rainfall-runoff events may increase their frequency and lead to a more frequent remobilization of these sediments. These phenomena have been observed after the flood mid-July 2021 in the Ahr and neighboring valleys.

Being aware of the widespread problem, monitoring of these contaminants started in 1998. Observations have shown since then a decrease in the concentration of most of the monitored contaminants, except Cu and Zn [1]. Within the last decades, new issues have emerged on soil and environmental contamination. On one hand, harmful chemical compounds have been banned. But many of them, such as DDT, may still persist in soils. The introduction of highly efficient and more complex agrochemicals may show non-desired interactions with the former ones, such as enhancing their mobilization [74, 75].

5 Microplastics

The research on microplastics in soils, this is plastic particles with a size <5 mm, has emerged late, mainly due to important methodological issues [76]. The author laments a severe lack of knowledge on quantitative evidence of their occurrence in soils, even though the knowledge on their existence in soils is not new. The sources and pathways of microplastics into the environment and soils have been identified by now [77–79], as well as their ubiquitous occurrence in soils [80, 81]. However, there is still only little empirical and quantitative research available on the immission of particular plastics into soils. The actual knowledge is based mainly on estimations and models [82]. These authors generated a first systematic approach for the classification and quantification of plastics into the environment. They identified two main types of sources: littering and plastics with intended use. In addition, there is still a strong inconsistency in definition and analytics of this environmental problem [83]. In general, particles smaller than 5 mm are defined as microplastics, those between 5 and 25 mm as mesoplastics, and those >25 mm as macroplastics. When regarding the smallest fraction, there is also an additional differentiation possible. Sources of microplastics may be primary: these are products used in cosmetics, cleaning products, etc., but also from abrasion of wheels and washing of synthetic tissues. Secondary ones are those coming from degradation and fragmentation of macroplastics in soils [77, 79, 83, 84]. The authors of the studies demonstrate also, that the pathways are similar variable as their sources [85]. First overall estimations indicate that the highest amount, by far, comes from tire abrasion [79, 82, 85], showing a high seasonal and geographical variability, but mainly close to the roads and highways. Only a small amount (and smaller than 10 μm) are transported airborne.

Anyhow, being Germany a highly industrialized country with intensive high-input agriculture, another important source of microplastics arriving in the soils is by application of sewage sludge, plasticulture and compost. Taking into account the land surfaces, a modeling approach [83] shows that the cumulated input into soils during the last decades reached 1.34 kg ha^{-1} (sewage sludge), 0.32 (compost), 0.07 (plasticulture). However, the spatial heterogeneity is considered as very high. Anyhow, there is experimental evidence of other input pathways, as areas show also detectable amounts of microplastics where these agricultural practices are absent [80, 81, 86] or demonstrating the transport as aerosols [79].

To understand the effects of microplastics in soils and the environment, there is still a lack in knowledge on their behavior in the environment. They can be defined as extremely persistent in the soils. And MP accumulate substantially within the upper cm of the soil. But there is evidence of vertical transport, e.g., by bioturbation [86]. However, MP are prone to be translocated within landscape compartments. Rainfall simulations have shown that small particles (between 53 and 300 μm) reach an enrichment ratio within the transported material higher than 3 [87]. Especially the coarse particles are transported with erosion processes. These results match with the results published previously which found higher concentration on grasslands and riparian vegetation, showing with this the transport by flooding [86]. They found there up to two pieces per kg soil, which is high above amounts detected in other land uses.

6 Conclusions

Nowadays, there exist several processes of soil degradation in Germany. Soil erosion has been since the Neolithic the most important process, but with very variable intensities. However, the widespread introduction of intensive agriculture has led to a shift in degradations processes, adding a level of complexity, which makes it difficult to find comprehensive soil protection and degradation mitigation strategies.

On one side, we could observe an intensification of wind erosion processes, related to land-use strategies, especially in the north. In addition, the introduction of efficient machinery has caused a substantial damage of soil structure, which requires now advanced management strategies to minimize this soil threat.

On the other side, competing land and soil use requirements have increased pressure on agricultural and forestry soils. Their complete destruction by spreading of settlements and infrastructure, and nowadays energy production technology is an issue, which is still not under control. Germany has issued ambitious plans to reduce surface consumption but has failed by now its own plans.

In addition, new soil threats, or better, the awareness about them, have emerged during the last decades. Chemical time bombs and residues of soil treatments may contribute to soil degradation in future, especially with the introduction of new technologies. In addition, the ubiquitous spreading of plastic particles, which have reached now all compartments of the ecosystems, even in remote areas, seem to add to the processes of soil degradation, which we still do not understand in any way. But the latter one is, as mentioned, not only a problem of soil degradation in Germany.

References

1. Marahrens S, Schmidt S, Frauenstein J et al (2015) Bodenzustand in Deutschland. https://www.umweltbundesamt.de/publikationen/bodenzustand-in-deutschland
2. Saggau P, Kuhwald M, Duttmann R (2019) Integrating soil compaction impacts of tramlines into soil erosion modelling: a field-scale approach. Soil Syst 3:51. https://doi.org/10.3390/soilsystems3030051
3. Kuhwald M, Saggau P, Augustin K (2020) Konflikte um Flächennutzung und Bodenfunktionen in Agrarlandschaften. In: Duttmann R, Kühne O, Weber F (eds) Landschaft als Prozess. Springer Fachmedien Wiesbaden, Wiesbaden, pp 657–688
4. Statistisches Bundesamt (2021) Bodenfläche insgesamt nach Nutzungsarten in Deutschland. https://www.destatis.de/DE/Themen/Branchen-Unternehmen/Landwirtschaft-Forstwirtschaft-Fischerei/Flaechennutzung/Tabellen/bodenflaeche-insgesamt.html. Accessed 9 Nov 2021
5. Statistisches Bundesamt (2021) Flächenindikator "Anstieg der Siedlungs- und Verkehrsfläche". https://www.destatis.de/DE/Themen/Branchen-Unternehmen/Landwirtschaft-Forstwirtschaft-Fischerei/Flaechennutzung/Tabellen/anstieg-suv2.html. Accessed 9 Nov 2021
6. Penn-Bressel G (2017) Flächenverbrauch durch Siedlungen und Verkehr (Trends) und Flächenrucksäcke von Komponenten deutscher Energiesysteme. In: Meinel G, Schumacher U, Schwarz S et al (eds) Flächennutzungsmonitoring, vol 73. Rhombos-Verlag; Sächsische Landesbibliothek – Staats- und Universitätsbibliothek Dresden, Berlin, Dresden, pp 31–40
7. BodSchätzG (2019) Gesetz zur Schätzung des landwirtschaftlichen Kulturbodens. http://www.gesetze-im-internet.de/bodsch_tzg_2008/BJNR317600007.html. Accessed 11 Nov 2021
8. Rothkegel W (1930) Handbuch der Schätzungslehre für Grundbesitzungen
9. Wollny E (1879) Forschungen auf dem Gebiete der Agrikultur-Physik, 1–20. Carl Winter's Universitätsbuchhandlung
10. Richter G (1965) Bodenerosion: Schäden und gefährdete Gebiete in der Bundesrepublik Deutschland: Gutachten im Auftrag des Bundesministeriums für Ernährung, Landwirtschaft und Forsten. Bundesanstalt für Landeskunde und Raumforschung, Selbstrverlag
11. Batjes NH, Bridges EM (1993) Soil vulnerability to pollution in Europe. Soil Use Manage 9: 25–29. https://doi.org/10.1111/j.1475-2743.1993.tb00923.x
12. Emadodin I, Reiss S, Mitusov AV et al (2009) Interdisciplinary and multidisciplinary approaches to the study of long-term soil degradation: a case study from Schleswig-Holstein, Germany. Land Degrad Dev 20:551–561. https://doi.org/10.1002/ldr.941
13. Techen A-K, Helming K (2017) Pressures on soil functions from soil management in Germany. A foresight review. Agron Sustain Dev 37:64. https://doi.org/10.1007/s13593-017-0473-3
14. Richter G (1980) Three years of plot measurements in vineyards of the Moselle region some preliminary results. Zeitschrift fur Geomorphologie, Supplementband 35:81–91

15. Richter G (1979) (Soil erosion in vine growing areas of the Mosel region: the results of quantitative research, 1974-77). Forschungstelle Bodenerosion der Universitat Trier, Mertesdorf, (Ruwertal) 3
16. Richter G (1991) Combating soil erosion in vineyards of the Mosel-region. Forschungsstelle Bodenerosion – Universitat Trier 10
17. Auerswald K, Fiener P, Dikau R (2009) Rates of sheet and rill erosion in Germany – a meta-analysis. Geomorphology 111:182–193. https://doi.org/10.1016/j.geomorph.2009.04.018
18. Verheijen F, Jones R, Rickson RJ et al (2009) Tolerable versus actual soil erosion rates in Europe. Earth Sci Rev 94:23–38. https://doi.org/10.1016/j.earscirev.2009.02.003
19. Alewell C, Egli M, Meusburger K (2015) An attempt to estimate tolerable soil erosion rates by matching soil formation with denudation in Alpine grasslands. J Soil Sediment 15:1383–1399. https://doi.org/10.1007/s11368-014-0920-6
20. Schwertmann U, Vogl W, Kainz M (1987) Bodenerosion durch Wasser. Ulmer Verlag. 64 p
21. Dreibrodt S, Bork H-R (2005) Historical soil erosion and landscape development at Lake Belau (North Germany) – a comparison of colluvial deposits and lake sediments. Zeitschrift fur Geomorphologie, Supplementband 139:101–128
22. Reiß S, Dreibrodt S, Lubos C et al (2009) Land use history and historical soil erosion at Albersdorf (northern Germany) – ceased agricultural land use after the pre-historical period. Catena 77:107–118
23. Bork H-R, Schmidtchen G (2001) Soils: development, destruction, and conservation in Germany. Geogr Rundsch 53:4–9
24. Dreibrodt S, Lubos C, Terhorst B et al (2010) Historical soil erosion by water in Germany: scales and archives, chronology, research perspectives. Quat Int 222:80–95. https://doi.org/10.1016/j.quaint.2009.06.014
25. Dotterweich M, Haberstroh J, Siegmüller A et al (2003) Frühgeschichtliche Boden-und Reliefentwicklung am Talrand der Regnitz bei Altendorf (Oberfranken). Die Erde 134:4
26. Schmidtchen G, Bork H-R, Dotterweich M (2001) A case of severe gully erosion in eastern Brandenburg (Germany). Petermanns Geogr Mitt 145:74–82
27. Bork H-R (1998) Landschaftsentwicklung in Mitteleuropa: Wirkungen des Menschen auf Landschaften
28. BGR (ed) (2016) Bodenatlas Deutschland. Schweizerbart Science Publishers, Stuttgart
29. BGR (2014) Potentielle Gefährdung. https://www.bgr.bund.de/DE/Themen/Boden/Ressourcenbewertung/Bodenerosion/Wasser/Karte_Erosionsgefahr_node.html;jsessionid=ED60CE5DA43B8C1BBC8792E1133B0BC5.1_cid321. Accessed 8 Nov 2021
30. DIN 19708 DIN 19708:2017-08, Bodenbeschaffenheit_ – Ermittlung der Erosionsgefährdung von Böden durch Wasser mit Hilfe der ABAG
31. Auerswald K, Schmidt F (1986) Atlas der Erosionsgefährdung in Bayern. Karten zum flächenhaften Bodenabtrag durch Regen. GLA-Fachberichte, München
32. Auerswald K (1998) Bodenerosion durch Wasser. In: Richter G (ed) Bodenerosion: Analyse und Bilanz eines Umweltproblems; mit 38 Tabellen. Wiss. Buchges, Darmstadt, pp 33–42
33. Stroosnijder L (2005) Measurement of erosion: is it possible? Catena 64:162–173. https://doi.org/10.1016/j.catena.2005.08.004
34. Bug J, Mosimann T (2012) Lineare Erosion in Niedersachsen–Ergebnisse einer elfjährigen Messreihe zu Ausmaß, kleinräumiger Verbreitung und Ursachen des Bodenabtrags. Die Bodenkultur 63:2–3
35. Richter G, Negendank JFW (1977) Soil erosion processes and their measurement in the German area of the Moselle river. Earth Surf Process 2:261–278. https://doi.org/10.1002/esp.3290020217
36. Stehling E, Schmidt RG (2017) Das Datenarchiv der Forschungsstelle Bodenerosion in Mertesdorf (Ruwertal): Eine Dokumentation über 25 Messjahre (1974-1999); Informationszusammenstellung zum Gebrauch der Daten-CD, vol 16, Trier
37. Hacisalihoglu S (2007) Determination of soil erosion in a steep hill slope with different land-use types: a case study in Mertesdorf (Ruwertal/Germany). J Environ Biol 28:433

38. Richter G (1991) Erosion control in vineyards of the Mosel-region, FRG. Soil erosion protection measures in Europe. Proc EC Workshop Freising 1988:149–156
39. Seeger M, Rodrigo-Comino J, Iserloh T et al (2019) Dynamics of runoff and soil erosion on abandoned steep vineyards in the Mosel area, Germany. Water 11:2596
40. Kirchhoff M, Rodrigo-Comino J, Seeger M et al (2017) Soil erosion in sloping vineyards under conventional and organic land use managements (Saar-mosel valley, Germany) Erosión del suelo en viñas cultivadas en pendiente bajo sistemas de gestión convencional y orgánica (valle de Saar-mosela, Glemania). Cuadernos de Investigacion Geografica 43:119–140
41. Rodrigo Comino J, Brings C, Lassu T et al (2015) Rainfall and human activity impacts on soil losses and rill erosion in vineyards (Ruwer Valley, Germany). Solid Earth 6:823–837. https://doi.org/10.5194/se-6-823-2015
42. Kirchhoff M, Rodrigo-Comino J, Seeger M et al (2017) Soil erosion in sloping vineyards under conventional and organic land use managements (Saar-Mosel Valley, Germany). Cuadernos de Investigación Geográfica 43:119–140
43. Seeger M, Dittrich F, Iserloh T et al (2020) Diversifying steep slope viticulture – towards a sustainable intensive agriculture? Proceedings 30:51. Multidisciplinary Digital Publishing Institute
44. Dittrich F, Iserloh T, Treseler C-H et al (2021) Crop diversification in viticulture with aromatic plants: effects of intercropping on grapevine productivity in a steep-slope vineyard in the Mosel area, Germany. Agriculture 11:95
45. Richter G (1998) Bodenerosion: Analyse und Bilanz eines Umweltproblems. Wissenschaftliche Buchgesellschaft
46. Fiener P, Wilken F (2021) 3.2 Bodenerosion in Mitteleuropa-Auswirkungen des Klima-und Landmanagementwandels
47. Welle D (2022) Sandsturm löst Massenunfall bei Rostock aus | DW | 09.04.2011. Accessed 18 Jan 2022
48. Mangler J (2015) Unfall nach Sandsturm auf der A19: Verursacherin verurteilt. WELT
49. Mal P, Hesse JW, Schmitz M et al (2015) Konservierende Bodenbearbeitung in Deutschland als Lösungsbeitrag gegen Bodenerosion. Journal für Kulturpflanzen 67:310–319
50. Deumlich D, Funk R, Frielinghaus M et al (2006) Basics of effective erosion control in German agriculture. Z Pflanzenernähr Bodenk 169:370–381. https://doi.org/10.1002/jpln.200621983
51. Nerger R (2020) Detection of changes in soil using the long-term soil monitoring network Boden-Dauerbeobachtung Schleswig-Holstein (BDF-SH), Germany
52. Goossens D, Gross J, Spaan W (2001) Aeolian dust dynamics in agricultural land areas in lower saxony, Germany. Earth Surf Process Landf 26:701–720
53. Marzen M, Porten M, Ries JB (2022) Quantification of dust emissions during tillage operations in steep slope vineyards in the Moselle area. Agriculture 12:100
54. Goossens D, Poesen J, Gross J et al (2000) Splash drift on light sandy soils: a field experiment. Agronomie 20:12. https://doi.org/10.1051/agro:2000126
55. Fister W, Schmidt R-G (2008) Concept of a single device for simultaneous simulation of wind and water erosion in the field. In: Gabriels D, Cornelis W (eds) Proceedings of conference on desertification, vol 13. Gent, Belgium, pp 106–113
56. Marzen M, Iserloh T, Fister W et al (2019) On-site water and wind erosion experiments reveal relative impact on total soil erosion. Geosciences 9:478
57. Marzen M, Iserloh T, Casper MC et al (2015) Quantification of particle detachment by rain splash and wind-driven rain splash. Catena 127:135–141. https://doi.org/10.1016/j.catena.2014.12.023
58. Richter G (ed) (1998) Bodenerosion: Analyse und Bilanz eines Umweltproblems; mit 38 Tabellen. Wiss. Buchges, Darmstadt
59. Ledermüller S, Fick J, Jacobs A (2021) Perception of the relevance of soil compaction and application of measures to prevent it among German farmers. Agronomy 11

60. Augustin K, Kuhwald M, Brunotte J et al (2020) Wheel load and wheel pass frequency as indicators for soil compaction risk: a four-year analysis of traffic intensity at field scale. Geosciences 10:292. https://doi.org/10.3390/geosciences10080292
61. Duttmann R, Schwanebeck M, Nolde M et al (2014) Predicting soil compaction risks related to field traffic during silage maize harvest. Soil Sci Soc Am J 78:408–421. https://doi.org/10.2136/sssaj2013.05.0198
62. Horn R, Fleige H (2009) Risk assessment of subsoil compaction for arable soils in Northwest Germany at farm scale. Appl Vis Soil Eval 102:201–208. https://doi.org/10.1016/j.still.2008.07.015
63. Pulido-Moncada M, Munkholm LJ, Schjønning P (2019) Wheel load, repeated wheeling, and traction effects on subsoil compaction in northern Europe. Applications of Visual Soil Evaluation 186:300–309. https://doi.org/10.1016/j.still.2018.11.005
64. Dambeck R, Skrybeck C, Thiemeyer H (2015) Bodenphysikalische Untersuchungen zur Bewertung der Bodenverdichtung durch Forstmaschineneinsatz auf Lössstandorten im Marxheimer Wald (Hofheim a. Ts)
65. Hümann M (2010) Auswirkungen von Tieflockerung auf erstaufgeforsteten Ackerflächen. AFZ 65, H. 5:8
66. Hümann M, Schüler G, Müller C et al (2011) Identification of runoff processes – the impact of different forest types and soil properties on runoff formation and floods. J Hydrol 409:637–649. https://doi.org/10.1016/j.jhydrol.2011.08.067
67. Mercier P, Aas G, Dengler J (2019) Effects of skid trails on understory vegetation in forests: a case study from northern Bavaria (Germany). For Ecol Manage 453:117579
68. Saggau P, Kuhwald M, Hamer WB et al (2022) Are compacted tramlines underestimated features in soil erosion modeling? A catchment-scale analysis using a process-based soil erosion model. Land Degrad Dev 33:452–469. https://doi.org/10.1002/ldr.4161
69. Alakukku L (1999) Subsoil compaction due to wheel traffic. AFSci 8:333–351. https://doi.org/10.23986/afsci.5634
70. Kuhwald M, Dörnhöfer K, Oppelt N et al (2018) Spatially explicit soil compaction risk assessment of arable soils at regional scale: the SaSCiA-model. Sustainability 10. https://doi.org/10.3390/su10051618
71. Augustin K, Kuhwald M, Brunotte J et al (2019) FiTraM: a model for automated spatial analyses of wheel load, soil stress and wheel pass frequency at field scale. Biosyst Eng 180:108–120. https://doi.org/10.1016/j.biosystemseng.2019.01.019
72. Völkel J (2000) Bodenbelastung durch Schwermetalle. In: Leibnitz-Institut für Länderkunde (ed) Nationalatlas Bundesrepublik Deutschland. Spektrum, Akad. Verl., Leipzig, Heidelberg [u. a.], pp 112–113
73. Lottermoser BG (2012) Effect of long-term irrigation with sewage effluent on the metal content of soils, Berlin, Germany. Environ Geochem Health 34:67–76. https://doi.org/10.1007/s10653-011-9391-5
74. Neitsch J, Schwack W, Weller P (2016) How do modern pesticide treatments influence the mobility of old incurred DDT contaminations in agricultural soils? J Agric Food Chem 64:7445–7451. https://doi.org/10.1021/acs.jafc.6b03168
75. Weller P, Neitsch J (2017) Führt der Einsatz moderner Pflanzenschutzmittel zur Mobilisierung alter DDT-Rückstände in landwirtschaftlichen Nutzflächen? Mitteilungen der Fachgruppe Umweltchemie und Ökotoxikologie 2017:40–42
76. Rillig MC (2012) Microplastic in terrestrial ecosystems and the soil? Environ Sci Technol 46:6453–6454. https://doi.org/10.1021/es302011r
77. Leifheit EF, Rillig MC (2020) Mikroplastik in landwirtschaftlichen Böden - eine versteckte Gefahr? Berichte über Landwirtschaft - Zeitschrift für Agrarpolitik und Landwirtschaft, Aktuelle Beiträge https://doi.org/10.12767/BUEL.V98I1.279
78. Bertling J, Bertling R, Hamann L (2018) Mikro- und Makroplastik. Kurzfassung der Konsortialstudie. UMSICHT, Ursachen, Mengen, Umweltschicksale, Wirkungen, Lösungsansätze, Empfehlungen

79. Sommer F, Dietze V, Baum A et al (2018) Tire abrasion as a major source of microplastics in the environment. Aerosol Air Qual Res 18:2014–2028. https://doi.org/10.4209/aaqr.2018.03.0099
80. Piehl S, Leibner A, Löder MGJ et al (2018) Identification and quantification of macro- and microplastics on an agricultural farmland. Sci Rep 8:17950. https://doi.org/10.1038/s41598-018-36172-y
81. Harms IK, Diekötter T, Troegel S et al (2021) Amount, distribution and composition of large microplastics in typical agricultural soils in northern Germany. Sci Total Environ 758:143615. https://doi.org/10.1016/j.scitotenv.2020.143615
82. Jepsen D, Zimmermann T, Spengler L et al (2020) Kunststoffe in der Umwelt – Erarbeitung einer Systematik für erste Schätzungen zum Verbleib von Abfällen und anderen Produkten aus Kunststoffen in verschiedenen Umweltmedien. Texte, 198/2020, Dessau
83. Brandes E, Henseler M, Kreins P (2021) Identifying hot-spots for microplastic contamination in agricultural soils-a spatial modelling approach for Germany. 16:104041. https://doi.org/10.1088/1748-9326/ac21e6
84. Bertling J, Zimmermann T, Rödig L (2021) Kunststoffe in der Umwelt: Emissionen in landwirtschaftlich genutzte Böden. Fraunhofer-Gesellschaft
85. Schneider I, Scholz K-N, Biegel-Engler A et al (2021) Kunststoffe in Böden, Dessau
86. Weber CJ, Opp C (2020) Spatial patterns of mesoplastics and coarse microplastics in floodplain soils as resulting from land use and fluvial processes. Environ Pollut 267:115390. https://doi.org/10.1016/j.envpol.2020.115390
87. Rehm R, Zeyer T, Schmidt A et al (2021) Soil erosion as transport pathway of microplastic from agriculture soils to aquatic ecosystems. Sci Total Environ 795:148774. https://doi.org/10.1016/j.scitotenv.2021.148774

Agricultural Land Degradation in Portugal and Greece

Carla S. S. Ferreira, António C. Duarte, Anne K. Boulet, Adélcia Veiga, Giorgos Maneas, and Zahra Kalantari

C. S. S. Ferreira (✉)
Department of Physical Geography and Bolin Centre for Climate Research, Stockholm University, Stockholm, Sweden

Navarino Environmental Observatory, Costa Navarino, Navarino Dunes Messinia, Greece

Research Centre for Natural Resources, Environment and Society (CERNAS), Polytechnic Institute of Coimbra, Coimbra Agriculture School, Coimbra, Portugal
e-mail: carla.ferreira@natgeo.su.se

A. C. Duarte
Research Centre for Natural Resources, Environment and Society (CERNAS), Polytechnic Institute of Coimbra, Coimbra Agriculture School, Coimbra, Portugal

School of Agriculture/Polytechnic Institute of Castelo Branco, Castelo Branco, Portugal
e-mail: acduarte@ipcb.pt

A. K. Boulet and A. Veiga
Research Centre for Natural Resources, Environment and Society (CERNAS), Polytechnic Institute of Coimbra, Coimbra Agriculture School, Coimbra, Portugal
e-mail: anne.karine@esac.pt; adelcia.veiga@esac.pt

G. Maneas
Department of Physical Geography and Bolin Centre for Climate Research, Stockholm University, Stockholm, Sweden

Navarino Environmental Observatory, Costa Navarino, Navarino Dunes Messinia, Greece
e-mail: giorgos.maneas@natgeo.su.se

Z. Kalantari
Department of Physical Geography and Bolin Centre for Climate Research, Stockholm University, Stockholm, Sweden

Navarino Environmental Observatory, Costa Navarino, Navarino Dunes Messinia, Greece

Department of Sustainable Development, Environmental Science and Engineering, KTH Royal Institute of Technology, Stockholm, Sweden
e-mail: zahrak@kth.se

Paulo Pereira, Miriam Muñoz-Rojas, Igor Bogunovic, and Wenwu Zhao (eds.),
Impact of Agriculture on Soil Degradation II: A European Perspective,
Hdb Env Chem (2023) 121: 105–138, DOI 10.1007/698_2022_950,

Published online: 8 February 2023

Contents

1 Introduction 107
2 Soil Erosion 108
  2.1 State of the Problem for Agricultural Land in Portugal and Greece 109
  2.2 Consequences of Soil Erosion 111
  2.3 Soil Conservation Practices 112
3 Soil Compaction 114
  3.1 Causes of Soil Compaction 114
  3.2 Current Status of the Problem in Portugal and Greece 115
  3.3 Environmental Consequences of Soil Compaction 117
  3.4 Examples of Management Practices to Prevent Soil Compaction 117
4 Soil Contamination 119
  4.1 Causes of Contamination in Agricultural Soils 119
  4.2 Soil Contamination with Heavy Metals and the Current Situation in Portugal and Greece 121
  4.3 Soil Contamination with Pesticides and the Current Situation in Portugal and Greece 122
  4.4 Agricultural Soil Contamination with Microplastics 123
5 Soil Salinity and Sodicity 124
  5.1 Causes of the Problem 124
  5.2 Salinity and Sodicity Situation in Portugal and Greece 125
  5.3 Consequences for Crops 126
6 Final Considerations 127
References 129

Abstract Agricultural land degradation is a global problem affecting food production and other ecosystem services worldwide such as water regulation. It is driven by unsustainable land use and management practices (e.g. intensive tillage, overuse of agrochemicals) and can be aggravated by future climate change. Land degradation is particularly problematic in arid and semi-arid areas of southern Europe, and distinct soil degradation processes impair agricultural areas in Portugal and Greece. This chapter aims to improve understanding of various degradation processes affecting agricultural land, including soil erosion, compaction, contamination, and salinity and sodicity. It summarises the scientific literature on the current status of these degradation processes in agricultural areas of Portugal and Greece and their main causes and consequences. Moreover, it provides examples of best management practices implemented to mitigate agricultural land degradation. Some degradation processes are relatively well documented (e.g. erosion), while knowledge of the spatial extent of others such as soil compaction is still limited. A better understanding of soil degradation processes and of the counter-impacts of improved agricultural management practices is critical to support decision-making and ensure long-term fertility and productivity, thereby maintaining the sustainability of agriculture.

Keywords Agricultural land degradation, Compaction, Contamination, Greece, Portugal, Salinity and sodicity, Soil erosion

1 Introduction

Agricultural land supports crop and livestock production, as well as multiple ecosystem services of vital importance for local and global societies, and thus sustainable management of agricultural land is of paramount importance [1]. However, the importance of agricultural land for food security and the multitude of environmental services it provides are not always recognised by the public, while many stakeholders (e.g. policy-makers, land managers) are poorly involved in management strategies [2]. This is leading to diversified forms of land degradation, which are often ignored due to their gradual development over time and impacts on soil resources that are often neglected [3]. Land degradation encompasses an array of biophysical processes which result in reductions in land quality and is defined by loss of production [4]. In recent decades, it has been exacerbated by various factors including human activities and climate change [5]. Land degradation is an active process in arid, semi-arid and dry sub-humid areas [6], such as those found in Portugal and Greece.

Soils in southern Europe have been identified as particularly vulnerable to soil degradation [7], and at high or very high risk of desertification [8], in part favoured by the abundance of shallow soils [9]. Typically, these soils have high erosion rates [10], low levels of soil organic matter [11] and salinisation problems [12] and are threatened by compaction, contamination and loss of biodiversity [13, 14]. Such degradation processes are particularly problematic for agricultural land, given the negative impacts on soil fertility and crop yields [15]. It is estimated that land degradation leads to annual losses of six million ha of productive land globally [16]. On agricultural land, degradation has been aggravated in recent decades by rapid land use changes and intensive management practices [2]. The negative impacts on production often lead to land abandonment or are partly overcome by compensation through artificial provision of nutrients and increased use of irrigation [17].

Portugal, in the western European Mediterranean region, occupies an area of about 92,212 km^2 and has a population of 12 million [18]. Like many other countries, Portugal has experienced negative changes in its cropping systems over recent decades. Rainfed cereal, the main land use until the middle of the twentieth century, gradually became unprofitable in most marginal areas of Portugal due to decreasing crop prices driven by opening up to international markets [19]. This contributed to abandonment of marginal, mountainous or semi-mountainous areas and a decrease in the area of arable land where low-intensity management practices were used [20]. Implementation of the European Union (EU) Common Agricultural Policy (CAP) in 1992 led to afforestation of former agricultural land, intensification of production practices in more fertile areas (mainly lowlands), accelerated abandonment of marginal land and the collapse of traditional farming systems [20]. From 1999 to 2009, the area of total arable land in mainland Portugal decreased by 30% (from 1.7 to 1.2 million ha), associated with a similar area increase in pasture land (from 1.3 to 1.7 million ha) [21]. Utilised Agricultural Area (UAA) in Portugal is currently 3.641.680 ha [22], mainly comprising permanent grassland (52%), followed by arable land (29%) and permanent crops (19%) [23]. Between 2010 and 2020, the most relevant harvested crops in Portugal were grapes, permanent crops, olives, cereals and maize [24]. The agriculture sector represents 5% of total employment [22].

Greece, in the eastern European Mediterranean, occupies an area of about 132,000 km^2 and has a population of 11 million, mainly residing in coastal regions and on the islands [25]. Agriculture in Greece is a vital sector in terms of economic activity and employment (11% of total employment), contributing one-third of total national exports [22]. The agriculture system is characterised by small or average-sized farms [26], mainly located in lowlands (57%) but also extending into mountainous or semi-mountainous areas (43%) [27]. The UUA in Greece is 3591.42 million ha, comprising 39% arable land, 41% permanent grassland and 20% permanent crops [23]. Nearly 45% of the total cultivated area in Greece is irrigated [27]. A significant part of the arable land is intensively cultivated with tree and/or annual crops [26] and the dominant crops include olive, permanent crops, grapes, cereal, wheat and spelt [24].

This chapter describes the current status of agricultural land degradation in Portugal and Greece, focusing on soil erosion, compaction, contamination and salinity and sodicity degradation processes as well as their causes, environmental consequences and possible solutions to mitigate these processes. Although land degradation is an age-old problem, there is a new urgency in addressing and managing it in order to guarantee food security and safety for a growing global population and to ensure sustainable development of the agriculture sector.

2 Soil Erosion

Soil erosion is one of the most widespread forms of soil degradation in Europe [28]. Although a natural process, it has been largely accelerated by human activities, especially those associated with land use and management and in particular those linked with intensive tillage and ploughing in agricultural areas [29]. In the EU, 24% of soils are affected by erosion rates higher than the soil formation rate, driven by weathering and pedogenesis, which is estimated to be on average 1.4 t ha^{-1} $year^{-1}$ [30]. Higher erosion rates (>2 t ha^{-1} $year^{-1}$) occur on most of Greece's territory and on a significant proportion of Portugal's territory (Fig. 1). The rainfall regime, associated with short but intensive storm events, is an important driver of soil erosion in both countries and across the Mediterranean region [31]. In Portugal, the main causes of soil erosion include inappropriate management practices, overgrazing, deforestation, land abandonment, wildfires and construction activities, but agricultural land uses are reported to generate the highest soil erosion rates [20]. In Greece, the high susceptibility to soil erosion is driven by the combined impact of agriculture and site-specific conditions such as high susceptibility of most soils to erosion, the predominance of mountains and hilly landscapes (80%, one of

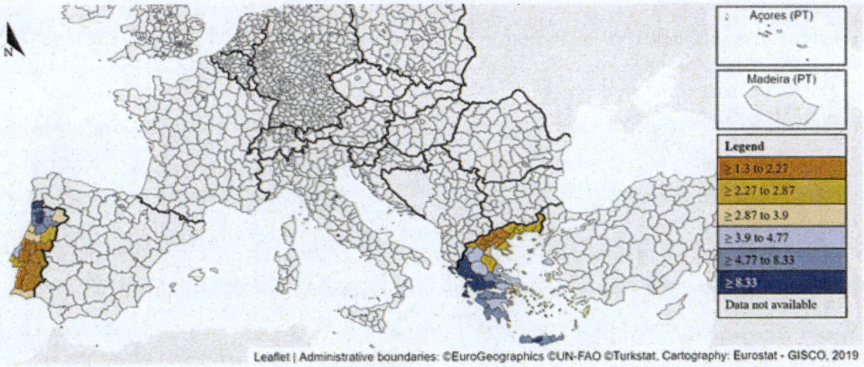

Fig. 1 Estimated water-driven soil erosion rates (t ha^{-1} $year^{-1}$) in NUTS 3 level administrative areas in Greek and Portuguese territories, 2016. Values based on the Revised Universal Soil Loss Equation (RUSLE) (adapted from [34])

the highest rates in Europe), erodible bedrock characterised by low permeability material favouring runoff (e.g. limestone, volcanic and transformed formation) and climate conditions [32, 33].

Between 2001 and 2012, there was a slight decrease in soil erosion in both Portugal and Greece, due to increased vegetation cover associated with farm abandonment caused by the global economic crisis [29]. This tendency became stronger in Portugal in 2010–2016, due to increasing use of soil conservation practices on arable land, but was reversed in Greece, where an increasing proportion of farmland is affected by high erosion rates due to unsustainable management practices [10].

2.1 *State of the Problem for Agricultural Land in Portugal and Greece*

Agricultural land is generally characterised by the highest erosion rates, due to intensive management practices and typical lack of vegetation cover [1, 20]. Intensive management is practised on about one-third of UAA in Europe and on a slightly higher percentage area in Portugal than Greece (Fig. 2). The average soil erosion rate for arable land in Portugal and Greece is ~2.75 t ha^{-1} $year^{-1}$ and ~ 2.90 t ha^{-1} $year^{-1}$, respectively [10]. The Portuguese rate for farmland is higher than the average soil erosion rate for all Portuguese land (~2.17 t ha^{-1} $year^{-1}$), but the Greek rate for farmland is lower than the Greek national average (~4.19 t ha^{-1} $year^{-1}$) [10]. It is estimated that severe soil erosion affects 5.8% and 11.8% of total agricultural area in Portugal and Greece, respectively [35].

Agricultural mechanisation has enabled great advances in crop management and productivity, but locally it can trigger soil compaction and decrease infiltration, favouring runoff and soil erosion [36]. To reduce soil compaction and improve

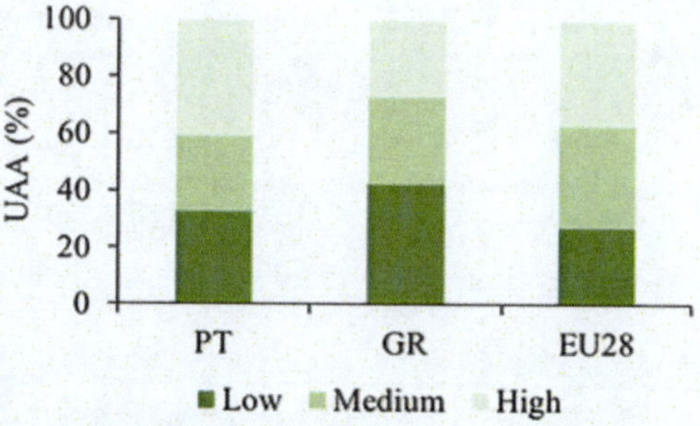

Fig. 2 Share of Utilised Agricultural Area (UAA) under low-, medium- and high-intensity management in Portugal (PT) and Greece (GR) in 2019, and average values for the 28 countries belonging to the European Union (EU28) [51]

Fig. 3 Soil erosion caused by the first rainstorms in a newly established kiwi plantation in central Portugal: (**a**) rill erosion in the upper part of the field, and (**b**) gully erosion in a downslope area (photo by Ferreira 2012)

soil conditions for root development, tillage and particularly ploughing (a highly intense form of tillage since it involves inverting the soil) has been intensively used in Portugal and Greece [37]. Studies performed in Portuguese vineyards report high sediment yields after tillage, even when little runoff is generated, since disturbed sediments are readily available to be transported [38]. Water is the most prevalent eroding agent, with sheet, rill and pipe erosion found in many farm fields in both Portugal and Greece. These can evolve into more severe forms such as gully erosion when fragile and bare soils are subject to intensive rainfall. Figure 3 shows the magnitude of soil erosion recorded in a small Portuguese kiwi plantation (1 ha) during the first storms after field preparation, which left the soil surface bare and totally exposed to erosive agents [39]. Other studies examining different land uses in Portugal have shown that soil transport by runoff peaks during autumn/winter coincides with the highest and most erosive rainfall [20]. This is associated with the typical seasonal pattern of the Mediterranean region, characterised by intense rainfall after a dry summer, with rainfall being concentrated in the autumn and winter months [40]. Gully erosion is particularly common in Greece and is responsible for erosion rates of up to 455 t ha^{-1} $year^{-1}$ [41].

Soil erosion on Portuguese and Greek farmland is also driven by other inappropriate farming techniques such as monoculture (which favours depletion of nutrients), limited use of organic fertilisers (which accelerates depletion of already limited organic matter), burning of crop residues (which leaves the soil surface bare and unprotected), cultivation on mountain slopes (e.g. for vineyards) and overgrazing [14, 33]. In intensively grazed areas, there is a significant reduction in vegetation cover and soil compaction due to trampling, which accelerates soil erosion [6]. Overgrazing is of particular importance in Greece since grazing land now comprises more than 40% of total land use in that country, following an increase in the number of grazing animals from 0.78 million in 1960 to 2.51 million in 2010, despite the estimated pasture carrying capacity being only ~0.65 million [6]. The increasing grazing pressure in Greece emerged particularly after the mid-1980s, driven by farm subsidies [6]. Around 71% of grassland in Greece is grazed for more than 10 months, so Greece has one of the highest grazing intensities in the EU in terms of duration and proportion [42].

Crop harvesting may also trigger soil losses, due to particles adhering to roots and tuber crops (e.g. carrots, potatoes). This form of soil loss depends on the type of crop and harvesting techniques (e.g. effectiveness and velocity of the technology used), as well as soil properties such as structure and texture and soil moisture content. Crop harvesting can make a relevant contribution to total soil losses (up to 28%), particularly in some regions of Greece due to the extent of tuber crops grown [28].

2.2 *Consequences of Soil Erosion*

Over time, soil erosion leads to several environmental impacts, at both on-site and off-site locations. The on-site reduction in soil depth caused by erosion restricts crop rooting space and the ability to store water. This may be particularly relevant in southern European countries typically characterised by high abundance of shallow soils [9], and in areas where water scarcity is a relevant problem during the driest periods [43]. Soil erosion affects biogeochemical processes, such as the soil carbon cycle, due to changes in nutrient mineralisation rates, and is often associated with a decreasing carbon sink [44]. These changes, coupled with losses of topsoil (the most fertile soil layer), cause on-site decreases in organic matter and nutrients and pose a threat to soil fertility and long-term productivity [45]. This can be especially relevant in southern European countries, where soils typically have low organic matter content (<2%) due to accelerated soil respiration and mineralisation driven by the regional climate (i.e. high soil temperature and moisture) [14]. Low organic matter content hinders the formation of stable soil aggregates that are resistant to the shearing forces of rainfall and runoff and reduces soil infiltration and water-holding capacity [33]. The depletion of nutrients and soil fertility caused by erosion and associated soil degradation lead to abandonment of agricultural land [20].

According to Bakker et al. [46], there is a 4.3% decline in crop yield for each 0.1 m of soil loss. Severe soil erosion in the EU is responsible for annual decreases of 3 and 0.6 million tonnes of wheat and maize, respectively, with the highest losses in countries such as Greece where extensive areas are occupied by these crops [35]. In 2010, it is estimated that the impact of soil erosion on Portuguese gross domestic product (GDP) was -€ 2.8 million but it reached -€12.6 million in Greece, based on the loss of crop productivity [35]. Apart from the relevant economic impacts of soil erosion, the loss of productivity also raises important concerns for food security and achievement of some of the United Nations Sustainable Development Goals (SDGs) [10].

Off-site impacts of soil erosion are linked to siltation of streams and reservoirs, reducing their drainage/storage capacity, which can increase the flood hazard [39]. It can also potentially impair water quality [38], leading to ecological disturbance of aquatic ecosystems and affecting recreational activities [29]. Soil erosion can also damage or destroy transportation networks such as roads and railways, and other public assets such as hydraulic infrastructure [47].

Sediments caused by erosion processes in agricultural areas may contribute to increasing concentrations of nutrients such as nitrogen [48], but also heavy metals [39] and pesticides [49] in the receiving water bodies. In Greece, several water bodies show moderate or high heavy metal contamination in sediments [26]. The impacts on water quality may impede achievement of some SDGs, e.g. SDG6: Clean water and sanitation and SDG14: Life below water.

2.3 *Soil Conservation Practices*

Soil erosion from agricultural land can be controlled by changing management practices and increasing the ground cover [20]. Increasing concerns regarding agricultural sustainability, partly triggered by EU guidelines and subsidies under the CAP and growing demand from consumers, have led to the adoption of improved management practices by farmers [14]. Most of these practices involve reduced-tillage intensity and/or enhanced soil cover. In 2016, arable land occupied 29% and 39% of Portuguese and Greek UAA, respectively [50]. On arable land, conventional tillage is widespread in Greece, being applied on 82% of the area, whereas in Portugal it is applied on 33% of the total area. In Portugal, 37% of arable land is managed using conservation tillage and 30% using no-tillage (Fig. 4a). In Portuguese vineyards located in the centre of the mainland, conservation tillage has led to runoff coefficients of 11–19%, which are much lower than those reported in nearby vineyards managed using conventional practices (42%) [38]. However, although conservation tillage has led to erosion rates (7–16 t ha^{-1} $year^{-1}$) of about half those recorded in conventional vineyards [38], it is not enough to prevent land degradation considering the estimated 1.4 t ha^{-1} $year^{-1}$ soil formation rate [30].

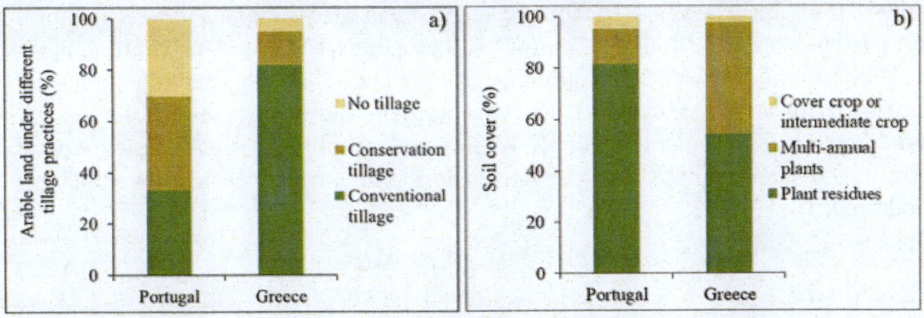

Fig. 4 Percentage cover of soil conservation practices in Portugal and Greece based on (**a**) tillage practices and (**b**) soil cover practices (data source: [51])

Soil conservation practices focusing on enhancing soil cover to protect against erosion largely involve leaving crop residues in the field (Fig. 4b). This is a widespread practice on arable land under soil conservation in Greece (81%), but less so in Portugal (44%). Grassland and intercropping (e.g. growing wheat together with clover-grass leys) are also used to reduce runoff and soil erosion, particularly in Portugal. In semi-arid areas, these measures are reported to reduce runoff by 34–89% and soil erosion by 45–94% [52]. Studies carried out in Mediterranean environments have shown that below 30% vegetation cover, soil erosion and runoff increase dramatically [53]. Unprotected soil surfaces may lead to erosion rates 100–1,000 times higher than in fields with permanent vegetation cover [20]. Vegetation protects the soil, e.g. by intercepting rainfall, which reduces the erosive power of impacting raindrops, and by reducing the volume of water reaching the soil surface, and is considered one of the basic approaches to control soil erosion [20].

Growing a cover crop is suggested to be one of the most promising measures to prevent erosion and soil degradation [54]. Cover crops are a specific form of mixed cropping in which a secondary crop is planted and grows after the main crop is harvested, thus protecting the soil between crop seasons, reducing runoff and soil erosion [37]. Field experiments performed in central Portugal have shown that leguminous cover crops reduce the risk of nutrient leaching during the rainy period and avoid groundwater pollution, by taking up and immobilising water-soluble nutrients in their roots and aerial parts [55]. Cover crops can also be used as green manure, for gradual release of nutrients to the subsequent main crop, which allows a reduction in at least 40% of nitrogen, 30% of phosphorus and 100% of potassium supplied by mineral fertilisers, while they also compete with weeds, limiting their growth [55].

Although vegetation cover can be a good strategy to prevent erosion, the dry conditions of the Mediterranean countries make it challenging to maintain soil cover. This leads to reduced vegetation cover during the summer and lower effectiveness in preventing soil erosion driven by the first autumn storms [56]. Other soil conservation practices such as application of plant residues (e.g. straw mulching, pruning wastes) can be more relevant to prevent land degradation in semi-arid and arid regions [57]. Incorporation of organic amendments (e.g. manure, green manure, compost) has been also found to improve soil properties and reduce erodibility, but it must be compliant with crop nutrient requirements and managed together with mineral fertilisation rates applied to avoid nutrient and heavy metal leaching and contamination of groundwater [58]. Widespread adoption of measures to reduce erosion and other degradation processes on agricultural land is critical for achieving food security [59].

3 Soil Compaction

Soil, a porous medium, tends to compress and increase its bulk density when subjected to pressure, and thus it can become compacted. The persistence of soil compaction, in surface and/or subsurface layers (i.e. below tillage depth, ~0.25 m), is one of the most important soil threats in Europe [60] and represents a great concern in some regions of Portugal and Greece [61].

3.1 Causes of Soil Compaction

Soil compaction is caused by exposure to pressure greater than the soil's strength, typically exerted by machinery and/or livestock [62]. The susceptibility of the soil to compaction depends on many factors, including natural and man-induced factors, or a combination of both if they occur simultaneously. Climate and soil properties such as soil texture, initial bulk density, water content and organic matter content largely determine the natural susceptibility of soil to compaction [63, 64]. Man-induced soil compaction is associated with land management [12], and tillage practices and high grazing intensity are among the land management practices with the greatest impacts on agricultural soil compaction [65]. In fire-affected areas in particular, traffic by heavy machinery is a major cause of soil compaction [66].

Although compaction is a major threat to soil quality, little has been done to reduce compaction of agricultural soils. Intensive mechanisation of crop production activities, particularly in developed countries, associated with high machine traffic in fields to perform multiple tillage operations and heavy machinery associated with powerful tractors are the main cause of soil compaction [67]. Despite the use of wider and more voluminous tyres, the changes in farm machinery have not been sufficient to compensate for the increasing weight of machinery and impacts on soil compaction [62]. Tillage, harvesting and application of chemicals and fertilisers are common operations affecting soil compaction on farms [1]. Tillage operations are widely reported to cause soil compaction at different depths [68]. In a modern drip-irrigation citric plantation located in the Mediterranean region, Cerdà et al. [69] found a rapid increase in bulk density from 1.05 to 1.33 g cm^{-3} over 13 years of machinery use. Schjønning et al. [70] reported increasing subsoil compaction, by a factor of 1.9, 3.0, 3.9 and 4.6 at 0.25, 0.50, 0.75 and 1.00 m depth in agricultural soils after 50 years of machinery use (e.g. mouldboard ploughing).

The pressure of animal traffic on the soil (overgrazing livestock) can result in soil compaction similar to that caused by machine traffic, depending on animal type, age and pressure exerted by the animals on the soil [63, 71, 72]. According to Taboada et al. [73], under the same set of soil conditions, the stress exerted by a cow in movement (98–192 kPa) can be twice the static pressure caused by a typical tractor (27–68 kPa), because of the angle of the force applied and the concentration of weight on a smaller area of hooves [71].

3.2 Current Status of the Problem in Portugal and Greece

Based on soil properties recorded in the European soil database, such as texture, relative proportion of fine particles, organic matter, level and type of aggregation, water regime and depth, and the dominant and secondary land use [74], Portuguese soils have lower susceptibility to soil compaction than Greek soils (Fig. 5). A relatively small part of mainland Portugal has high or very high susceptibility to soil compaction, whereas Greece has a large area with high susceptibility and a large area with medium susceptibility. Soils with lower susceptibility to compaction, mainly in the north of both Portugal and Greece, are mostly located in forestry and some are associated with extensive grazing systems. However, in the south of Portugal and centre and south of Greece, some regions of high and very high susceptibility of soil compaction are under intensive agricultural systems, most of them managed by conventional practices and without soil conservation measures, which can exacerbate the problem.

Increasing use of irrigation leads to higher soil moisture content, which favours compaction [12]. In Portugal, several irrigation projects (e.g. Alqueva which includes 69 dams, reservoirs and weirs over an extension of 120,000 ha, and considered the largest irrigation project in Europe) have been implemented in recent decades to enhance productivity in rainfed areas (Fig. 5b). Agricultural areas have expanded with the plantation of, e.g., new olive and almond orchards, managed under intensive practices. Thus, in the medium to long term, if improved soil management practices are not implemented, these areas can be affected by soil compaction and other soil degradation processes, with negative impacts on land resources and ecosystem sustainability. In Greece, the area under intensive farming is less extensive than in Portugal (Fig. 5b), but the soil is more susceptible to compaction than in Portugal (Fig. 5a). In Greece, around 71% of grassland is occupied by animals for 10 months of the year, which is one of the most intensive grazing rates of all European countries, both in duration and intensity [76].

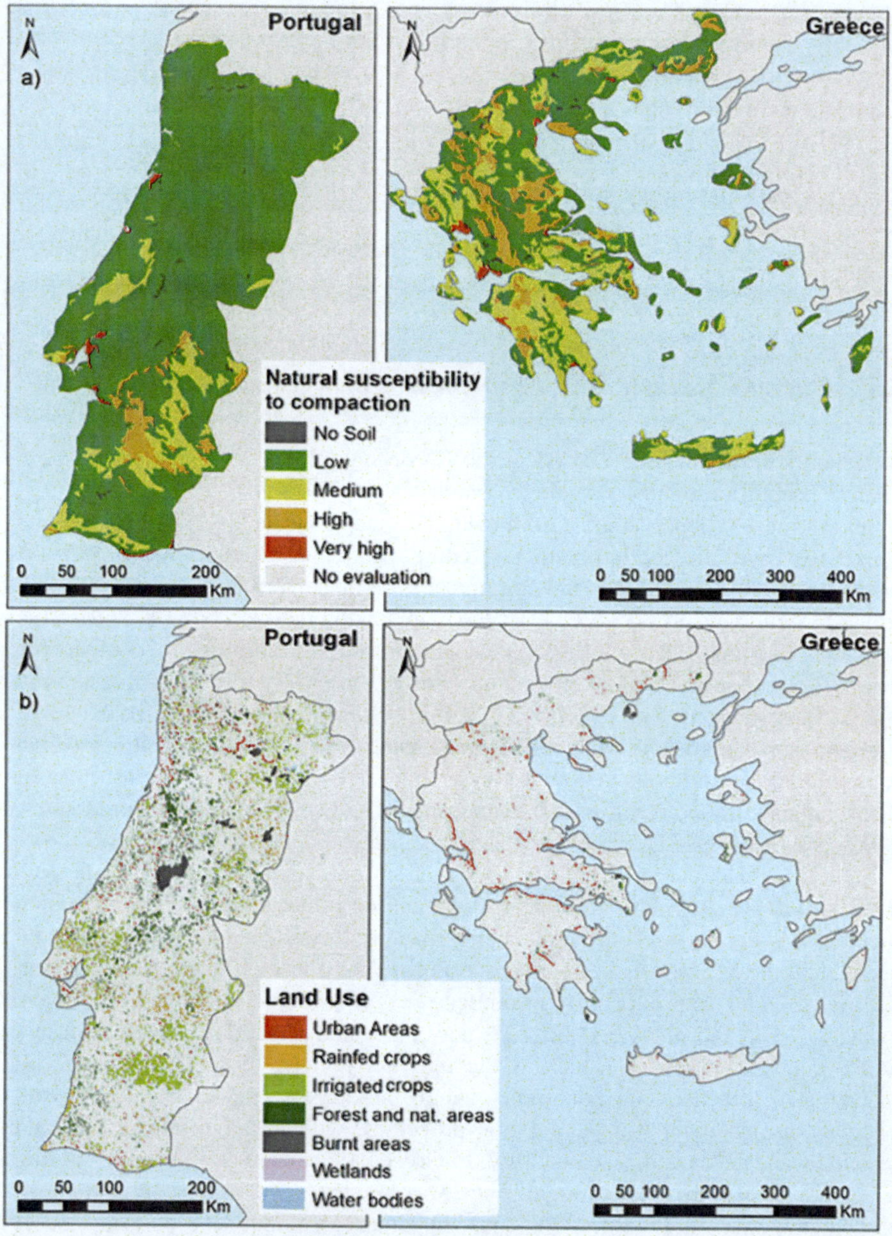

Fig. 5 (**a**) Natural susceptibility of soils in Portugal and Greece to compaction, and (**b**) change in land use between 2012 and 2018 (adapted from [74, 75])

3.3 Environmental Consequences of Soil Compaction

Soil compaction causes several environmental consequences on global farmlands, including those in Portugal and Greece. The compaction process primarily reduces the size and volume of pores, particularly macropores, decreasing the bulk density and negatively affecting several essential soil properties, such as hydraulic conductivity and infiltration capacity, which increases surface runoff and contributes to waterlogging during precipitation events [67]. With the application of compressive stress the soil can easily approach saturation. The compression of soil under saturated conditions is known as consolidation [77]. The reduction in soil water storage capacity driven by compaction also has consequences in terms of depletion of water resources [6]. In addition, increasing surface runoff from compacted soils, especially from soil surfaces lacking a protective cover, such as those where overgrazing is permitted, leads to a reduction in vegetation, increases erosion rates and, in extreme precipitation events, may trigger landslides and flooding [14]. A complete review of the effects of grazing animals on soil physical properties can be found in [78].

Some pollutants, like phosphorus and pesticides, are preferentially attached to fine soil particles and easily transported during erosion, which can have negative impacts on aquatic ecosystems, resulting, e.g., in eutrophication [79]. The transport of sediments to the natural drainage network, where the flow loses energy and deposits material, increases sedimentation effects, reducing the water storage capacity of dams and the flow capacity of natural streams (Fig. 6). This in turn increases flood susceptibility [39].

Soil compaction (in unsaturated soil) reduces diffusion of oxygen into soil, affecting microbial activity, and increases penetration resistance for roots and thus root growth [70]. The constraints imposed by soil compaction on the availability of air and water, and on uptake of nutrients by plants, lead to decreasing yield capacity [80]. In Europe, soil compaction leads to crop yield reductions ranging from 2.5–15% [15] to 25–50% [81].

3.4 Examples of Management Practices to Prevent Soil Compaction

Conservation agriculture is considered to be a good solution to prevent soil compaction. It includes a combination of several agricultural practices based on minimal soil disturbance, e.g. in seedbed preparation, retention of crop residues on the soil surface and use of combined management operations to minimise machine traffic in fields [82]. In Portugal, long-term research has shown that conservation agriculture can increase soil organic matter and improve aggregate stability and porosity throughout the soil profile, helping to prevent soil compaction [83]. Adoption of no-tillage or reduced-tillage systems, besides preventing soil compaction, also prevents mineralisation of stable components of soil organic matter and thus contributes

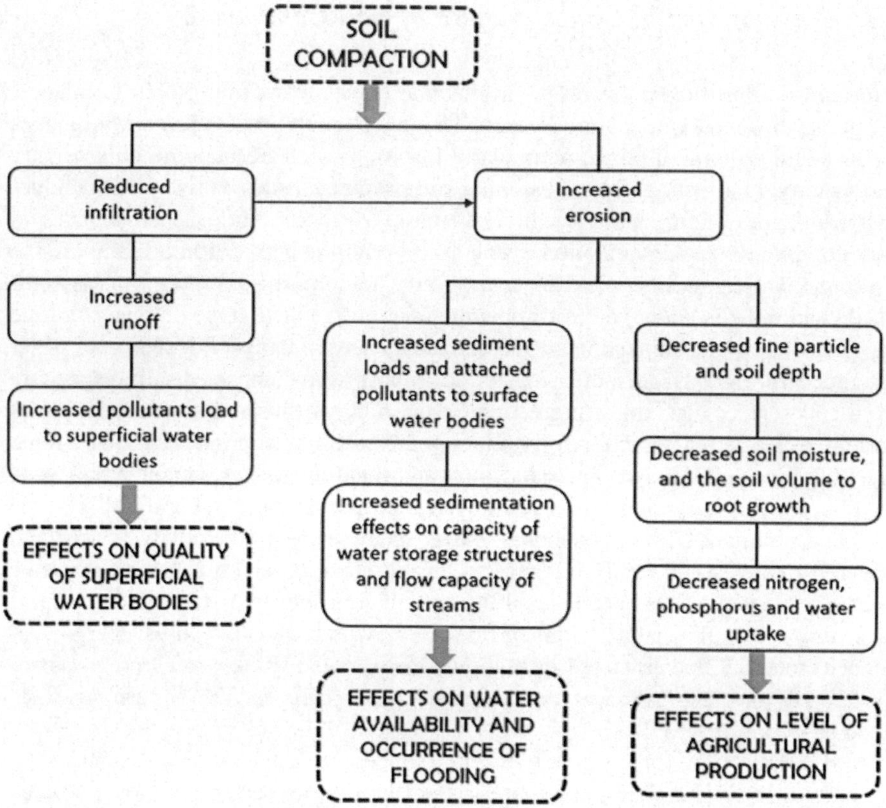

Fig. 6 Environmental impacts of soil compaction on soil on water resources, aquatic ecosystems and crop productivity

to sequestration of organic carbon in the soil [82]. As shown in Fig. 4, the Greek arable land managed under no-tillage or conservation tillage practices is nonrepresentative. In contrast, these management practices are implemented in the majority of Portuguese arable land. In fact, Portugal is the European country showing the greatest decrease in conventional tillage practices [23].

In grazing livestock systems, maintenance of vegetation cover on the soil lessens the direct impact of rainfall drops and decreases water erosion, reducing the effect of external forces on the soil. In conditions of extensive grazing, some studies report benefits of animal manure in increasing soil organic matter concentration and resilience to soil compaction [77].

4 Soil Contamination

Soil contamination refers to the presence of physical, chemical and/or biological substances causing temporary or permanent loss of one or several soil functions [64]. In recent decades, soil contamination has become one of the major concerns for developed and developing countries, particularly in agricultural soils since it impairs food safety [84]. Soil contamination affects, e.g., crop yields by lowering the activity of soil biota and reducing biodiversity, with negative impacts on nutrient cycling and soil structure. The changes in soil physical properties driven by inappropriate management practices enhance the susceptibility to degradation processes such as soil erosion, which contributes to the spread of contaminants into downslope areas [85]. Soil contamination in agricultural areas may pose threats to food safety by providing a pathway for human exposure to pollutants such as pesticides and heavy metals [84]. A study by [86] revealed that more than 29% of food samples analysed across the EU contained multiple pesticide residues. Global concerns have led more than 170 countries within the United Nations Environmental Assembly (UNEA-3) to express their willingness to reduce soil contamination [87].

4.1 Causes of Contamination in Agricultural Soils

The intensification of agricultural practices has been associated with increasing use of agrochemicals to enhance nutrient availability for crops and to control pests and diseases, in order to improve crop yields, but this is a major cause of soil contamination [49, 68].

Overuse of fertilisers, through excessive inputs of synthetic fertilisers, inappropriate management of organic amendments (e.g. manure, sewage sludge) and/or intensive livestock density, has been identified as a major driver of soil contamination [87, 88]. Nitrogen-based fertilisers are among the most widely consumed and manufactured fertilisers worldwide [51]. In Portugal and Greece, average sales of nitrogen fertilisers between 2010 and 2019 represented 67% and 74% of all fertiliser sales (nitrogen, phosphorus and potassium), respectively (Fig. 7). Although Portugal has about half the Greek level of sales of synthetic fertilisers, the rate of application of inorganic nitrogen and phosphorus fertiliser per unit area of cropland is higher, although still below the European average in both countries (Fig. 8). A decreasing tendency in the rate of fertilisation was seen from 1990 to 2009, especially in terms of phosphorus inorganic fertiliser, possibly due to increasing application restrictions and the development of the global economic crisis. However, in recent years the fertilisation rate seems to have increased slightly (Fig. 8).

Fertilisation rates higher than plant nutrient requirements can lead to nutrient imbalances in the soil, with negative impacts on the ability of plants to absorb some of the necessary nutrients (e.g. calcium and magnesium), causing nutrient deficiency in plant tissues or even toxic levels, which increases plant vulnerability to pests and

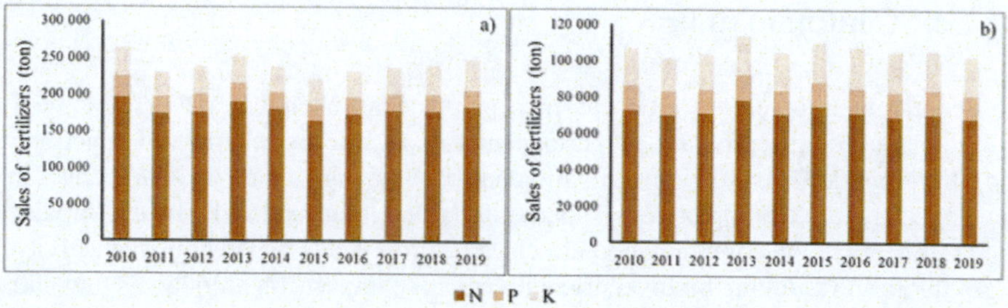

Fig. 7 Total sales of inorganic fertilisers 2010–2019 in (**a**) Portugal and (**b**) Greece (data source: [51])

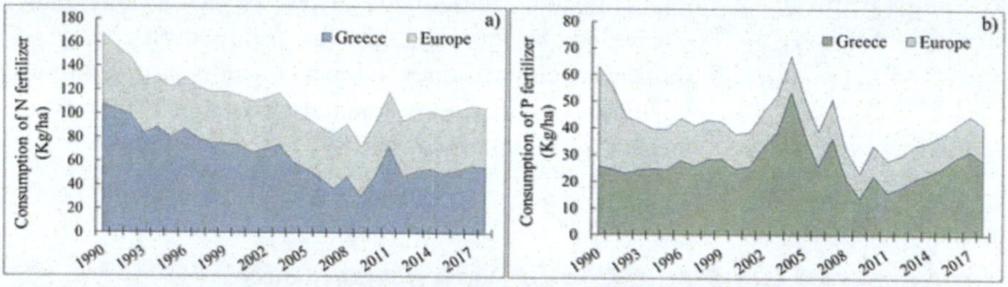

Fig. 8 Ratio of (**a**) nitrogen and (**b**) phosphorus mineral fertiliser in total fertiliser consumption and area of cropland (including arable land and permanent crops) in Portugal, Greece and Europe in the period 1990–2018 (data source: [89])

diseases and decreases crop productivity [90]. Inefficient management of nitrogen and phosphorus within agricultural fields may also lead to eutrophication and acidification of surface waters and contamination of groundwater [38, 88, 91]. In mainland Portugal, the risk of groundwater contamination driven by nitrogen fertilisation in agriculture, considering the spatial extent of the agricultural fertilisation hazard and site-specific aquifer vulnerability, increased in extent from 8,800 to 82,679 ha of the territory between 1999 and 2009 [21]. This was despite action programmes and mandatory restriction measures associated with fertilisation rates and animal stocking rates in Nitrate Vulnerable Zones delineated in 2004 (covering 4.4% of the territory), following the European Nitrates Directive, and the abandonment of grazing practices, particularly in southern regions [21].

Inappropriate management of manure and other organic amendments such as sewage sludge, which are widely used due to their potential to improve soil properties and functions, can also contribute to over-fertilisation and soil contamination [92, 93]. Soil contamination due to organic amendments has been associated with heavy metals, pathogens and veterinary antibiotic residues [94–96].

Given the positive impacts of sewage sludge on soil properties (e.g. aggregate stability, organic matter content and biodiversity) [68] and their increasing availability driven by a growing population, agricultural land can be a useful recipient of this type of solid waste. In 2016, 13.89 and 21.53 thousand tons of sewage sludge were applied to agricultural areas in Portugal and Greece, respectively, corresponding to ~12% and ~ 18% of national sludge production, respectively [97]. In order to prevent soil contamination driven by sewage sludge, the European Council Directive 86/278/EEC regulates sludge application in agriculture. The implementation of legislation regulating application of sewage sludge in Portugal, which involves regular soil monitoring and a complex permit system, led to a ~ 57% decrease in land application between 2012 and 2016, whereas in Greece a slight increasing tendency has been observed [97].

4.2 *Soil Contamination with Heavy Metals and the Current Situation in Portugal and Greece*

Heavy metals are among the most toxic, persistent and complex non-biodegradable contaminants reported in agricultural soils [98]. They are easily accumulated in tissues and living organisms, representing a threat to the health and well-being of animals, plants and humans [99]. Although heavy metals may occur naturally in soils, depending on bedrock properties, concentrations above background levels are likely to be found in soils receiving intensive applications of synthetic fertilisers, pesticides, manure and sewage sludge [100, 101].

In general, both Portugal and Greece have higher heavy metal concentrations in agricultural soil than the average values across Europe (Table 1). This is particularly the case in terms of copper (Cu), zinc (Zn) and lead (Pb), with concentrations about twice the European average. This may be due to the high fertiliser doses required to

Table 1 Baseline concentrations (mg kg^{-1}) of heavy metals (mean and standard deviation (STD)) in soils in Portugal, Greece and Europe, based on pan-European studies (adapted from ESDA database: esdac.jrc.ec.europa.eu)

		Portugal	Greece	Europe
As	Mean	13.45	10.18	6.89
	STD	5.05	3.41	1.68
Cd	Mean	0.20	0.29	0.11
	STD	0.10	0.11	0.66
Cr	Mean	46.73	26.30	23.77
	STD	23.21	7.42	5.17
Cu	Mean	22.20	26.97	14.83
	STD	6.06	10.29	3.70
Hg	Mean	0.05	0.05	0.05
	STD	0.02	0.01	0.01
Ni	Mean	20.38	56.31	20.43
	STD	6.78	44.22	6.95
Pb	Mean	31.66	24.78	19.06
	STD	14.96	9.18	5.16
Zn	Mean	71.28	72.37	58.88
	STD	28.95	23.86	14.26

overcome the relatively low soil organic matter content and low nutrient availability [38] and high use of pesticides to mitigate the high climate-related susceptibility to pests and diseases [102].

High soil copper concentrations are often driven by extensive use of copper-based fungicides [100], particularly in olive and wine-producing regions [75], and by application of sewage sludge [101]. In fact, vineyards in Central Portugal have been reported to have copper, lead and nickel (Ni) concentrations exceeding the legal thresholds [68]. In Greece, particularly high concentrations of nickel have been reported (Table 1) in the Atalanti [103] and Thessaloniki [104] regions. These high concentrations have been recorded on maize farms, together with high concentrations of chromium (Cr), molybdenum (Mo), selenium (Se) and antimony (Sb) [105]. Heavy metals can remain in the environment for many years, and thus impair water resources [58]. In Greece, heavy metal concentrations exceeding sediment quality guidelines have been reported in many regions since the 1970s [26].

Relatively high accumulation of heavy metals in agricultural Mediterranean soils is also favoured by relatively low leaching rates under low precipitation surplus [106]. To mitigate the impacts of high heavy metal concentrations, the FAO recommends remediation of contaminated agricultural soils and implementation of sustainable soil management practices [87].

4.3 Soil Contamination with Pesticides and the Current Situation in Portugal and Greece

In agricultural soils, contamination with pesticides is a major concern due to their high toxicity and persistence [85] and their global use [87]. In Europe, almost 500 active substances are approved for use [49] but less than 0.1% of the pesticides applied to crops reach the target pest, raising relevant environmental concerns [85]. Furthermore, pesticides may induce pest resistance and leave pesticide residues in crops, affecting food safety [49]. Within the EU, 1.7% of food samples analysed in 2015 revealed contamination with pesticides, whereas in Portugal the value reached 2.8% [107].

In Portugal, the average application rate of pesticides per unit cropland area is slightly lower than the European average, but more than twice the Greek rate (Fig. 9a). However, between 2011 and 2018 the use of pesticides in Portugal decreased by ~41%, whereas in Greece it increased by ~21% (Fig. 9b).

A study by [109] found that more than 80% of EU agricultural topsoil is contaminated with pesticide residues and that Portugal is one of the countries with more widespread contamination, whereas Greece is among the countries with the lowest contamination rates. In Portugal, the highest occurrence of pesticide residues is mostly reported in vineyards, where at least four different compounds have been detected [109]. Residues of glyphosate and their first-order metabolites are the main

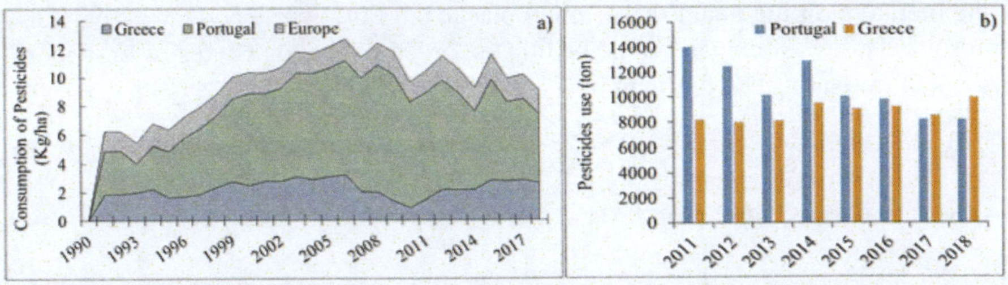

Fig. 9 a) Consumption of pesticides per unit cropland area (including arable land and land under permanent crops) in Greece, Portugal and Europe 1990–2018, and b) changes over time in the use of pesticides in Portugal and Greece during the period 2011–2018 (data source: [108])

active substances reported in Portuguese agricultural soils [110]. In Guadiana River Basin, extending over ~60 km^2 in southern Portugal, which is a biodiversity hotspot highly impacted by agriculture, the most abundant pesticides detected in soil are bentazone and 2,4-D, the most ubiquitous are terbuthylazine and terbutryn [49]. Moreover, 18 out of 38 pesticides detected are no longer approved in Europe and five of these are included in the list of priority substances for elimination. The highest concentrations of pesticides are in intensively irrigated agricultural plots such as olive groves and vineyards [49]. These results highlight the need to implement actions for sustainable use of pesticides in agricultural areas of Portugal.

In Greece, herbicide residues have been reported in soils on maize and tomato farms [111], while traceable levels of insecticides have been detected in soil samples from peach orchards [112]. The accumulation of pesticide residues and metabolites depends on their mobility in the soil, determined by pesticide chemical properties and several biophysical properties of the soil [113].

4.4 *Agricultural Soil Contamination with Microplastics*

Microplastics are one of the substances classified as 'chemicals of emerging concern' [15], which have been increasingly studied and widely detected in different environments [114]. In agricultural soils, small fragments of plastics have raised concerns because of their numerous and uncontrolled sources, although their distribution and abundance are affected by soil texture [115]. In general, microplastics in agricultural soils are mostly associated with practices relating to fertilisation, such as application of sewage sludge [116] and manure [117, 118], and mulching [119].

Soil contamination with microplastics has also been associated with other chemical pollution in agricultural soils. For example, polyethylene films used in agriculture may be a potential vector of several pesticide residues [119, 120]. In Greece, it is estimated that total plastic pesticide packaging waste amounts to 0.028 kg 1,000 m^{-2} farm^{-1} year^{-1} [121]. Despite increasing concern about the environmental impacts of microplastics in agriculture, there is a lack of information in the literature on the quantities of microplastics in Portuguese and Greek agricultural soils. This research gap should be filled in order to develop strategies to support farmers in mitigating the problem.

5 Soil Salinity and Sodicity

Salinisation refers to accumulation of soluble salts in the soil and comprises one of the main soil degradation processes in Europe [122]. Salinisation is associated with accumulation of salts, such as magnesium, potassium, calcium, carbonates and chorine, leading to electrical conductivity of the saturated soil extract of more than 4 dS m^{-1} at 25°C [64]. When the soil is affected by accumulation of exchangeable sodium, the problem is called sodicity. In sodic soils, the percentage of exchangeable sodium is above 15 [123]. In Europe, salinisation and soil sodisation affect about 30.7 M ha, which represents 3.3%, of all saline and sodic soils worldwide [124]. Soil salt accumulation affects most plants, leading to decreasing crop productivity and, in extreme cases, to land desertification [12].

5.1 Causes of the Problem

Salt-affected soils may occur naturally (primary salinisation) or may be caused by human-induced processes involving incorporation of highly water-soluble salts into the soil (secondary salinisation). Natural salinisation is of marine origin, driven by the direct action of the tides in coastal regions, deposition of marine salts transported by the wind, transfer of saline water to areas with limited drainage or ascending capillary flow of groundwater due to high evapotranspiration rate in arid and semi-arid areas [125]. Sodisation results mainly from the meteorisation of rocks with minerals rich in sodium [126]. The global extent of primary salinisation is estimated to be slightly under 1 billion ha and it mainly affects coastal areas [127]. This problem is particularly apparent in coastal southern Europe.

Human-induced salt accumulation is often recorded in agricultural soils receiving insufficient precipitation and/or with low hydraulic conductivity and weak drainage capacity, which restricts the leaching of salts. It is associated with inadequate management practices such as (1) intensive use of fertilisers or corrective agents, particularly in conditions of limited leaching; (2) inappropriate irrigation practices including insufficient water provisioning, overexploitation of coastal groundwater aquifers causing seawater intrusion and use of water rich in soluble salts (e.g. wastewater) and (3) application of saline products of industrial origin [128]. Secondary salinisation affects ~77 M ha globally, of which 58% are located in irrigated areas [127]. It is estimated that 20% of irrigated area worldwide is salt-affected [129]. Irrigation on drylands sustains crop production, but high evapotranspiration rates of water may lead to accumulation of salts in the root zone if not enough water is supplied to leach the salt beyond the root zone. Waterlogging without adequate drainage has also become a serious cause of soil salinisation [12].

5.2 Salinity and Sodicity Situation in Portugal and Greece

In Portugal, an area of 150,000 ha is estimated to be affected by salt accumulation, two-thirds of which is caused by primary or natural salinisation in low-lying areas affected by tides, riverbanks and estuaries in coastal areas, and only one-third is of human-related origin, associated with inadequate irrigation and drainage practices and irrigation with poor-quality water [130]. In agricultural areas, moderate or high soluble salt accumulation occurs in some soil layers during certain periods of the year, namely in summer and/or autumn. It is found in some irrigated agricultural areas located in semi-arid and arid regions in the interior of the country, such as the Alentejo. Here, the dry climate conditions co-occur with soils presenting deficient internal drainage due to low permeability of the clay B-horizon, which impedes leaching of soluble salts [130].

In Greece, salt-affected soils are found in a large variety of climatic, topographical and soil conditions, but particularly in the western coastal region, the marine Mediterranean Ionian islands and the Mediterranean mainland zone, including the south-eastern region (Aegean) and up to Thessaly [6]. Saline soils develop in particular in areas with alluvial parent material (lake or wind deposits), and thus are mainly found on plains or gentle sloping areas or in concavities (Fig. 10), and in

Fig. 10 Soil salinisation in an arable field in Messinia, Greece (photo by Maneas 2021)

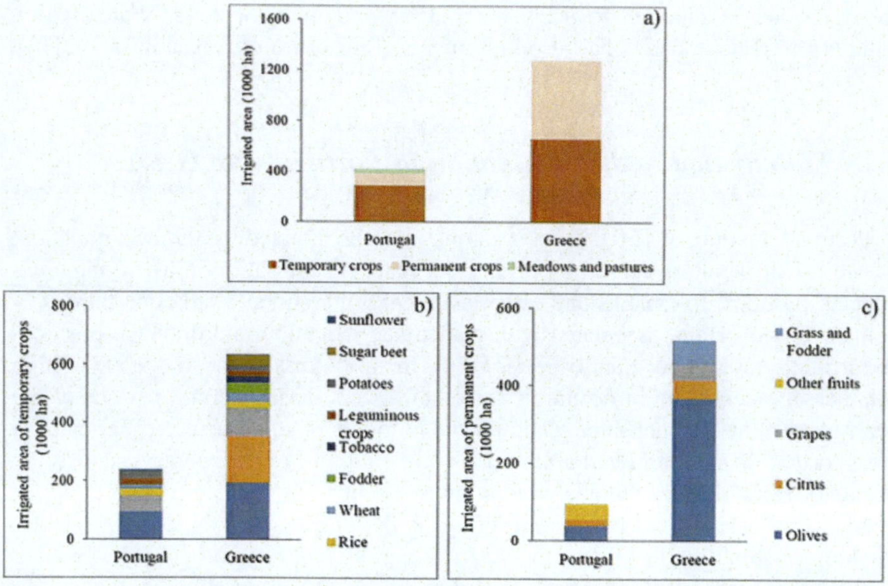

Fig. 11 (**a**) Extent of irrigated area in Portugal and Greece occupied by the main crop types in 2017, and corresponding irrigated area of (**b**) temporary and (**c**) permanent crops (data source: [102])

winter-flooded valleys surrounding by mountains. Saltwater intrusion into coastal agricultural areas and intensive agricultural activities are the main causes of the bad/poor chemical status of some water bodies in Greece [27].

Although salinisation in agricultural soils in both Portugal and Greece is not a major concern due to the relatively limited spatial extent, predicted climate change in coming decades, namely predicted increases in temperature, will increase the area requiring irrigation, and thus an increase in salt-affected areas and soil degradation can be expected. Figure 11 shows the current extent of irrigated agricultural areas and associated crops in Portugal and Greece.

5.3 *Consequences for Crops*

The accumulation of salts in the soil leads to imbalances in plant nutrition and may cause plant toxicity, due to excessive absorption of some ions and changes in regulatory mechanisms, such as increasing osmotic pressure driven by lower water availability in the root zone [131]. In soils with high sodium concentration, typically associated with higher pH, there is a decrease in the solubility of some macronutrient (e.g. Ca, Mg, P) and micronutrients (e.g. Fe, Mn, Zn), and in their availability to plants [130]. An excess of exchangeable sodium can also cause dispersion of clay particles, leading to disruption of aggregate stability, decreased permeability to water

Table 2 Sensitivity of different crop species to excess salt and sodium in soil. T: tolerant; MT: moderately tolerant; MS: moderately sensitive; S: sensitive (adapted from [133–136])

Crop	Salt		Sodium
Temporary crops	Maize	MS	S
	Cotton	S	S-MS
	Vegetables	MS	
	Rice	S	T
	Wheat	T-MT	MT
	Fodder	T-MT	T
	Leguminous crops	S-MS-MT	S
	Potatoes	MS	
	Sugar beet	T	T
	Sunflower	MT	
Permanent crops	Olives	MT	S
	Citrus	S	S
	Grapes	MS	S
	Other fruits	S	S
	Grass and forage crops	MT-MS	T

and air, greater resistance to root growth and greater susceptibility to surface crusting, which impedes the emergence of seedlings [126].

The impacts of salinity or sodicity on soil properties and plants are often associated with decreasing crop yields [64]. But different crops show variable tolerance to salt accumulation in the soil (Table 2). In terms of crop species, wheat and sugar beet are more tolerant than maize and vegetable crops, whereas rice and cotton are more sensitive. In terms of sodium excess tolerance, rice appears to be tolerant, wheat only moderately tolerant and legume crops such as cabbage and lentil are relatively sensitive to excess exchangeable sodium (Table 2). Besides, individual plants demonstrate variable tolerance to salt depending on the growing stage, but are generally more sensitive during germination [132].

6 Final Considerations

Land degradation, a deleterious form of disturbance driven by human-induced processes and affected by climate conditions, is a major concern worldwide. It is a particular concern in agricultural areas due to its impacts in decreasing soil fertility and productivity, threatening global food security. Land degradation affects a great area of agricultural land globally, including large areas in Portugal and Greece. In these countries, agricultural land is exposed to individual degradation processes or a combination of these. Intensive management practices and rather limited vegetation cover result in high erosion rates in both countries, although with slightly higher values recorded in Greece due to more extensive hilly terrain, higher erodibility of the soil resulting from bedrock properties and larger areas with overgrazing. Intensive grazing and mechanisation of agriculture (e.g. tillage) cause significant levels of soil compaction, aggravated by inappropriate irrigation practices (e.g. furrow irrigation), not only at the soil surface but also in the subsoil. Although the real status of the problem in Greece and Portugal is largely unknown, the natural susceptibility of the soil to compaction is considerably greater in Greece than in Portugal. However, highly and very highly susceptible areas should receive particular management attention in both countries. Intensive agricultural practices are often associated with overuse of agrochemicals, with concentrations of heavy metals (e.g. Zn, Cr and Pb) and pesticide residues in the Portuguese and Greek soils being considerably higher than average values across Europe. Although there are relatively few studies of soil contamination, some research performed in Portugal and Greece has highlighted the impact of unsuitable agricultural management practices on degradation of water resources, including groundwater and surface waters. Overuse of mineral fertilisers has also led to increasing problems with salinity and sodicity in agricultural soils, particularly in coastal areas, due to excessive groundwater abstraction. This problem is more pronounced in Greece than in Portugal due to the longer coastline in Greece. Other land degradation processes, such as loss of soil organic matter and biodiversity, are also concerns in southern European countries and should be considered when addressing agricultural land degradation in Portugal and Greece. However, these processes were outside the scope of the present review.

The degradation processes affecting agricultural land also result in several on-site and off-site problems, threatening the productivity of cropping systems and impairing the quality of the environment, with relevant consequences for society and the economy. Increasing recognition of these problems has generated awareness of the need for improved agricultural land management practices to assure the sustainability of agriculture and ensure land degradation neutrality by 2030. Some soil conservation practices have been implemented in Portugal and Greece (e.g. no-tillage, cover crops), but on a rather limited proportion of total arable land. Approaches to improve soil and crop management are required to prevent and/or mitigate soil degradation and enable water saving in agricultural fields. Building dialogue between scientists, policy-makers, decision-makers and farmers is of utmost importance to ensure effective adaptation and implementation of management practices to resolve agricultural land degradation and achieve food security.

Acknowledgments This research was supported by Navarino Environmental Observatory, a project funded by Formas, 2017-00608, and another project funded by the Portuguese Science and Technology Foundation, PTDC/EEI-ROB/2459/2021.

References

1. Bai Z, Caspari T, Gonzalez MR, Batjes NH, Mader P, Bunemann EK, Goede R, Brussaard L, Xu M, Ferreira CSS, Reintam E, Fan H, Mihelic R, Glavan M, Tóth Z (2018) Effects of agricultural management practices on soil quality: a review of long-term experiments for Europe and China. Agric Ecosyst Environ 265:1–7. https://doi.org/10.1016/j.agee.2018.05.028
2. Bado VB, Bationo A (2018) Integrated management of soil fertility and land resources in sub-Saharan Africa: involving local communities. Adv Agron 150:1–33. https://doi.org/10.1016/bs.agron.2018.02.001
3. Vlek P, Le QB, Tamene L (2008) Land decline in land-rich Africa – a creeping disaster in the making. CGIAR Science Council Secretariat, Roma, Italy, 55 p
4. FAO (Food and Agriculture Organization of the United Nations) (2000) Guide diagnostic participatif des contraintes et des potentialit_es pour la gestion des sols et des _el_ements nutritifs des plantes. FAO, Land and Water Development Division, Rome
5. Bourlion N, Ferrer R (2018) The Mediterranean region's development and trends: framework aspects. In: FAO and Plan Bleu. 2018. State of Mediterranean Forests 2018. Food and Agriculture Organization of the United Nations, Rome and Plan Bleu, Marseille. Chapter 1, pp 2–15
6. Kosmas C, Detsis V, Karamesouti M, Kounalaki K, Vassiliou P, Salvait L (2015) Exploring long-term impact of grazing management on land degradation in the socio-ecological system of Asteroussia Mountains, Greece. Land 4:541–559. https://doi.org/10.3390/land4030541
7. Lahmar R, Ruellan A (2007) Soil degradation in the Mediterranean region and cooperative strategies. Cah Agric 16(4):318–323. https://doi.org/10.1684/agr.2007.0119
8. Prăvălie R, Patriche C, Bandoca G (2017) Quantification of land degradation sensitivity areas in southern and central Southeastern Europe. New results based on improving DISMED methodology with new climate data. Catena 158:309–320. https://doi.org/10.1016/j.catena.2017.07.006
9. Lagacherie P, Álvaro-Fuentes J, Annabi M, Bernoux M, Bouarfa S, Douaoui A, Grunberger O, Hammani A, Montanarella L, Mrabet R, Sabir M, Raclot D (2018) Managing Mediterranean soil resources under global change: expected trends and mitigation strategies. Reg Environ Chang 18:663–675. https://doi.org/10.1007/s10113-017-1239-9
10. Panagos P, Ballabio C, Poesen J, Lugato E, Scarpa L, Montanarella L, Borrelli P (2020) A soil erosion indicator for supporting agricultural, environmental and climate policies in the European Union. Remote Sens (Basel) 12:1365. https://doi.org/10.3390/rs12091365
11. Aguilera E, Lassaletta L, Saz-Cobena A, Garnier J, Vallejo A (2013) The potential of organic fertilizers and water management to reduce N2O emissions in Mediterranean climate cropping systems. A review. Agric Ecosyst Environ 164:32–52. https://doi.org/10.1016/j.agee.2012.09.006
12. Stolte J, Tesfai M, Oygarden L, Kvaemo S, Keizer J, Verheijen F, Panagos P, Ballabio C, Hessel R (2016) Soil threats in Europe. Status, methods, drivers and effects on ecosystem services. A review report, deliverable 2.1 of the RECARE project. European Union. https://doi.org/10.2788/828742
13. EC (2006) Proposal from the Commission to the Council, the European Parliament, the European Economic and Social Committee and the Committee of the Regions for a Directive of the European Parliament and of the Council establishing a framework for the protection of soil and amending. Directive 2004/35/EC. COM (2006) 232 final. Brussels, European Commission
14. Ferreira CSS, Seifollahi-Aghmiuni S, Destouni G, Ghajarnia N, Kalantari Z (2022) Soil degradation in the European Mediterranean region: processes, status and consequences. Sci Total Environ 805:150106. https://doi.org/10.1016/j.scitotenv.2021.150106
15. EEA (2019) The European environment state and outlook 2020. Knowledge for transition to a sustainable Europe. European Environmental Agency. https://doi.org/10.2800/96749

16. UNDP/GEF (2004) Reclaiming the land sustaining livelihoods: lessons for the future. United Nations Development Fund/Global Environmental Facility, November 2004
17. Martín-Ortega PN, García-Montero LG, del Rio S, Penas A, Marchetti M, Lasserre B, Ozdemir E, Robredo FG, Pascual C, Guerrero CC, Alberdi I, Canellas I, Guerrero X, Hernández L, Jauregui MM, Miguel AS, Vallejo R, Sibelet N, Rivas-Marínez S (2018) Importance of Mediterranean forests. In FAO and Plan Bleu. State of Mediterranean Forests 2018. Food and Agriculture Organization of the United Nations, Rome and Plan Bleu, Marseille. pp 31–50
18. World Bank (2021) Land area. https://data.worldbank.org/indicator/AG.LND.TOTL.K2?locations=PT
19. Pinto-Correia T, Mascarenhas J (1999) Contribution to the extensification/intensification debate: new trends in the Portuguese montado. Landscape Urban Plann 46:125e131. https://doi.org/10.1016/S0169-2046(99)00036-5
20. Nunes AN, Almeida AC, Coelho COA (2011) Impacts of land use and cover type on runoff and soil erosion in a marginalarea of Portugal. Appl Geogr 31:687–699. https://doi.org/10.1016/j.apgeog.2010.12.006
21. Serra J, Cameira M, Cordovil C, Hutchings N (2021) Development of a groundwater contamination index based on the agricultural hazard and aquifer vulnerability: application to Portugal. Sci Total Environ 772:145032. https://doi.org/10.1016/j.scitotenv.2021.145032
22. EC (2021) CAP indicators-employment, 2019 https://agridata.ec.europa.eu/extensions/IndicatorsSectorial/EmploymentByEconomicActivity.html
23. Eurostat (2016) Main land types in UAA for Greece and Portugal, by NUTS2 regions. Available at: https://ec.europa.eu/eurostat/databrowser/view/tai05/default/table?lang=en (Downloaded on 13/5/2021)
24. FAOSTAT (2020) Agricultural practices. Available at: https://ec.europa.eu/eurostat/databrowser/view/ef_mp_prac/default/table?lang=en (Downloaded on 13/5/2021)
25. RBMPs (2015) 1st river basin management plans, Greece https://eceuropaeu/environment/water/participation/map_mc/countries/greece_enhtm Accessed 1 Sept 2020 (in Greek)
26. Karaouzas I, Kapetanaki N, Mentzafou A, Kanellopoulos T, Skoulikidis N (2021) Heavy metal contamination status in Greek surface waters: a review with application and evaluation of pollution indices. Chemosphere 263:128192. https://doi.org/10.1016/j.chemosphere.2020.128192
27. Kourgialas (2021) A critical review of water resources in Greece: the key role of agricultural adaptation to climate-water effects. Sci Total Environ 775:145857. https://doi.org/10.1016/j.scitotenv.2021.145857
28. Panagos P, Borrelli P, Poesen J (2019) Soil loss due to crop harvesting in the European Union: a first estimation of an underrated geomorphic process. Sci Total Environ 664:487–498. https://doi.org/10.1016/j.scitotenv.2019.02.009
29. Borrelli P, Robinson DA, Panagos P, Lugato E, Yang JE, Alewell C, Wuepper D, Montanarella L, Ballabio C (2020) Land use and climate change impacts on global soil erosion by water (2015-2070). Proc Natl Acad Sci U S A 117(36):21994–22001. https://doi.org/10.1073/pnas.2001403117
30. Verheijen RGA, Jones RJA, Rickson RJ, Smith CJ (2009) Tolerable versus actual soil erosion rates in Europe. Earth Sci Rev 94(1–4):23–38. https://doi.org/10.1016/j.earscirev.2009.02.003
31. Peña-Angulo D, Nadal-Romero E, Gonzalez-Hidalgo JC, Albaladejo J, Andreu V, Bagarello V, Barhi H, Batalla RJ, Bernal S, Gienes R, Campo J, Campo-Bescós MA, Canatario-Duarte A, Cantón Y, Casali J, Castillo V, Cerdà A, Cheggour A, Cid P, Cortesi N, Desir G, Díaz-Pereira E, Espigares T, Estrany J, Fernández-Raga M, Ferreira CSS, Ferro V, Gallart F, Giménez R, Gemeno E, Gómez JA, Gómez-Gutiérrez A, Gómez-Macpherson H, González-Pelayo O, Hueso-González P, Kairis O, Karatzas GP, Klotz S, Kosmas C, Lana-Renault N, Lasanta T, Latron J, Lázaro R, Le Bissonnais Y, Le Bouteiller C, Licciardello F, López-Tarazón JA, Lucía A, Marín C, Marqués MJ, Martínez-Fernández J, Martínez-mena M, Martínez-Murillo JF, Mateos L, Mathys N, Merino-Martín L,

Moreno-e las Heras M, Moustakas N, Nicolau JM, Novara A, Pampalone V, Raclot D, Rodríguez-Blanco ML, Rodrigo-Comino J, Romero-Díaz A, Roose E, Rubio JL, Ruiz-Sinoga JD, Schnabel S, Senciales-González JM, Simonneaux V, Solé-Benet A, Taguas EV, Taboada-Castro MM, Taboada-Castro MT, Todisco F, Úbeda X, Varouchakis EA, Vericat D, Wittenberg L, Zabaleta A, Zorn M (2019) Spatial variability of the relationships of runoff and sediment yield with weather types throughout the Mediterranean basin. J Hydrol 571:390–405. https://doi.org/10.1016/j.jhydrol.2019.01.059

32. Theocharopoulos SP, Florou H, Walling DE, Kalantzakos H, Christou M, Tountas P, Nikolaou T (2003) Soil erosion and deposition rates in a cultivated catchment area in central Greece, estimated using the 137Cs technique. Soil Tillage Res 69:153–162. https://doi.org/10.1016/S0167-1987(02)00136-8
33. Efthimiou (2020) The new assessment of soil erodibility in Greece. Soil Tillage Res 204: 104720. https://doi.org/10.1016/j.still.2020.104720
34. Eurostat (2021) Estimated soil erosion by water, by erosion level, land cover and NUTS 3 regions. Available at: https://ec.europa.eu/eurostat/databrowser/view/aei_pr_soiler/default/map?lang=en (Downloaded on 13/5/2021)
35. Panagos P, Ballabio C, Lugato E, Jones A, Borrelli P, Scarpa S, Orgiazzi A, Montanarella L (2018) Potential sources of anthropogenic copper inputs to European agricultural soils. Sustainability 10:2380. https://doi.org/10.3390/su10072380
36. Chevigny E, Quiquerez A, Petit C, Curmi P (2014) Lithology, landscape structure and management practice changes: key factors patterning vineyard soil erosion at metrescale spatial resolution. Catena 121:354–364. https://doi.org/10.1016/j.catena.2014.05.022
37. Barão L, Alaoui A, Ferreira CSS, Basch G, Schwilch G, Geissen V, Sukkel W, Lemesle J, Garcia-Orenes F, Morugán-Coronado A, Mataix-Solera J, Kosmas C, Glavan M, Pintar M, Tóth B, Hermann T, Vizitiu OP, Lipiec J, Reintam E, Xu M, Jiaying D, Fan H, Wang F (2019) Assessment of promising agricultural management practices. Sci Total Environ 649:610–619. https://doi.org/10.1016/j.scitotenv.2018.08.257
38. Ferreira CSS, Keizer JJ, Santos LMB, Serpa D, Silva V, Cerqueira M, Ferreira AJD, Abrantes N (2018) Runoff, sediment and nutrient exports from a Mediterranean vineyard under integrated production: an experiment at plot scale. Agric Ecosyst Environ 256:184–193. https://doi.org/10.1016/j.agee.2018.01.015
39. Ferreira CSS, Walsh RPD, Blake WH, Kikuchi R, Ferreira AJD (2017) Temporal dynamics of sediment sources in an urbanizing Mediterranean catchment. Land Degrad Dev 28(8): 2354–2369. https://doi.org/10.1002/ldr.2765
40. Imeson AC (1990) Climate fluctuations and soil erosion under Mediterranean conditions. Technical Report. International University, Valencia, Spain
41. Poesen J, Vanwalleghem T, De Vente J, Knapen A, Verstraeten G, Martinez-Casasnova JA (2006) Gully erosion in Europe. In: Boardman J, Poesen J (eds) Soil erosion in Europe. Wiley, Chichester, pp 515–536
42. Panagos P, Imeson A, Meusburger K, Borrelli P, Poesen J, Alewell C (2016) Soil conservation in Europe: wish or reality? Land Degrad Dev 27:1547–1551. https://doi.org/10.1002/ldr.2538
43. Yves R, Koutroulis A, Samniego L, Vicente-Serrano SM, Volaire F, Boone A, Page ML, Llasat MC, Albergel C, Burak S, Cailleret M, Kalin KC, Davi H, Dupuy J-L, Greve P, Grillakis M, Hanich L, Jarlan L, Polcher J (2020) Challenges for drought assessment in the Mediterranean region under future climate scenarios. Earth Sci Rev 210:103348. https://doi.org/10.1016/j.earscirev.2020.103348
44. Lugato E, Smith P, Borrelli P, Panagos P, Ballabio C, Orgiazzi A, Fernandez-Ugalde O, Montanarella L, Jones A (2018) Soil erosion is unlikely to drive a future carbon sink in Europe. Sci Adv 4:eaau3523. https://doi.org/10.1126/sciadv.aau3523
45. Alewell C, Ringeval B, Ballabio C, Robinson DA, Panagos P, Borrelli P (2020) Global phosphorus shortage will be aggravated by soil erosion. Nat Commun 11:4546. https://doi.org/10.1038/s41467-020-18326-7

46. Bakker MM, Govers G, Jones RA, Rounsevell MDA (2007) The effect of soil erosion on Europe's crop yields. Ecosystems 10:1209–1219. https://doi.org/10.1007/s10021-007-9090-3
47. Kalantari Z, Ferreira CSS, Koutsouris AJ, Ahmer AK, Cerdà A, Destouni G (2019) Assessing flood probability for transportation infrastructure based on catchment characteristics, sediment connectivity and remotely sensed soil moisture. Sci Total Environ 661:393–406. https://doi.org/10.1016/j.scitotenv.2019.01.009
48. Serpa D, Nunes J, Santos J, Sampaio E, Jacinto R, Veiga S, Lima J, Moreira M, Corte-Real J, Keizer J, Abrantes N (2015) Impacts of climate and land use changes on the hydrological and erosion processes of two contrasting Mediterranean catchments. Sci Total Environ 538:64–77. https://doi.org/10.1016/j.scitotenv.2015.08.033
49. Palma P, Fialho S, Lima A, Catarino A, Costa MJ, Barbieri MV, Monllor-Alcaraz LS, Postigo C, Lopez de Alda M (2021) Occurrence and risk assessment of pesticides in a Mediterranean Basin with strong agricultural pressure (Guadiana Basin: southern of Portugal). Sci Total Environ 794:148703. https://doi.org/10.1016/j.scitotenv.2021.148703
50. Eurostat (2016) Farming intensity in Greece and Portugal at a country level. Available at: https://agridata.ec.europa.eu/extensions/IndicatorsEnvironmental/FarmingIntensity.html (Downloaded on 13/5/2021)
51. Eurostat (2016) Utilized Agricultural Area (UAA) in Greece and Portugal, by NUTS2 regions. Available at: https://ec.europa.eu/eurostat/databrowser/view/EF_M_FARMLEG__custom_948812/default/table?lang=en (Downloaded on 13/5/2021)
52. Zuazo V, Pleguezuelo C (2008) Soil-erosion and runoff prevention by plant covers. A review. Agron Sustain Dev 28(1):65–86. ffhal-00886458, Springer Verlag/EDP Sciences/INRA
53. Gimeno-García E, Andreu V, Rubio JL (2007) Influence of vegetation recovery on water erosion at short and medium-term after experimental fires in a Mediterranean shrub land. Catena 69:150e160. https://doi.org/10.1016/j.catena.2006.05.003
54. Alaoui A, Barão L, Ferreira CSS, Schwilch G, Basch G, Garcia-Orenes F, Morugan A, Mataix-Solera J, Kosmas C, Glavan M, Tóth B, Hermann T, Vizitiu OP, Lipiec J, Frac M, Reintam E, Xu M, Di J, Fan H, Sukkle W, Lemesle J, Geissen V, Fleskens L (2020) Impacts of agricultural management practices on soil quality in Europe and China using a visual soil assessment methodology. Agron J:1–16. https://doi.org/10.1002/agj2.20216
55. Boulet AK, Alarcão C, Ferreira C, Kalantari Z, Veiga A, Campos L, Ferreira A, Hessel R Agro-ecological services delivered by legume cover crops grown in succession with grain corn crops in the Mediterranean region. Open Agric 6:609–626. https://doi.org/10.1515/opag-2021-0041
56. Ruiz Sinoga JD, Martinez-Murillo JF (2009) Effects of soil surface components on soil hydrological behaviour in a dry Mediterranean environment (southern Spain). Geomorphology 108:234–245. https://doi.org/10.1016/j.geomorph.2009.01.012
57. Keesstra S, Pereira P, Novara A, Brevik EC, Azorin-Molina C, Parras-Alcántara L, Jordán A, Cerdà A (2016) Effects of soil management techniques on soil erosion in apricot orchards. Sci Total Environ 551–552:357–366. https://doi.org/10.1016/j.scitotenv.2016.01.182
58. Megremi I, Vasilatos C, Vassilakis E, Economou-Eliopoulos M (2019) Spatial diversity of Cr distribuion in soil and groundwater sites in relation with land use management in a Mediterranean region: the case of C. Evia and Assopos-Thiva Basins, Greece. Sci Total Environ 651: 656–667. https://doi.org/10.1016/j.scitotenv.2018.09.186
59. Evans D, Janes-Bassett V, Borrelli P, Chenu C, Ferreira CSS, Griffiths R, Kalantari Z, Keesstra S, Lal R, Panagos P, Robinson D, Seifollahi-Aghmiuni S, Smith P, Steenhuis T, Thomas A, Visser S (2021) Sustainable futures over the next decade are rooted in soil science. Eur J Soil Sci 1-16. https://doi.org/10.1111/ejss.13145
60. EEA (2012) The State of soil in Europe. European Enviornmental Agency and Joint Research Centre. Report EUR 25186 EN
61. JRC (2008) European Soil Data Centre. Joint Research Centre, European Commission. https://esdac.jrc.ec.europa.eu/themes/soil-compaction. Accessed 15 Oct 2020

62. Garcia-Tomillo A, Figueiredo T, Almeida A, Rodrigues J, Dafonte J, Paz-González A, Nunes J, Hernandez Z (2017) Comparing effects of tillage treatments performed with animal traction on soil physical properties and soil electrical resistivity: preliminary experimental results. Open Agric 2:317–328
63. Greenwood KL, McKenzie BM (2001) Grazing effects on soil physical properties and the consequences for pastures a review. Aust J Exp Agric 41:1231–1250. https://doi.org/10.1071/EA00102
64. FAO (Food and Agriculture Organization of the United Nations) (2015) World fertilizer trends and outlook to 2018. Food Agriculture Organization United Nations, Rome, Italy
65. Alaoui A, Diserens E (2018) Mapping soil compaction – a review. Curr Opin Environ Sci Health 5:60–66. https://doi.org/10.1016/j.coesh.2018.05.003
66. Prats SA, Malvar MC, Coelho CA, Wagenbrenner JW (2019) Hydrologic and erosion responses to compaction and added surface cover in post-fire logged areas: isolating splash, interrill and rill erosion. J Hydrol 575(2019):408–419. https://doi.org/10.1016/j.jhydrol.2019.05.038
67. Bluett C, Tullberg JN, McPhee JE, Antille DL (2019) Soil tillage research: why still focus on soil compaction? Soil Tillage Res 194:104282. https://doi.org/10.1016/J.STILL.2019.05.028
68. Ferreira CSS, Veiga A, Caetano A, Gonzalez-Pelayo O, Karine-Boulet A, Abrantes N, Keizer J, Ferreira AJD (2020) Assessment of the impact of distinct vineyard management practices on soil physico-chemical properties. Air Soil Water Res 13:1–13. https://doi.org/10.1177/1178622120944 8
69. Cerdà A, Novara A, Moradi E (2021) Long-term non-sustainable soil erosion rates and soil compaction in drip-irrigated citrus plantation in eastern Iberian Peninsula. Sci Total Environ 787:147549. https://doi.org/10.1016/j.scitotenv.2021.147549
70. Schjønning P, van den Akker JH, Keller T, Greve MH, Lamandé M, Simojoki A, Stettler M, Arvidsson J, Breuning-Madsen H (2015) Driver-pressure-state-impact-response (DPSIR) analysis and risk assessment for soil compaction a European perspective. In: Sparks DL (ed) Advances in agronomy, pp 183–237
71. Abdalla M, Hastings A, Chadwick DR, Jones DL, Evans CD, Jones MB, Rees RM, Smith P (2018) Critical review of the impacts of grazing intensity on soil organic carbon storage and other soil quality indicators in extensively managed grasslands. Agric Ecosyst Environ 253: 62–81
72. Monaghan RM, Laurenson S, Dalley DE, Orchiston TS (2017) Grazing strategies for reducing contaminant losses to water from forage crop fields grazed by cattle during winter. N Z J Agric Res 60:333–348. https://doi.org/10.1080/00288233.2017.1345763
73. Taboada MA, Rubio G, Chaneton EJ (2011) Grazing impacts on soil physical, chemical, and ecological properties in forage production systems. In: Hatfield JL, Sauer TJ (eds) Soil management: building a stable base for agriculture. SSSA, Madison, pp 301–320
74. European Commission (2008) Map for Europe of natural susceptibility of soils to compaction, European Commission – Joint Research Centre. Available from ESDAC.jrc.ec.europa.eu
75. EC (2020) Producing 69% of the world's production, the EU is the largest producer of olive oil. European Commission. https://ec.europa.eu/info/news/producing-69-worlds-production-eu-largest-producer-olive-oil-2020-feb-04_en
76. Nikolaos E (2020) The new assessment of soil erodibility in Greece. Soil Tillage Res 204: 104720. https://doi.org/10.1016/j.still.2020.104720
77. Blanco-Canqui H, Benjamin JG (2013) Impacts of soil organic carbon on soil physical behavior. In: Logsdon S, Berli M, Horn R (eds) Quantifying and modeling soil structure dynamics, advanced agricultural system model, vol 3. SSSA, Madison, pp 11–40
78. Drewry JJ, Cameron KC, Buchan GD (2008) Pasture yield and soil physical property responses to soil compaction from treading and grazing – a review. Aust J Soil Res 46(3): 237–256. https://doi.org/10.1071/SR07125

79. Carretta L, Tarolli P, Cardinali A, Nasta P, Romano N, Masin R (2021) Evaluation of runoff and soil erosion under conventional tillage and no-till management: a case study in northeast Italy. Catena 197:104972. https://doi.org/10.1016/j.catena.2020.104972
80. Hallett P, Balana B, Towers W, Moxey A, Chamen T (2012) Studies to inform policy development with respect to soil degradation sub project A: cost curve for mitigation of soil compaction Defra project SP1305
81. Eswaran H, Lal R, Reich PF (2001) Land degradation: an overview. Natural resources conservation service soils, United States Department of Agriculture. https://www.nrcs.usda.gov/wps/portal/nrcs/detail/soils/use/?cid=nrcs142p2_054028. Accessed 13 Sept 2020
82. Lal R (2010) A dual response of conservation agriculture to climate change: reducing CO2 emissions and improving the carbon sink. Proceedings of the European congress on conservation agriculture. Towards agro-environmental climate and energetic sustainability. Madrid, Spain, pp 3–19
83. Carvalho M, Lourenço E (2014) Conservation agriculture – a Portuguese case study. J Agro Crop Sci. ISSN 0931-2250
84. FAO (Food and Agriculture Organization of the United Nations) (2018) Soil pollution: a hidden reality. Global Soil partnership. Rome. http://www.db.zs-intern.de/uploads/1525681402-SoilPollution.pdf
85. Silva V, Mol HGJ, Zomer P, Tienstra M, Ritsema CJ, Geissen V (2019) Pesticide residues in European agricultural soils – a hidden reality unfolded. Sci Total Environ 653:1532–1545. https://doi.org/10.1016/j.scitotenv.2018.10.441
86. EFSA (European Food Safety Authority) (2020) The 2018 European Union report on pesticide residues in food. EFSA EFSA J 18(4):6057. https://doi.org/10.2903/j.efsa.2020.6057
87. Rodríguez-Eugenio McLaughlin M, Penneock D (2018) Soil pollution: a hidden reality, Education Chemistry, Rome. https://doi.org/10.5124/jkma.1998.41.10.1032
88. Cameira MR, Rolim J, Valente F, Mesquita M, Dragosits U, Cordovil CMS (2021) Translating the agricultural N surplus hazard into groundwater pollution risk: implications for effectiveness of mitigation measures in nitrate vulnerable zones. Agric Ecosyst Environ 306:107204. https://doi.org/10.1016/j.agee.2020.107204
89. FAOSTAT (2021) Ratio between the totals by nutrient of agricultural use of chemical or mineral fertilizers, and the area of cropland (sum of arable land and permanent crops) at national and European level reported in the FAOSTAT for the period 1990 to 2018. Available at: http://www.fao.org/faostat/en/#data/EF (Downloaded on 13/5/2021)
90. Albornoz F (2016) Crop responses to nitrogen overfertilization: a review. Sci Hortic 205:79–83. https://doi.org/10.1016/j.scienta.2016.04.026
91. Peng Y, Zhang M, Lee SY (2017) Food availability and predation risk drive the distributional patterns of two pulmonate gastropods in a mangrove-saltmarsh transitional habitat. Mar Environ Res 130:21–29. https://doi.org/10.1016/j.marenvres.2017.07.005
92. Carbonell G, Imperial RM, Torrijos M, Delgado M, Rodriguez JA (2011) Effects of municipal solid waste compost and mineral fertilizer amendments on soil properties and heavy metals distribution im maize plants (Zea mays L.). Chemosphere 85:1614–1623. https://doi.org/10.1016/j.chemosphere.2011.08.025
93. Faissal A, Ouazzani N, Parrado JR, Dary M, Manyani H, Morgado BR, Barragán MD, Mandi L (2017) Impact of fertilization by natural manure on the microbial quality of soil: molecular approach. Saudi J Biol Sci 24:1437–1443. https://doi.org/10.1016/j.sjbs.2017.01.005
94. Verlicchi P, Zambelo E (2015) Pharmaceuticals and personal care products in untreated and treated sewage sludge: occurrence and environmental risk in the case of application on soil-acritical review. Sci Total Environ 538:750–767
95. Bloem E, Albihn A, Elving J, Hermann L, Lehmann L, Sarvi M, Schaaf T, Schick J, Turtola E, Ylivainio K (2017) Contamination of organic nutrient sources with potentially toxic elements, antibiotics and pathogen microorganisms in relation to P fertilizer potential and treatment options for the production of sustainable fertilizers: a review. Sci Total Environ 607–608:225–242. https://doi.org/10.1016/j.scitotenv.2017.06.274

96. Urra J, Alkorta I, Lanzén A, Mijangos I, Garbisu C (2019) The application of fresh and composted horse and chicken manure affects soil quality, microbial composition and antibiotic resistance. Appl Soil Ecol 135:73–84. https://doi.org/10.1016/j.apsoil.2018.11.005
97. Eurostat (2016) Tillage practices agro-environmental indicator. Accessed at: https://ec.europa.eu/eurostat/statistics-explained/index.php?title=Agri-environmental_indicator_-_tillage_practices
98. Xia Y, Luo H, Li D, Chen Z, Yang S, Liu Z, Yang T, Gai C (2020) Efficient immobilization of toxic heavy metals in multi-contaminated agricultural soils by amino-functionalized hydrochar: performance, plant responses and immobilization mechanisms. Environ Pollut 261:114217. https://doi.org/10.1016/j.envpol.2020.114217
99. Guittonny-Philippe A, Masotti V, Hohener P, Boudenne J-L, Viglione J, Laffont-Schwob I (2014) Constructed wetlands to reduce metal pollution from industrial catchments in aquatic Mediterranean ecosystems: a review to overcome obstacles and suggest potential solutions. Environ Int 64:1–16. https://doi.org/10.1016/j.envint.2013.11.016
100. Ballabio C, Panagos P, Lugato E, Huang J-H, Orgiazzi A, Jones A, Fernández-Ugalde O, Borrelli P, Montanarella L (2018) Copper distribution in European tospoils: an assessment based on LUCAS soil survey. Sci Total Environ 15(636):282–298. https://doi.org/10.1016/j.scitotenv.2018.04.268
101. Panagos P, Standardi G, Borrelli P, Lugato E, Montanarella L, Bosello F (2018) Cost of agricultural productivity loss due to soil erosion in the European Union: from direct cost evaluation approaches to the use of macroeconomic models. Land Degrad Dev 29:471–484. https://doi.org/10.1002/ldr.2879
102. FAO (Food and Agriculture Organization of the United Nations) (2011) The state of the world's land and water resources for food and agriculture – managing systems at risk. Food Agriculture Organization United Nations and Earthscan, Rome, Italy
103. Kanellopoulos C, Mitropoulos P (2015) Geochemistry of serpentine agricultural soil and associated groundwater chemistry and vegetation in the area of Atalanti, Greece. J Geochem Explor 158:22–33. https://doi.org/10.1016/j.gexplo.2015.06.013
104. Tzempelikou E, Zeri C, Iliakis S, Paraskevopoulou V (2021) Cd, Co, Cu, Ni, Pb, Zn in coastal and transitional waters of Greece and assessment of background concentrations: results from 6 years implementation of the water framework directive. Sci Total Environ 774:145177. https://doi.org/10.1016/j.scitotenv.2021.145177
105. Antoniadis V, Golia EE, Wang S, Shaheen JR (2019) Soil and maize contamination by trave elements and associated health risk assessment in the industrial area of Volos, Greece. Environ Int 124:79–88. https://doi.org/10.1016/j.envint.2018.12.053
106. EC (2020) Caring for soil is caring for life. European Commission https://doi.org/10.2777/918775
107. EFSA (European Food Safety Authority) (2017) The 2015 European Union report on pesticide residues in food. EFSA J 15. https://doi.org/10.2903/j.efsa.2017.4791
108. FAOSTAT (2021) Use of pesticides per area of cropland (which is the sum of arable land and land under permanent crops) at national and European level reported in the FAOSTAT for the period 1990 to 201. Available at: http://www.fao.org/faostat/en/#data/EF (Downloaded on 13/5/2021)
109. Silva V, Montanarella L, Jones A, Fernández-Ugalde O, Mol HGJ, Ritsema CJ, Geissen V (2018) Distribution of glyphosate and aminomethylphosphonic acid (AMPA) in agricultural topsoils of the European Union. Sci Total Environ 621:1352–1359. https://doi.org/10.1016/j.scitotenv.2017.10.093
110. Geissen V, Silva V, Lwanga EH, Beriot N, Oostindie K, Bin Z, Pyne E, Busink S, Zomer P, Mol H, Ritsema CJ (2021) Cocktails of pesticide residues in conventional and organic farming systems in Europe – legacy of the past and turning point for the future. Environ Pollut 278. https://doi.org/10.1016/j.envpol.2021.116827
111. Karasali H, Marousopoulou A, Macgera K (2016) Pesticide residue concentration in soil following conventional and low-imnput crop management in a Medoterraneam agro-

ecosytem, in Central Greece. Sci Total Environ 541:130–142. https://doi.org/10.1016/j.scitotenv.2015.09.016

112. Danis T, Karagiozoglou DT, Tsakiris IN, Alegakis AK, Tsatsakis AM (2011) Evaluation of pesticides redisues in Grek peaches during 2002-2007 after the implementation of integrated crop management. Food Chem 126:97–103. https://doi.org/10.1016/j.foodchem.2010.10.083
113. Kleveno J, Loague K, Green R (1992) Evaluation of a pesticide mobility index: impact of recharge variation and soil profile heterogeneity. J Contam Hydrol 11:83–99. https://doi.org/10.1016/0169-7722(92)90035-D
114. Dahl M, Bergman S, Björk M, Diaz-Almela E, Granberg M, Gullström M, Leiva-Dueñas C, Magnusson K, Marco-Méndez C, Piñeiro-Juncal N, Mateo MÁ (2021) A temporal record of microplastic pollution in Mediterranean seagrass soils. Environ Pollut 273. https://doi.org/10.1016/j.envpol.2021.116451
115. Li X, Chen L, Mei Q, Dong B, Dai X, Ding G, Zeng EY (2018) Microplastics in sewage sludge from the wastewater treatment plants in China. Water Res 142:75–85. https://doi.org/10.1016/j.watres.2018.05.034
116. van den Berg P, Huerta-Lwanga E, Corradini F, Geissen V (2020) Sewage sludge application as a vehicle for microplastics in eastern Spanish agricultural soils. Environ Pollut 261. https://doi.org/10.1016/j.envpol.2020.114198
117. Harms IK, Diekötter T, Troegel S, Lenz M (2021) Amount, distribution and composition of large microplastics in typical agricultural soils in northern Germany. Sci Total Environ 758. https://doi.org/10.1016/j.scitotenv.2020.143615
118. Yang J, Li R, Zhou Q, Li L, Li Y, Tu C, Zhao X, Xiong K, Christie P, Luo Y (2021) Abundance and morphology of microplastics in an agricultural soil following long-term repeated application of pig manure. Environ Pollut 272. https://doi.org/10.1016/j.envpol.2020.116028
119. Lan T, Wang T, Cao F, Yu C, Chu Q, Wang F (2021) A comparative study on the adsorption behavior of pesticides by pristine and aged microplastics from agricultural polyethylene soil films. Ecotoxicol Environ Saf 209. https://doi.org/10.1016/j.ecoenv.2020.111781
120. Wang K, Ma Z, Zhang X, Ma J, Zhang L, Zheng J (2020) Effects of vegetation on the distribution of soil water in gully edges in a semi-arid region. Catena 195:104719. https://doi.org/10.1016/j.catena.2020.104719
121. Garbounis G, Komilis D (2021) A modeling methodology to predict the generation of wasted plastic pesticide containers: an application to Greece. Waste Manag 131:177–186. https://doi.org/10.1016/j.wasman.2021.06.005
122. CEC (2006) Thematic strategy for soil protection. COM(2006)231final. Commission of the European Communities, Brussels 22.9.2006
123. Sentís IP (2014) Advances in the prognosis of soil sodicity under dryland and irrigated conditions. Int Soil Water Cons Res 2(4):50–63. https://doi.org/10.1016/S2095-6339(15)30058-7
124. Rengasamy P (2006) World salinisation with emphasis on Australia. J Exp Bot Plants 57(5): 1017–1023. https://doi.org/10.1093/jxb/erj108
125. Tóth G, Montanarella L, Rusco E (2008) Threats to soil quality in Europe. EUR:23438. https://esdac.jrc.ec.europa.eu/images/eusoils_old/Library/Themes/Salinization/Resources/salinisation.pdf
126. Keren R (2000) Salinity. In: Sumner ME (ed) Handbook of soil science. CRC Press, Boca Raton, pp G3–G25
127. Cherlet M, Hutchinson C, Reynolds J, Hill J, Sommer S, von Maltitz G (2018) World atlas of desertification. Publication Office of the European Union, Luxembourg
128. Ghassemi F, Jakeman AJ, Nix HA (1995) Salinisation of land and water resources: human causes, extent, management and case studies. University of New South Wales Press Ltd, Sydney. 526 p
129. UN (2017) Global land outlook. Secretariat of the United Nations to Combat Desertification. https://knowledge.unccd.int/publication/full-report. Accessed 10 June 2021

130. Gonçalves MC, Martins JC, Ramos TB (2015) A salinização do solo em Portugal. Causas, extensão e soluções Revista de Ciências Agrárias 38(4). https://doi.org/10.19084/RCA15140
131. Singh A (2021) Soil salinization management for sustainable development: a review. J Environ Manage 277:111383. https://doi.org/10.1016/j.jenvman.2020.111383
132. Lo'ay AA, El-Ezz SFA (2021) Performance of 'Flame seedless' grapevines grown on different rootstocks in response to soil salinity stress. Sci Hortic 275:109704. https://doi.org/10.1016/j.scienta.2020.109704
133. Abrol IP, Yadav JSP, Massoud FI (1988) Salt-affected soils and their management. FAO soils bulletin, vol 39. FAO, Rome
134. Pearson GA (1960) Tolerance of crops to exchangeable sodium, agriculture information bulletin no. 216 Agricultural Research Service, United States Department of Agriculture
135. Gupta SK, Sharma SK (1990) Response of crops to high exchangeable sodium percentage. Irrig Sci 11:173–179. https://doi.org/10.1007/BF00189455
136. Maas EV (1993) Testing crops for salinity tolerance proceedings workshop on adaptation of plants to soil stresses. In: Maranville JW, Baligar BV, Duncan RR, Yohe JM (eds) INTSORMIL Pub No. 94-2, Univ of Ne, Lincoln, pp 234–247, 4 Aug 1993

Agricultural Soil Degradation in Hungary

M. Birkás and I. Dekemati

Contents

1 Introduction 140
2 Soil Compaction 141
3 Water and Wind Erosion 143
4 Flooding and Water Logging 145
5 Landslides 147
6 Soil Contamination 147
7 Soil Sealing 148
8 Salinity 149
9 Acidification 150
10 Agrochemical Use 151
11 Conclusions 154
References 155

Abstract Soil degradation is a serious phenomenon both in Hungary and worldwide. Although it may be induced by natural causes, human activity, especially in the past two centuries has contributed more significantly to soil degradation. There are many known consequences of soil degradation. Although in the short-term moderating the rate of deterioration can be acceptable, the impacts of climate change on soil degradation seem to be a real hindering factor to crop production in the longer term. In Hungary, soil compaction, water and wind erosion, and water logging may present difficulties in the future. Based on negative experiences in crop production in dry and wet seasons, soil management should be used to prevent or alleviate soil compaction. Landslides, soil contamination, salinity, acidification, and agrochemical use are likely to be kept within limits by complying with national and EU regulatory requirements. Limiting soil sealing will be however a difficult issue in the future.

M. Birkás (✉) and I. Dekemati
Department of Agronomy, Institute of Crop Production Sciences, Hungarian University of Agriculture and Life Sciences, Gödöllő, Hungary
e-mail: Birkas.Marta@uni-mate.hu; Dekemati.Igor@uni-mate.hu

Paulo Pereira, Miriam Muñoz-Rojas, Igor Bogunovic, and Wenwu Zhao (eds.),
Impact of Agriculture on Soil Degradation II: A European Perspective,
Hdb Env Chem (2023) 121: 139–158, DOI 10.1007/698_2022_949,

Published online: 8 February 2023

Keywords Acidification, Agrochemicals, Landslide, Salinity, Soil compaction, Soil contamination, Soil sealing, Water and wind erosion, Water logging

1 Introduction

The total area of Hungary is 9,303,000 ha, of which 57.1% (or 5,310,000 ha) is agricultural land, containing 46.4% or 4,318,000 ha as cultivated land. The topsoil textures of Hungarian soils can be characterized as follows: sand 15%, sandy loam 12%, loam 47%, and loamy clay and clay 26% [1]. Soil tillage is sufficient for crop cultivation on 59% of the soils, while the remaining 41% needs additional cultivation management. Várallyay [2, 3] stated that approximately 34.8% of the soils are sensitive to degradation and compaction (e.g., Solonetz, Gleysols, and Vertisols), 13.9% are non-sensitive (Calcisols), 23.0% are slightly sensitive (Arenosols, Cambisols, Histosols), and 28.3% have moderate sensitivity (Luvisols, Chernozems) (Fig. 1).

The Hungarian climate is continental, although extreme phenomena have occurred more frequently in the last three decades. The average annual precipitation ranges from 800 mm in the west to 500 mm in the east. During the past decade,

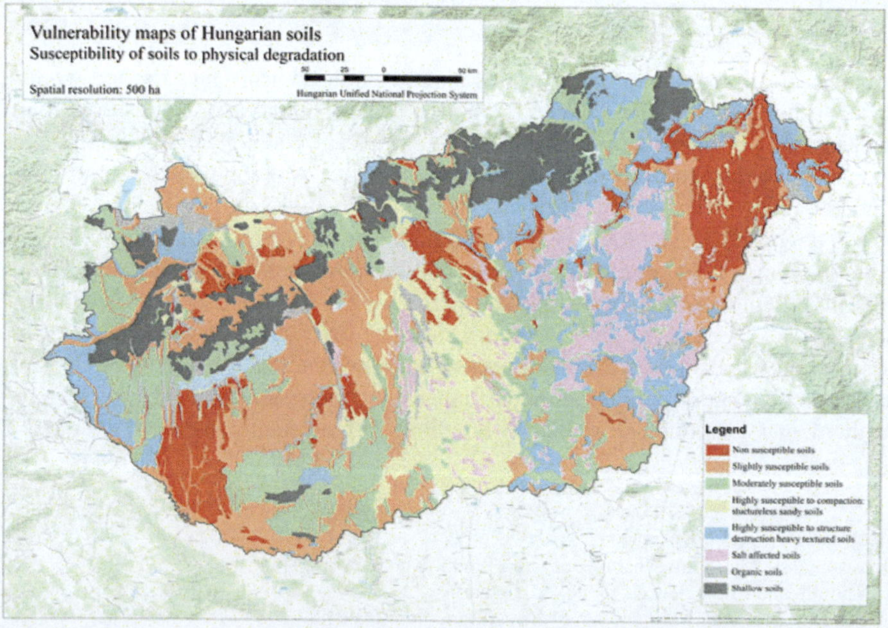

Fig. 1 Vulnerability of Hungarian soils to soil compaction and structural destruction. *Source: Reproduced with permission from* [3]

2 years were dry, 1 year was rainy and 7 years – due to the alternation of the dry and rainy periods – were extreme.

The major types of soil degradation in terms of the total land area of Hungary are erosion (24%), physical degradation (14%), acidification (12%), salinization/alkalization (8%), and extreme drying and water logging (5%) [4]. Overgrazing is not an assessed type of soil damage in Hungary, because the number of grazing livestock has steadily declined in Hungary in recent decades. In addition to these types of soil degradation, this chapter also discusses soil compaction, landslides, soil contamination, and soil sealing.

2 Soil Compaction

Regarding soil compaction, this chapter is focused on tillage-induced soil compaction, with respect to its formation, prevention, and alleviation, in relation to farming activities.

In classical Hungarian soil management literature, human-induced compaction is a soil quality problem frequently referred to as compacted structure; compacted, dense layer; crust of furrow bottom; dense crust below the tilled layer; smeared layer or other similar terms.

Soil compaction is a consequence of natural processes and human activity (farming). Farming-induced soil compaction originates from traffic and tillage operations. Tillage-induced mechanical stress, in other words, pan compaction, is a side effect of the parts of tools used in tillage (e.g. ploughshare, or disk plates), and occurs at the depth of tillage. The plough pan typically forms in the ploughed layer, depending on the most frequent ploughing depth, which ranges between 22 and 35 cm [5]. The disk pan forms underneath the most frequent disking depth, i.e., between 12 and 20 cm [6, 7].

Plough pan compaction is a term that has been used since the 1920s, while disk-pan compaction was first observed, evaluated, and described by Birkás [6] in Hungarian and later by Birkás et al. [8] in English language. According to Nyiri [9], the area of soils affected by natural and human-induced compaction covered 1.4 million hectares. Later, 1.82 million hectares of compacted areas were estimated on the basis of farm soil monitoring [10]. Table 1 presents measurements of farming-induced compaction carried out between 1976 and 2019, in 8 time-periods.

Types of tillage-induced soil compaction were grouped according to the depth of occurrence by Birkás et al. [10]. Seven typical types of compactions caused by incorrect tillage practices were identified, i.e., compaction below 55 cm, between 35–45 cm, 28–32, 22–25, 18–22 cm, 12–15 cm, and below 10 cm. To date, there are no similar categorizations carried out in other Hungarian publications. These measurements were conducted on a total area of 28,360 ha under the following crops: winter wheat (25%), oilseed rape (20%), maize (20%), sugar beet (5%), sunflower (15%); soybean (3%), and others (alfalfa, peas, winter, and spring barley 12%).

Table 1 Soil compaction observed on 28,360 ha of land during eight examination periods in Hungary (1976–2019). Soil layer data are presented in percentage values (%). Source [10]

Depth of loosened soil layer (cm)	1 1976–1987	2 1988–1990	3 1991–1997	4 1998–2001	5 2002–2007	6 2008–2010	7 2011–2013	8 2014–2019
≥55	14	4	1	0	11	9	17	3
35–45	22	12	6	2	21	26	33	27
28–32	44	47	42	36	30	34	31	22
22–25	14	22	23	14	21	16	11	22
18–22	6	10	16	22	12	10	7	9
12–15	0	3	7	14	5	5	0.6	12
≤10	0	2	5	12	0	0	0.4	5
Total area (ha)	2,420	2,860	2,580	1,860	4,690	2,870	3,410	7,670

Non-compacted soil layers to a depth of 28–55 cm had “favorable looseness” (bulk density not exceeding 1.38–1.48 t m^{-3} and 2.5–2.8 MPa penetration resistance), which was 52% of the monitored soil area. Less climate damage was found in deeply loosened soils than in soils that had shallow tillage. Soil looseness to a depth of 22–25 cm was acceptable for crop production in favorable seasons; however, it was suboptimal for crop production in years with extreme weather conditions. The damages on soil compacted below 18 cm were more severe than in soils with a deeper loosened layer. This type of soil condition covers about 1.12 million ha, where disk tillage is frequently used. The need of alleviation can be explained by the wider application of subsoiling (on 25–30% of total cultivated land).

Both the area of the loosened layer and its depth fluctuated strongly during the eight monitored time periods. The reasons for these variations were the shortcomings of tillage practices and the use of heavy machinery. Nowadays, because of climate change, the settling effect of heavy rains on soils must be added to this assessment. The adverse effects of soil compaction on crop production are well known from literature [10–13]. In recent years, the restriction of normal water movement has become the main cause of soil compaction. Soils degraded by any form of compaction have severely suffered from drought damage and persistent water stagnation above the surface and pan layers. The prevention of compaction is a general expectation, and management solutions are adapted to mitigate it depending on the local features of crop production areas [11].

New data from 2019 show that level and extent of soil susceptibility to physical degradation requires more attention in future landscape management (Fig. 2).

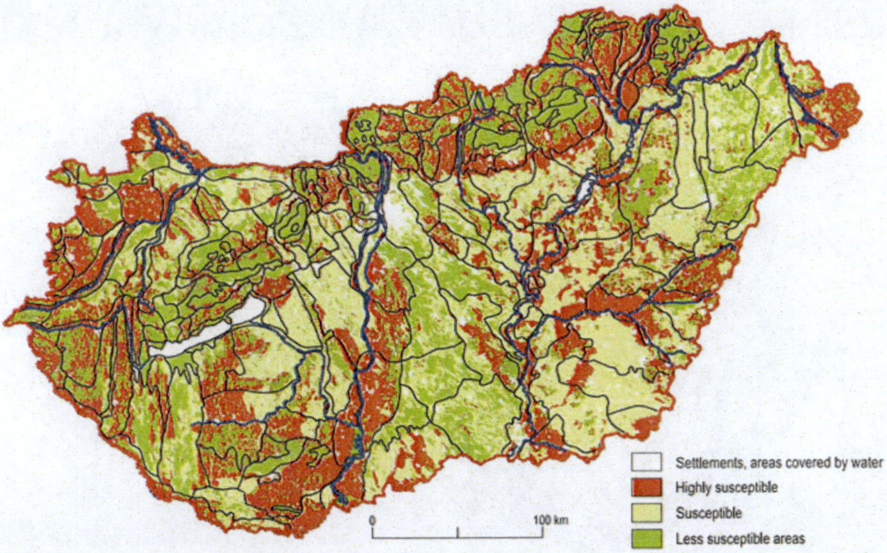

Fig. 2 Areas susceptible to physical degradation in Hungary. White = settlements, areas covered by water; red = highly susceptible; yellow = susceptible; green = less susceptible areas. *Source: Reproduced with permission from* [14]

3 Water and Wind Erosion

Erosion, which is a form of physical degradation, is a major threat to soil resources and may impair their ability to provide a range of ecosystem goods and services [15]. Two types of erosions endanger soils, i.e., water and wind. Some of the natural factors that influence water erosion intensity are rainfall intensity, soil erodibility, length of slope, slope steepness, and the exposure of the sloped soil. Wind erosion occurs when strong winds remove the light-textured soil and transport it to unexpected areas.

Both types of soil erosion cause damages (soil loss or sedimentation) on-site and off-site. In the Hungarian context, water and wind erosion phenomena were exacerbated in the last two decades, due to climate phenomena and improper farm management [16]. According to Michéli et al. [4], the size of the eroded area reached 35.3% of cultivated soils in Hungary. Similar data were published by Kertész and Centeri [17], outlining that over one-third of croplands were eroded (2.3 million ha) (Fig. 3). Overall, 25% of the total area of Hungary is affected by water erosion and 16% by wind erosion with two-thirds of the area covered by loose sediment susceptible to soil erosion [18]. As agricultural activity has been extended to the hilly regions in Hungary, the risk of soil erosion has also expanded and both sheet and gully erosion occurred [19]. Kertész and Křeček [19] outlined that water erosion on arable land is especially hazardous to large arable fields created mainly in the 1960s and 1970s. Among the many consequences of water erosion damages, the two

Fig. 3 Soil erosion risk map of Hungary. The legend categories (1–9) correspond to the following tons per ha per year values: 1 = 0.0–0.5; 2 = 0.5–1.0; 3 = 1.0–1.5; 4 = 1.5–2.0; 5 = 2.0–5.0; 6 = 5.0–8.0; 7 = 8.0–11.0; 8 = 11.0–100.0; 9 = > 100.0 *Source: Reproduced with permission from* [20]

most serious ones are the loss of organic matter-rich topsoil and sedimentation, and the decrease in moisture-storing capacity and biodiversity [4].

Rusco et al. [21] listed the on-site and off-site damages due to soil erosion in Europe. On-site damages are loss of organic matter, soil structure degradation, soil surface compaction, reduction of water penetration, supply reduction at the water table, surface erosion, nutrient removal, increase of coarse elements, rill and gully generation, plant uprooting, and reduction of soil productivity. Off-site damages include floods, water pollution, infrastructure burial, and obstruction of drainage networks, changes in watercourse shape and water eutrophication. Twenty percent of Hungary is covered by windblown sand, and high wind erosion risk endangers 10% of the surface [19]. Wind erosion risk is expected to increase in Hungary, due to the growing frequency of windy and dry periods, mainly during the spring and summer seasons. Farming is an additional influencing factor, via soil tillage and inappropriate crop residue management. Removing the protective tree lines from the sides of fields to enlarging the production areas has also increased wind damages (Fig. 4) [22].

The tolerable rate of soil erosion loss is <1.0 t ha/year [15]. However, Klik and Rosner [23] found that mean long-term annual erosion loss rates for conventional tillage ranged between 8.6 and 33.2 t ha^{-1}, for mulch tillage between 3.6 and

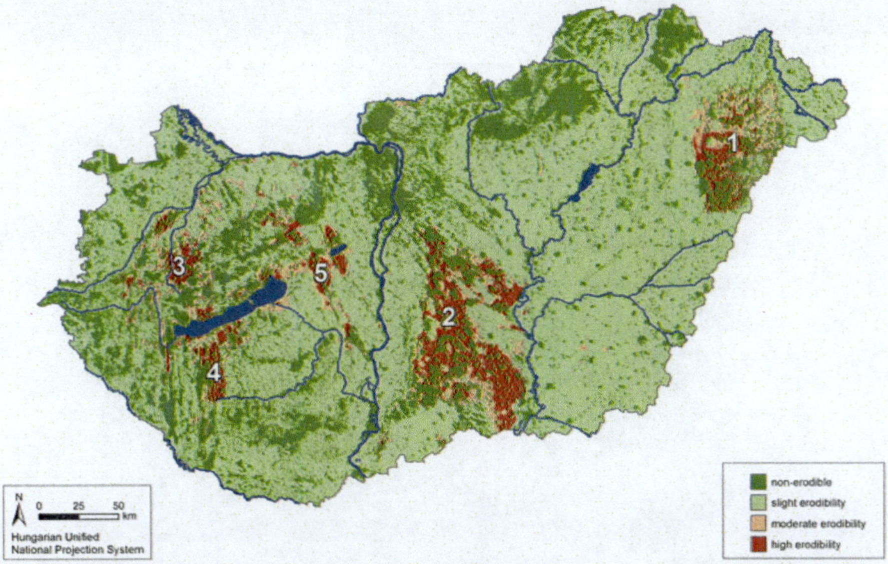

Fig. 4 Categorized wind erosion susceptibility map of Hungarian soils. The five distinct areas with typically higher wind erosion risk are numbered: Nyírség (1), Danube-Tisza Interfluve (2), Glacis in the foreground of the Transdanubian Mountains (3), Inner-Somogy (4), Transdanubian loess region (5). *Source: Reproduced with permission from* [22]

5.3 t ha^{-1} and for no-till between 1.9 and 3.0 t ha^{-1}. The long-term results of this study showed two main benefits with the use of mulch and no-till in well-drained soils, i.e., improved soil aggregate stability and water contents, and reduction in soil and nutrient losses.

Hungary has a relatively large agricultural area where wind erosion interacts with water erosion, displaying a complexity of fluvio-aeolian processes [16]. Avoidance of damages is a complex task that generally consists of site adopted land use, soil conservation tillage, and soil preservation crop residue management. In order to reduce soil loss, land use for soil conservation in endangered areas benefits from higher surface cover as well as lower flow velocities [23, 24].

4 Flooding and Water Logging

The Carpathian basin includes all of Hungary, parts of Slovakia, the Czech Republic, Romania, and Croatia, stretching out of the EU into Serbia and Ukraine). Its geological formation, geographic location, and climate threats is continuously exposed to water-related phenomena, such as water surplus and deficit [25]. The Hungarian area that is endangered by excess water covers about 4.4 million ha, which is 47% of the country's territory [26]. Figure 5 shows soil vulnerability in

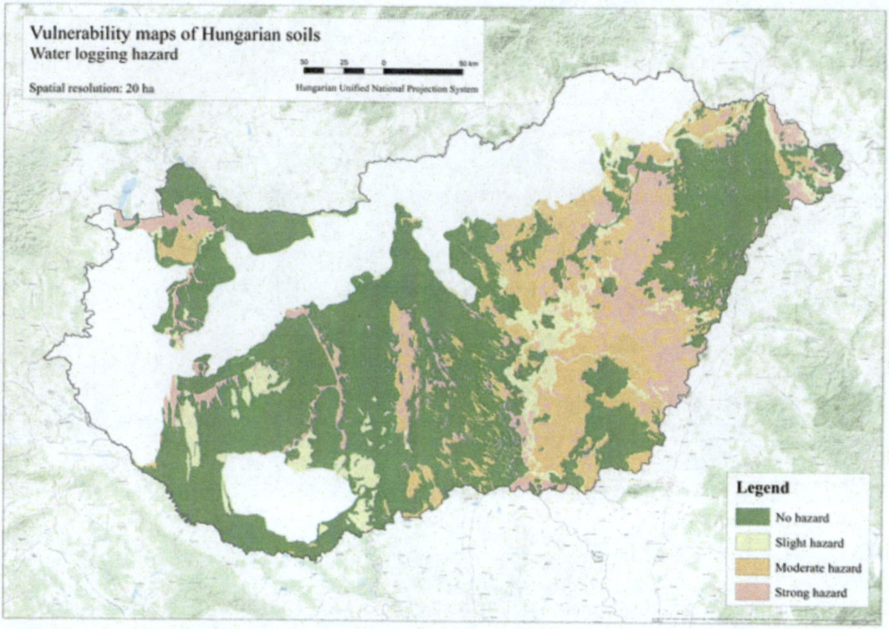

Fig. 5 Vulnerability of Hungarian soils to water logging hazard. *Source: Reproduced with permission from* [3]

Hungarian relation [3]. In particular, high exposures to water hazards occur in periods following long-term rainfall activity. Natural factors that influence inland excess water and have been closely investigated are hydrometeorological (e.g., abundant rains), topographic (e.g., heterogeneous micro-relief), and hydrogeological (e.g., impermeability of soil horizons). Regulation of the river Tisza (1846–1908) has been considered the most effective solution for preventing flooding for decades. The objective of the regulation was to reduce the Great Plain's risk of being flooded and to create a well-defined channel for the river. In addition to floods, inland waters also pose a serious challenge and the hazards caused by them have risen significantly due to climate extremes. Removing the water surplus from flooded areas is still difficult to manage on 2.7 million hectares.

Surplus water in water-logged fields is a factor that practically prevents tillage, so it needs to be drained and stored in a safe and reliable way [25]. The alleviation of damage caused by permanent inundation, traffic-induced damage, and the damage caused by tillage carried out under the force of necessity, should be developed into an annual program. This remediation program's tasks pertaining to soils may include organic matter conservation, the use of tools that avoid damage to wet soils, and the provision of financial assistance for the purchasing of such tools.

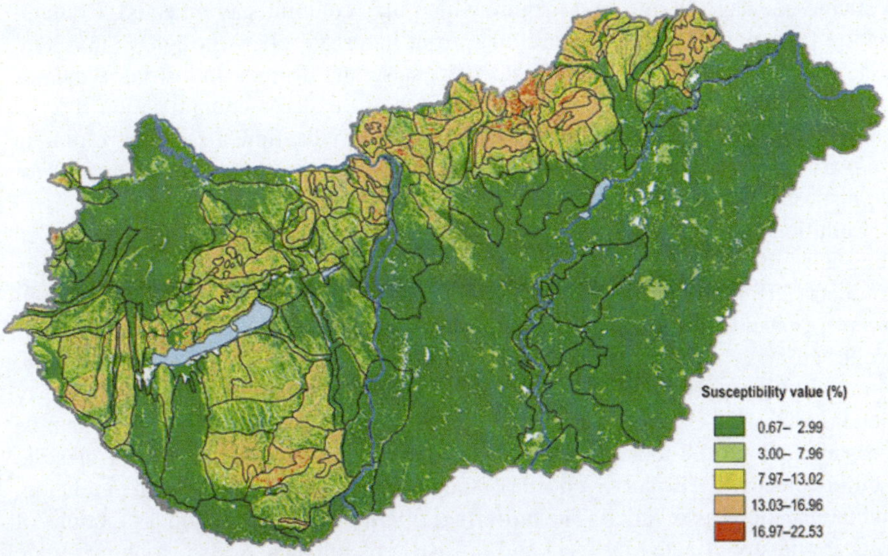

Fig. 6 Landslide susceptibility choropleth map of Hungary showing the boundaries of micro-regions affected by mass movements (boundaries after the National Atlas of Hungary, 2018, 126–127.) *Source: Reproduced with permission from* [27]

5 Landslides

Landslides are widespread in Hungary because of the wide availability of unconsolidated sediments (Fig. 6). They occur on hillslopes, piedmonts, mountains, and basins. Most landslides develop on Oligo-Mio-Pliocene clays, sands, and marls, with Pleistocene paleosols of high clay content covered by loess mantles of different thicknesses [19].

Soil structure, relief values, sliced landslides, and intensive rains are the main reasons for landslide damages [27]. Valleys of rivers on riverbanks and big lakeshores are endangered areas [19]. Landslide investigations are underway, and the investigation of hazard sites has been completed, but the size of the endangered area has not been estimated yet [19].

6 Soil Contamination

Soil contamination from non-point (diffuse) and point sources causes damage to several soil functions and the contamination of groundwater and surface water [19]. Kertész and Křeček [19] noted that the main non-point sources of soil contamination are the depositions from runoff, surface waters, and from eroded soil and atmospheric depositions (acidification and eutrophication compounds, fertilizers, pesticides, sewage sludge, and manure which may contain heavy metals). Contamination from point sources can originate from industrial plants no longer in operation, municipal and industrial waste disposals, and former industrial accidents [28]. Operating industrial plants and mining sites are also a contamination risk for soils and groundwater [19]. Overall, rapid industrialization and urbanization in Hungary have caused contamination of soils, as soils in these areas are used for the disposal of waste products [19].

Pollution with heavy metals and radioactive nuclides has only local character, so large territories of the country are suitable for producing environmentally healthy products [29]. However, soil and water contamination by chemicals (toxic microelements, pesticides, chlorinated aliphatic hydrocarbons, petroleum derivatives, polycyclic aromatic hydrocarbons, and synthetic detergents) is likely permanent. Muranyi [30] reported that the obtained results following the monitoring of heavy metal pollution in Hungary were within harmful limits according to the soil sample measurements in the topsoil of about 100,000 agricultural fields covering close to 5 million hectares. The surveyed area of the contaminated soils was small in relation, and the locations also varied. The buffering, filtering, and transforming functions of soils are mostly affected by local and diffuse contaminates. Soils can absorb toxic metals without harm up to a critical point but overloading the buffering capacity can result in a release of the substance back to the environment [31]. Stefanovits [32] stated three basic factors affecting soil buffering capacity: clay content, humus content, and presence or absence of carbonates. He outlined that soil buffering capacity alleviates or eliminates human-induced damages - such as physical, chemical, and biological impacts - to some extent. Publications cited above state that the level of soil contamination is still under levels considered to be dangerous.

7 Soil Sealing

Soil sealing is the isolation of soil by an impervious material from the atmosphere, hydrosphere, and biosphere [19]. Kertész and Křeček [19] stated that sealed surfaces are lost to agricultural land uses and limit or hamper ecological soil functions, including carbon storage and habitat for unique biota. In addition, they noted that the total sealed area in Hungary in 2015 covered 3.98% of the country's surface at the same time, 5–7 thousand hectares of land are lost from agricultural production every year. As they compared the percentage of sealed areas in Hungary with that of other European countries, they concluded that the value may be below average. The proportion of sealed areas is expected to increase steadily, but at the same time the proportion of areas that could be reorganized for agricultural activity is quite high (e.g., areas previously used for livestock or former farm yards and buildings).

8 Salinity

Salt-induced soil degradation is a major drawback to optimal functioning of soils in arid and semiarid regions of the Hungarian Great Plain. Salt-affected soils cover about 600 thousand hectares in Hungary. Considering the saline nature of soil and the areas degraded by salinization (Solonetzic meadow, deeply saline Chernozems, among others), according to Micheli et al. [4] salinization occurs in more than one million hectares (about 10% of the total area) in Hungary. Larger parts of saline areas are under nature protection due to their special, unique value. Micheli et al. [4] described that salinity and/or alkalinity and their consequences are significant limiting factors on soil fertility in the Hungarian Great Plain. They noted that the "natural" solonchaks, solonchak-solonetzes meadow solonetzes and solonetzic meadows soils are Gleyic Solonchak and Gleyic Solonetz in World References Base for Soils.

Saline areas are present in 67 natural micro-regions of the country. The largest saline areas are in the micro-regions Hortobágy (93,900 ha), Tiszafüred–Kunhegyes Plain (35,900 ha), Csongrád Plain (29,600 ha), Szolnok–Túr Plain (27,700 ha), Bihar Plain (27,300 ha) and Dévaványa Plain (26,800 ha) [14].

Salinization occurs in salt-affected soils and it causes a local risk if temporary water logging occurs and brings excess salt from deeper layers to the surface. The other cause for salinization is improper irrigation management. According to data about secondary salinization (Fig. 7) with respect to water depth [14], a 1 m rise of the level of groundwater would only affect a smaller area (34,000 ha), compared to a

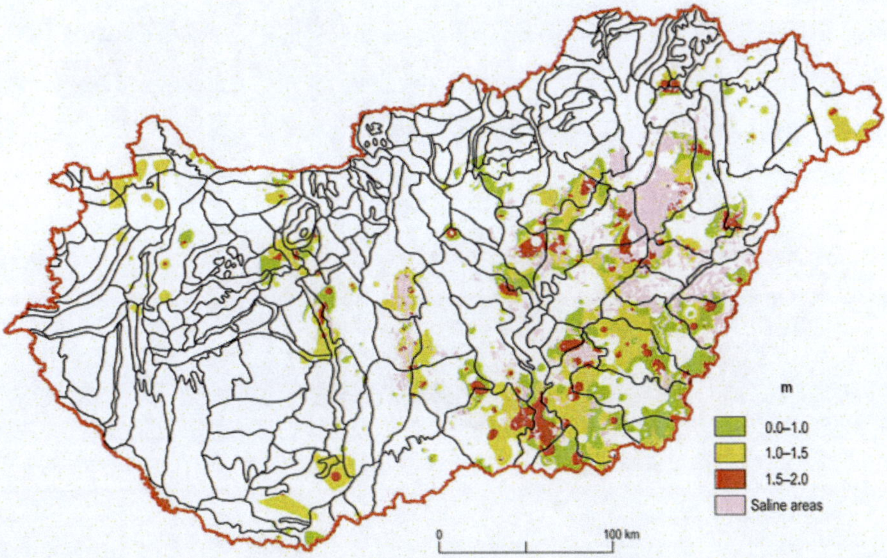

Fig. 7 Extent of areas susceptible to secondary salinization as a function of potential rise of water level. *Source: Reproduced with permission from* [14]

rise by 1.5 or 2 m, where the size of the areas affected by secondary salinization would be greater than 145,000 ha, and 235,700 ha respectively. Micheli et al. [4] noted that the presence of shallow (or easily and rapidly rising) groundwater with high levels of sodium ion compositions represents a potential hazard of further development of the salinization-alkalization processes in extensive areas of the Hungarian Great Plain.

9 Acidification

Soil acidification is one of the most significant degradation processes in Hungary. Forty-six percent of the total area is susceptible to acidification, of which 14% is highly susceptible, 5% is susceptible, 23% is moderately susceptible, and 4% is slightly susceptible [30]. Micheli et al. [4] described that 12% of Hungarian soils are strongly acidic and 43% are slightly acidic. As they noted, acidic soils are mainly found in West and South Transdanubia, the Transdanubia Mountains, the North Hungarian Mountains, and the alluvial regions of the Tisza and Rába rivers (Fig. 8).

According to Micheli et al. [4], the major natural cause of soil acidification is the leaching of neutral or alkaline weathering products including atmospheric deposition (SO_x, NO_x). Agricultural practices (harvesting crops, applying acidifying fertilizers)

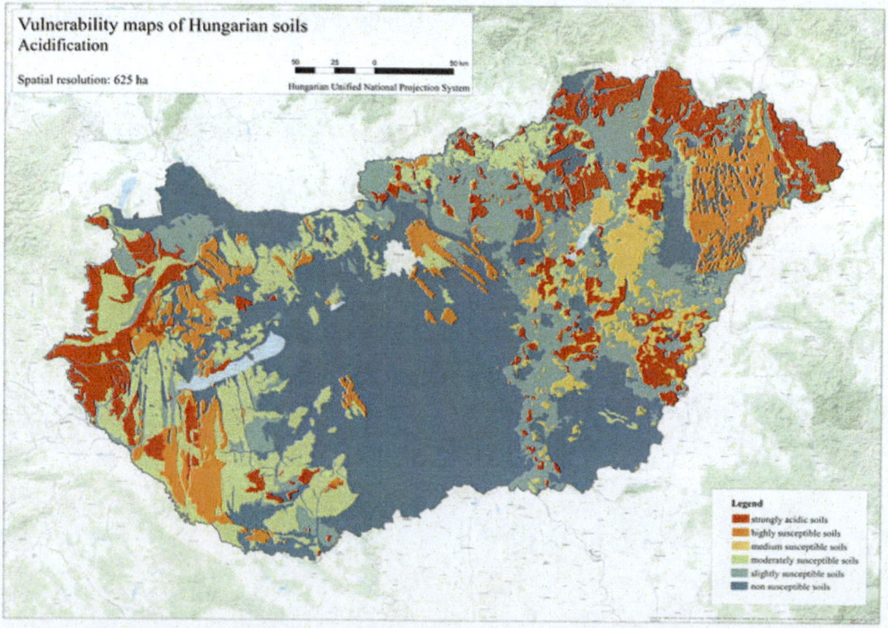

Fig. 8 Vulnerability of Hungarian soils to acidification. *Source: Reproduced with permission from* [3]

are also important influencing factors. The improvement of acidic soils with liming and lime fertilization is a long process. Looking to the future, acidification would need to be alleviated at 10% of acidic soils per year, and maintenance liming would have to be carried out every 10 years [33].

10 Agrochemical Use

The efficiency of crop production is influenced by several factors, and yield for each crop variety and region mainly depends on nutrient supply. The use of agrochemicals (e.g., growth regulators, pesticides, and fertilizers) has a positive effect on yield and increased growth, thereby providing stability for agricultural production [34]. Thus, agrochemistry is an integral part of the world's current agricultural production systems. Intensive agriculture practices rely on the widespread use of agrochemicals to improve crop productivity by controlling harmful pests, pathogens, and unwanted weeds (Fig. 9). Given that the world's population is growing, and agricultural land is shrinking, it is justified that there is an increased use of agrochemicals with the aim of increasing yields to meet food demand. Hungary has been

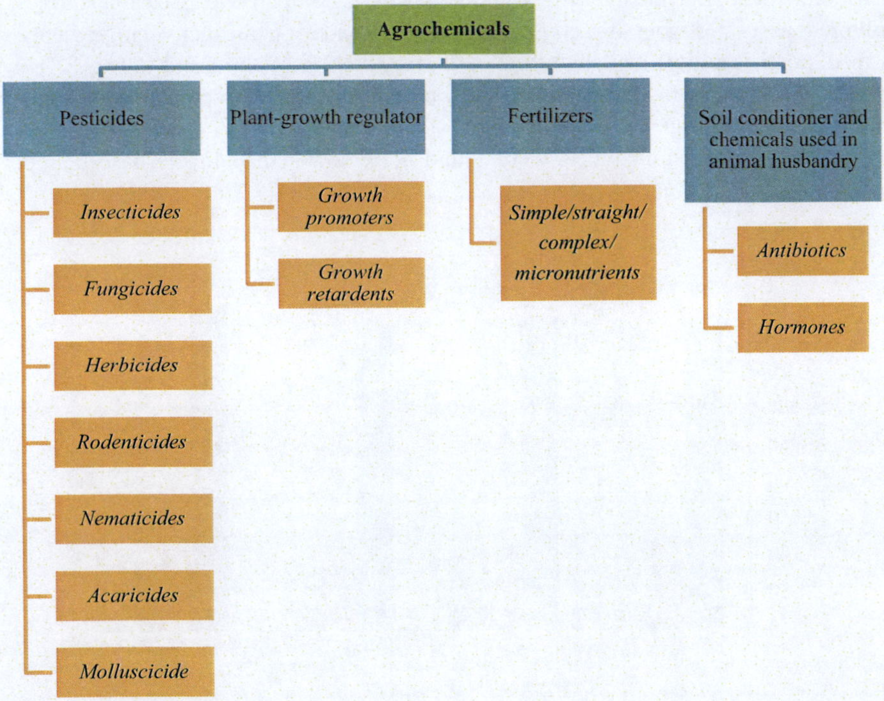

Fig. 9 Classification of agrochemicals (*Adapted from*: Impact of agrochemicals on soil health) *Source: Reproduced with permission from* [34]

recognized for centuries as a significant and world-renowned agricultural producer due to its significant natural resources. According to the latest data from KSH [35], in 2019 it produced a total of 15.6 million tons of cereals, of which there was 5.37 million tons of wheat (*Triticum aestivum* L.), and 8.23 million tons of maize (*Zea mays* L.).

In Hungary, the amount of fertilizer usage is closely related to the changes in the structure of production units. In the 1950s, about one million small family farms were forced to merge into a few thousand large farms or cooperatives (collectivization) which became state property. The objectives of this merge were to intensify and strengthen agriculture (e.g., field size and the use of agrochemicals have increased) [36]. However, in 1989, the socialist regime collapsed which was also of great influence on agriculture, causing the use of agrochemicals to decline drastically (e.g., an 80% decrease in fertilizer use) [37].

Compared to the 1980s, the use of nitrogen fertilizer in the early 1990s decreased by 75%, and the use of phosphorus and potassium by 95%, a trend that had not changed much until 2000. However, it is important to point out that the use of organic manure has declined significantly since the early 1970s, while fertilizer use per hectare has doubled [38].

According to the latest data from AKI [39], the total volume of fertilizer sales increased by 7% in 2020 compared to the average of the last 5 years. Fertilizer distributors in 2020 achieved a 4.2% increase in sales compared to 2019 and sold 1.9 million tons of fertilizer directly to farmers. Considering the active ingredients, farmers purchased 650 thousand tons in 2020, which is 20 thousand tons (+3.1%) higher NPK (Nitrogen, Phosphorus, and Potassium) fertilizers compared to 2019 (Fig. 10). Also, nitrogen as an active ingredient increased by 6.6%, while phosphorus levels and potassium levels were 3 and 4.1% lower than in the base period, respectively.

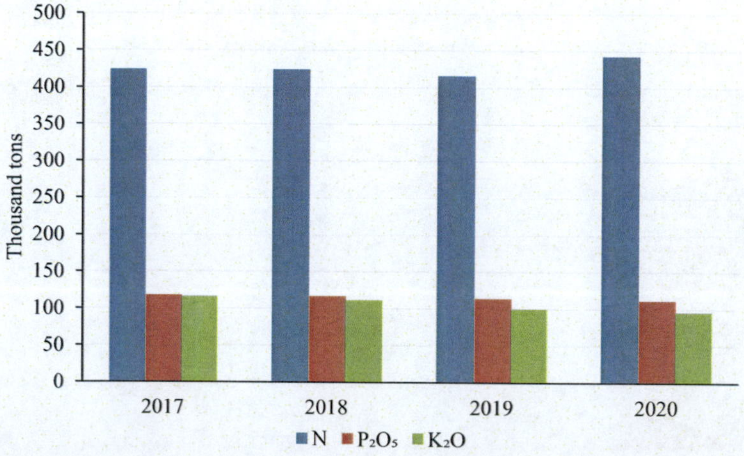

Fig. 10 Nitrogen, phosphorus, and potassium fertilizers (thousands of tons) sold to farmers in Hungary. Amounts based on the active ingredient (2017–2020) *Source: Reproduced with permission from* [39]

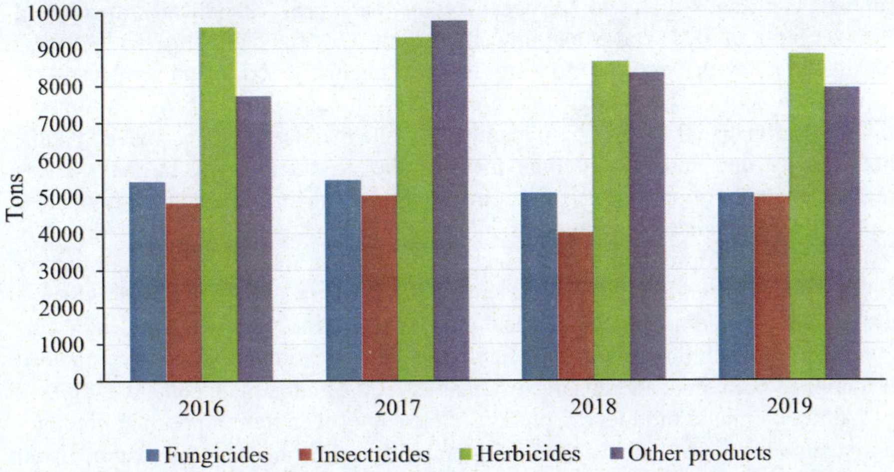

Fig. 11 Development of pesticide product sales by product group (2016–2019). *Source: Reproduced with permission from* [41]

As for consumption, it is important to highlight that the annual fertilizer use and NPK ratio significantly fluctuate, because in many cases the use of fertilizer does not depend on actual demand but depends on the current financial situation of the farms. In addition to the above problem, climate change also contributes to fluctuating yield averages [40].

In the field of agrochemical inputs, in addition to fertilizers and fertilization, pesticides occupy a large part and play an important role in intensive production.

Over the past 40 years, data shows that the highest pesticide use was in the 1980s and 1990s, although the lowest was in 2000 (with a total of 10.9 thousand tons). The use of pesticides on an annual basis is mainly determined by the weather of a given vegetation period, more precisely by the size of the infections, the period of the appearance of pests, and the expected damage. The latest data shows that the volume of pesticides used in 2019 increased by 2.7% compared to 2018. Within this, the volume of insecticides marketed increased the most (+23.6%) compared to the previous year (Fig. 11). On the other hand, it is very important to highlight that in 2019, herbicides were the top consumed pesticide with 8,840 tons [41].

Considering the ratio of pesticide consumption, agriculture is the dominant land use in Europe, accounting for almost half of the total area of the EU-27. Its effects are therefore far-reaching and affect areas outside agricultural production [42]. Two-thirds of the entire territory of Hungary is under agricultural production, and according to agricultural land per capita, it is among the first in European countries [43]. The impact of excessive and uncontrolled application of agrochemicals is thus huge because the proportion of intensive cultivation branches (arable, garden, orchard, and vineyard) is the highest in Hungary after Denmark (54.2%). In the European Union and Central and Eastern European countries, this proportion is only around 30–40% [44]. Therefore, Firbank [45] points out that the intensification of agriculture presents the largest challenge to nature conservation, and thus at the same time a major threat to European biodiversity [46]. Specifically, increasing fertilizer input in fields and expanding arable land at the scale of entire landscapes is considered a key driver of biodiversity loss and declining ecosystem services [47].

With the help of technology and digitalization in agriculture, there is an opportunity to rationalize the use of agrochemicals, while maintaining the achieved yields, their quality and safety, and thus increase the productivity of farmers. These guidelines are an important part of an integrated pest management strategy and the application of precision technologies.

Integrated crop production is a management system that produces high-quality food and other products while preserving and improving soil fertility and a diverse environment while respecting ethical and social expectations. Thus with this approach, agrotechnical, biological, and chemical methods are carefully balanced to minimize pollutants and ensure sustainable yet profitable management [48].

Within integrated production, precision farming is even more important in reducing the amount of pesticide used. Takács and Takács [49] highlighted that precision technology prioritizes the rational use of agrochemical inputs, thus ensuring an increase in economic efficiency. Several studies have confirmed the justification for the application of precision technology [50]. Studies conducted in England have shown that by applying precision agricultural technologies input costs can be reduced by 63% [51].

11 Conclusions

Hungarian soils are exceptionally favorable for agricultural production within Europe. In addition to natural influences and climate phenomena, farming practices can have also a great impact on soil quality. The application of the suggested changes of cultivated land use, coupled with rational agricultural and environmental policies are proposed in hope of limiting the degradation processes and promoting the quality of Hungarian soils.

We conclude by stating a few observations regarding soil degradation problems in Hungary:

- Increased presence of farming-induced compaction, the occurrence of water and wind erosion, floods, and water logging are current problems and also a critical issue in the future. Moreover, flooding events, including water logging and landslides will be connected with climate extremes in the future. Complex prevention methods for these issues are to be applied in endangered areas, which requires deeper knowledge. The proportion of sealed areas is expected to increase steadily, however, some areas may be reused for agricultural production.
- Salt-affected soil is found in nature-protected areas for the most part and to a lesser extent in agricultural areas induced by irrigation (secondary salinization). Acidification would need to be alleviated at 10% of acidic soils per year, and maintenance liming would have to be carried out every 10 years. The use of

agrochemicals, including growth regulators, pesticides, and fertilizers has a positive effect on yield, thereby providing stability for agricultural production. However, to avoid environmental pollution, an integrated chemical management strategy is to be applied using precision technologies.

– Hungarian landscapes - some of the most valuable in Europe- are relatively well maintained with a low level of soil contamination. Thus, the above-described processes should be taken into consideration and be addressed accordingly. Soils are the foundation of agriculture, with an important role in the EU Farm to Fork Strategy [52]. Therefore, all degradation processes represent important environmental problems, that must be further investigated and analyzed. Results of these investigations could then be used as the basis of a well-established policy.

References

1. Stefanovits P (1981) Talajtan (*Soil science*). Budapest, Mezőgazdasági Kiadó (in Hungarian)
2. Várallyay G (1989) Soil degradation processes and their control in Hungary. Land Degrad Rehabil 1:171–178
3. Várallyay G (2011) Soil quality: a step back in land evaluation. In: Tóth G, Németh T (eds) Land quality and land use information in the European Union, pp 21–33
4. Michéli E, Várallyay Gy, Pásztor L, Szabó J (2003) Land degradation in Hungary. In: Jones RJA, Montanarella L (eds) Land degradation in central and Eastern Europe. European Soil Bureau Research Report, vol 10, 20688 EN Ispra, Italy, pp 198–206
5. Birkás M (2000) Soil compaction situation in Hungary; Consequences and possibilities of the alleviation. DSc Theses, Gödöllő (in Hungarian)
6. Birkás M (1987) Agronomical factors influencing the quality of soil tillage. PhD thesis, Gödöllő (in Hungarian)
7. Chen Y, Tessier S (1997) Techniques to diagnose plow and disk pans. Can Agric Eng 39(2): 143–147
8. Birkás M, Antal J, Dorogi I (1989) Conventional and reduced tillage in Hungary – a review. Soil Tillage Res 13:233–252
9. Nyiri L (1993) Soil structure and influencing possibilities. In: Nyiri L (ed) Földműveléstan, Mezőgazda Kiadó, Budapest, Hungary, pp 66–69 (in Hungarian)
10. Birkás M, Kisic I, Bottlik L, Jolánkai M, Mesic M, Kalmár T (2009) Subsoil compaction as a climate damage indicator. Agric Conspec Sci 74(2):1–7
11. Birkás M, Jolánkai M, Gyuricza C, Percze A (2004) Tillage effects on compaction, earthworms and other soil quality indicators in Hungary. Soil Tillage Res 78:185–196
12. Bogunovic I, Pereira P, Kisic I, Sajko K, Sraka M (2018) Tillage management impacts on soil compaction, erosion and crop yield in Stagnosols (Croatia). Catena 160:376–384. https://doi.org/10.1016/j.catena.2017.10.009
13. Van Ouwerkerk C, Soane BD (1994) Conclusions and recommendations for further research on soil compaction in crop production. In: Soane BD, Van Ouwerkerk C (eds) Soil compaction in crop production. Elsevier, pp 627–642
14. Madarász B (2019) Salinization, secondary salinization. In: Kertész Á (ed) Landscape degradation in Hungary. Budapest Geographical Institute, MTA RCAES, pp 43–49
15. Verheijen FGA, Jones RJA, Rickson RJ, Smith CJ (2009) Tolerable versus actual soil erosion rates in Europe. Earth Sci Rev 94(1–4):23–38

16. Négyesi G, Lóki J, Buró B, Bertalan-Balázs B, Pásztor L (2019) Wind erosion researches in Hungary – past, present and future possibilities. Hung Geogr Bull 68(3):223–240. https://doi.org/10.15201/hungeobull.68.3.2
17. Kertész Á, Centeri C (2006) Hungary. In: Boardman J, Poesen J (eds) Soil erosion in Europe. Wiley, Chichester, pp 139–153
18. Kertész Á, Jakab G (2010) Gully erosion in Hungary, review and case study. Procedia Soc Behav Sci 19:693–670
19. Kertész Á, Křeček J (2019) Landscape degradation in the world and in Hungary. Hung Geogr Bull 68(3):201–221. https://doi.org/10.15201/hungeobull.68.3.1
20. Pásztor L, Waltner I, Centeri C, Takács K, Laborczi A (2015) Soil erosion risk map of Hungary. http://www.mta-taki.hu/hu/osztalyok/gis-labor/agrotopo
21. Rusco E, Montanarella L, Bosco C (2008) Soil erosion: a main threats to the soils in Europe. In: Toth G, Montanarella L, Rusco E (eds) Threats to soil quality in Europe. JRC, Italy, EUR 23438 EN, pp 37–46
22. Pásztor L, Négyesi G, Laborczi A, Kovács T, László E, Bihari Z (2016) Integrated spatial assessment of wind erosion risk in Hungary. Nat Hazards Earth Syst Sci 16:2421–2432
23. Klik A, Rosner J (2020) Long-term experience with conservation tillage practices in Austria: impacts on soil erosion processes. Soil Tillage Res 203:104669
24. Kisic I, Bogunovic I, Zgorelec Z, Bilandzija D (2017) Effects of soil erosion by water under different tillage treatments on distribution of soil chemical parameters. Soil Water Res 12(4): 36–43. https://doi.org/10.17221/25/2017-SWR
25. Halbac-Cotoara-Zamfir R, Günal H, Birkas M, Rusu T, Brejea R (2015) Successful and unsuccessful stories in restoring despoiled and degraded lands in Eastern Europe. Adv Environ Biol 9(23):368–376
26. Kuti L, Kerék B, Vatai J (2006) Problem and prognosis of excess water inundation based on agrogeological factors. Carpathian J Earth Environ Sci 1(1):5–18
27. Józsa E, Lóczy D, Soldati M, Drăguţ LD, Szabó J (2019) Distribution of landslides reconstructed from inventory data and estimation of landslide susceptibility in Hungary. Hung Geogr Bull 3:255–268. https://doi.org/10.15201/hungeobull.68.3.4
28. EEA (2003) Europe's environment – the third assessment. Environmental Assessment Report No. 10. European Environment Agency, Copenhagen
29. Günal H, Korucu T, Birkas M, Özgöz E, Halbac-Cotoara-Zamfir R (2015) Threats to sustainability of soil functions in central and Southeast Europe. Sustainability 7:2161–2188. https://doi.org/10.3390/su7022161
30. Murányi A (2000) Quality and contamination of agricultural soils in Hungary as indicated by environmental monitoring and risk assessment. In: Soil quality, sustainable agriculture and environmental security in central and Eastern Europe. Springer, Heidelberg, pp 61–77
31. Gentile AR (2000) Soil degradation in Europe. In soil degradation in central and Eastern Europe: the assessment of the status of human-induced degradation; United Nations Environment Programme (UNEP), and ISRIC—World Soil Information, Wageningen, The Netherland, pp 68–89
32. Stefanovits P (2005) Environmental buffering and loading capacity of soils. In: Balogh M, Péterfi B (eds) The importance of soils in the 21st century. MTA Társadalomkutató Központ Budapest, pp 373–400 (in Hungarian)
33. Blaskó L (2005) Soil improvement in present and future. In: Balogh M, Péterfi B (eds) The importance of soils in the 21st century. MTA Társadalomkutató Központ, Budapest, pp 267–290. (in Hungarian)
34. Asit M, Binoy S, Sanchita M, Meththika V, Ashok KP, Madhab CM (2020) In: Majeti NVP (ed) Agrochemicals detection, treatment and remediation. Chapter 7 – impact of agrochemicals on soil health. Butterworth-Heinemann, pp 161–187
35. KSH (2021) (Hungarian Central Statistical Office) https://www.ksh.hu/mezogazdasag. Accessed 1 Apr 2021

36. Ángyán J, Tardy J, Vajnáné-Madarassy A (eds) (2003) Védett és érzékeny természeti területek mezőgazdálkodásának alapjai, Mezőgazda, Budapest (in Hungarian)
37. Liira J, Aavik T, Parrest O, Zobel M (2008) Agricultural sector, rural environment and biodiversity in the central and eastern European EU member states. AGD Landscape Environ 2:46–64
38. Németh T (2002) A tápanyag-utánpótlás jelenlegi helyzete, időszerű kérdései. In: Győri Z, Jávor A (eds) Az agrokémia időszerű kérdései. ISBN: 963 9274 39 9
39. AKI (2021) Műtrágya-értékesítés mezőgazdasági termelőknek 2020. Agrárközgazdasági Intézet (Ed. Demeter Edit), ISSN 1418 2130 Online Hungarian: https://asir.aki.gov.hu Accessed 21 Mar 2021
40. Gockler L (2016) Műtrágya- és szervestrágya-felhasználás hazánkban Mezőgazdasági Technika 10:40–43
41. AKI (2019) Növényvédő szerek értékesítése (2019) NAIK Agrárgazdasági Kutatóintézet (Ed. Demeter Edit), ISSN 1418 2130 Online Hungarian: Agrárstatisztikai Információs Rendszer (ASIR) – Agrárközgazdasági Intézet (gov.hu). Accessed 18th Jan 2021
42. Green ER, Cornell JS, Scharlemann WPJ, Balmford A (2005) Farming and the fate of wild nature. Science 307(5709):550–555. https://doi.org/10.1126/science.1106049
43. Szabó P (1994) Soil degradation in Hungary. Farm land erosion. In: Wicherek S (ed) Temperate plains environment and hills, pp 563–569
44. Magda R (2001) A magyarországi természeti erőforrások gazdaságtana és hasznosítása, Mezőgazda Kiadó. (in Hungarian) p 167
45. Firbank LG (2005) Striking a new balance between agricultural production and biodiversity. Ann Appl Biol 146(2):163–175. https://doi.org/10.1111/j.1744-7348.2005.040078.x
46. Báldi A, Faragó S (2007) Long-term changes of farmland game populations in a post-socialist country (Hungary). Agric Ecosyst Environ 118:307–311. https://doi.org/10.1016/j.agee.2006.05.021
47. Zhao HZ, Hui C, Ouyang F, Liu J-H, Guan X-Q, He D-H, Ge F (2013) Effects of inter-annual landscape change on interactions between cereal aphids and their natural enemies. Basic Appl Ecol 14:472–789. https://doi.org/10.1016/j.baae.2013.06.002
48. Boller EF, Avilla J, Joerg E, Malavolta C, Wijnands FG, Esbjerg P (2004) Integrated production principles and technical guidelines. IOBC/WPRS Bull 22(4). ISBN: 9290671084
49. Takács-György K, Takács I (2011) Use of agrochemicals – environmental, social and economic impacts of alternative farming strategies: precision Weed management. 1st world sustainability forum, 1–30 Nov 2011, pp 2–16
50. Rider TW, Vogel JW, Dille JA, Dhuyvetter KC, Kastens TL (2006) An economic evaluation of site-specific herbicide application. Precis Agric 7:379–392. https://doi.org/10.1007/s11119-006-9012-y
51. DEFRA (2013) Farm practices survey autumn 2012 – England. Department for Environment, Food and Rural Affairs (DEFRA), p 41
52. Montanarella L, Panagos P (2021) The relevance of sustainable soil management within the European Green Deal. Land Use Policy 100:104950. https://doi.org/10.1016/j.landusepol.2020.104950

Agricultural Land Degradation in Iceland

Isabel C. Barrio and Ólafur Arnalds

Contents

1 Introduction 160
2 Environmental Conditions 160
3 Changes in Iceland Since Human Settlement 162
4 Past and Present of Icelandic Agriculture 163
5 A Framework Model for Land Degradation in Iceland (Ice-LaCoRe) 166
6 Present Condition of the Land 169
7 Agriculture, Land Degradation, and Climate: The Future? 171
8 Final Considerations 173
References 174

Abstract Iceland is located just south of the Arctic Circle. Its cold climate, volcanic origin, erodible soils, and relative isolation make it very sensitive to human impact. Humans arrived in Iceland ~1,150 years ago, bringing with them their pastoral ways of life, which had large impacts on Iceland's subarctic ecosystems. The relatively short history of human land use and the good documentation of this period provide a unique opportunity to study the drivers of land degradation related to human land use practices and how the interactions between society, economy, and the natural environment have changed over time. Centuries of agricultural use in Iceland under marginal natural conditions have caused severe and large-scale land degradation, which is a main environmental concern still today. A framework model for land condition response in Iceland (Ice-LaCoRe) helps separate the underlying reasons, drivers, processes, states, and consequences of land degradation, where decoupling mechanisms disrupt the cycle and favour inaction in dealing with the poor state of the land. Recognition of the poor state of Icelandic ecosystems and the identification of such decoupling mechanisms is a critical step to the effective implementation of sustainable land uses and to prevent further land degradation.

I. C. Barrio (✉) and Ó. Arnalds
Agricultural University of Iceland, Hvanneyri, Iceland
e-mail: isabel@lbhi.is

Paulo Pereira, Miriam Muñoz-Rojas, Igor Bogunovic, and Wenwu Zhao (eds.),
Impact of Agriculture on Soil Degradation II: A European Perspective,
Hdb Env Chem (2023) 121: 159–178, DOI 10.1007/698_2022_920,

Published online: 2 December 2022

Keywords Extensive sheep grazing, Land condition response model, Marginal agriculture, Soil erosion, Subarctic

1 Introduction

Land degradation and extensive soil erosion are main environmental concerns in Iceland. The location of Iceland, its volcanic origin and relative isolation make it very sensitive to human impacts. The country was settled after 870 AD by people of Scandinavian and Celtic origin, who arrived to a fragile subarctic environment that had not been subjected to grazing animals other than birds and insects before the arrival of man [1]. After settlement, large-scale grazing and woodland clearing, under cold climatic conditions and frequent volcanic activity, resulted in dramatic ecosystem degradation throughout much of the country [2]. In this chapter, we discuss the nature and extent of land degradation in Iceland and why agricultural uses have had such pronounced effect. We focus on soil erosion and overgrazing, as these are the main aspects regarding agricultural land degradation in Iceland. Soil erosion in Iceland is mostly associated with open land or rangelands rather than cultivated fields [3]. Other aspects, like slash-and-burn agriculture, soil contamination due to overuse of agrochemicals, salinity, or microplastics, that are pressing issues in other countries around the world, are not a main concern in Iceland.

2 Environmental Conditions

Iceland is a 103,000 km^2 volcanic island in the North Atlantic Ocean, just south of the Arctic Circle. Despite its northern latitude, Iceland has a mild maritime climate thanks to the Gulf Stream that brings warm waters to the North Atlantic Ocean [4]. Winters are mild with average monthly temperatures for the whole country often near −5°C, while summer average monthly temperatures are moderated by oceanic influence and are often around 7°C [5]. Annual precipitation varies considerably and there is a marked precipitation gradient across the country, ranging broadly from 400 mm in North Iceland to 2,000 mm in South Iceland [6]. In addition, the weather in Iceland is characterized by strong winds and frequent weather fluctuations [7].

Topography plays an important role in determining the environmental conditions across Iceland. About 60% of the surface area of the island lies above 400 m a.s.l. in mountainous areas and higher elevation plateaus. This creates a sharp contrast between the highlands and the surrounding lowland areas along the coast. These distinct areas also differ in the predominant land uses: most of the Icelandic lowlands, especially the areas below 200 m a.s.l. are well suited for a variety of agricultural uses, including dairy production and sheep grazing, while the highlands have a subarctic character with sporadic permafrost areas [8] and are mostly used as communal summer grazing areas.

An important feature of Iceland is that it is an active volcanic island. The North-Atlantic rift-zone cuts through the country, from the northeast to the southwest. Volcanic ash deposition and aeolian redistribution of unstable glacial sediments made of basaltic volcanic glass make up the parent materials of the soils, which are largely classified as Andosols [3]. These soils are fertile, with high pH and favourable nutrient availability, but are mostly devoid of lattice-clay minerals that normally enhance cohesion. Allophane and ferrihydrite are the main clay particles, which tend to form stable silt-sized aggregates. Aeolian processes also add silty sediments to the soils and both of these factors make Icelandic Andosols silty and susceptible to erosion by wind and water. Bare soils are prone to intense frost action due to the rapid hydraulic conductivity of the silty soils and frequent freeze–thaw cycles caused by the combination of the subarctic climate and oceanic influences [3]. The susceptibility of Icelandic soils to erosion has had an important bearing for the dramatic degradation processes that have taken place since human settlement.

Glaciers have played a major role in shaping both landscapes and biodiversity in Iceland. During the Pleistocene, repeated Quaternary glaciations extirpated species within Iceland and allowed only brief windows for colonization during interglacial periods. In comparison with continental areas at similar latitudes, Iceland has low number of plant species, with a small subset of the species found in North Western Europe and lacks endemic species [9]. There are no native mammalian herbivores in Iceland. The two species of large herbivores present nowadays, sheep and reindeer, are both introduced. Sheep were introduced as livestock by the Norse settlers in the ninth century; reindeer were introduced from Norway in the 1700s and are now restricted to some parts of East Iceland [10]. The only native herbivores in Iceland are birds (ptarmigan, migratory geese, and swans) and invertebrates. Several species of migratory geese use Iceland as a stopover in their migration to their breeding areas in the Arctic, but several species breed in Iceland, including the pink-footed goose (*Anser brachyrhynchus*) that breeds in large colonies in the highlands. The only native mammal in Iceland is the Arctic fox (*Alopex lagopus*) [11]. The American mink (*Mustela vison*) was brought to Iceland in the early 1930s for mink farming; escaped animals established in the wild since at least 1937 and are now present throughout the country, mainly in the lowlands [12]. In contrast to other terrestrial systems at high latitudes, Iceland has no native small mammalian herbivores either, like lemmings or voles, except for the introduced wood mouse (*Apodemus sylvaticus*) mostly in lowland areas [13]. In turn, Iceland hosts internationally important breeding populations of many bird species, particularly waders [14], that extensively use the agricultural lands and wetlands in the lowlands [15, 16].

3 Changes in Iceland Since Human Settlement

Iceland was settled in the ninth century by people of Scandinavian and Celtic origin, with a rapid population increase into the eleventh century. The population remained between 40,000 and 80,000 through much of the Middle Ages until the nineteenth century [17]. The current population is about 360,000 people, which yields an average population density of 3.6 inhabitants/km^2 (European Commission 2021). When only the areas below 200 m a.s.l. are considered, which represent around 25% of the country and comprise the main urban and agricultural areas, human population density is around 13 inhabitants/km^2. This population density is still much lower than the average of the European countries with the lowest population density (Finland 18.2 inhabitants/km^2, or Norway 17.3 inhabitants/km^2; [18]). Despite this low population density, humans have had pervasive impacts on Icelandic ecosystems.

Iceland provides a unique opportunity to try to disentangle the influence of different drivers of landscape change, because human influence can be clearly tracked in the palaeoenvironmental record and historical narratives [19]. Multiple sources, including archaeological and historical records, provide considerable knowledge on environmental changes in Iceland. For instance, analyses of pollen assemblages preserved in lake sediments and peat [20], macrofossil record and sediment analyses [21, 22] provide insights into pre-settlement vegetation. During the Holocene, birch (*Betula pubescens*) woodlands were common in Iceland, but their extent fluctuated with variations in climate [20, 23, 24]. At the time of Settlement (around 870 AD) a large proportion of the country was covered with vegetation [25], and birch woodlands were widespread. Estimates of the cover of birch forests at the time of settlement range between 8% [24] and 40% of the country [26], although some authors place this number around 24% [27]. Currently, birch woodlands occupy about 1.5% of Iceland [28].

Pre-settlement vegetation was dominated by birch and willow shrublands in drier areas at lower elevations together with a variety of wetland systems [20]. Highland systems included dwarf shrub heathlands dominated by *Betula nana*, *Salix* spp., and *Juniperus communis*, wetlands and sparsely vegetated rocky and sandy surfaces [22]. Barren areas characterized the highest elevations and occurred also in locations subjected to periodic volcanic disturbance and flooding by glacial rivers [29]. Without the influence of humans and large herbivores, pre-settlement vegetation patterns were likely determined by topography and substrate properties, and responded mainly to fluctuations in climate [24] and aeolian processes [22] and, on shorter time scales, to volcanism.

Pronounced changes have occurred over the past 1,150 years since the settlement. The collapse of many ecosystems in large areas has left barren desert surfaces behind, while other systems where soil and vegetation cover remained have been profoundly altered [1] (Fig. 1).

Fig. 1 Present day vegetation in Iceland ranges from well vegetated areas to barren deserts. (**a**) Birch forests occupied much larger areas before human settlement around 870 AD. (**b**) Many wetland areas in the lowlands have been drained for agricultural purposes. (**c**) Extensive areas with sparse vegetation cover across Iceland are likely the result of human impacts. (**d**) Barren and poorly vegetated areas cover about 40% of Iceland. (Photographs by Ólafur Arnalds [46])

4 Past and Present of Icelandic Agriculture

Despite the relatively mild conditions, the climate of Iceland is considered marginal for agriculture [30]. Temperature is more limiting than precipitation for plant cultivation and winter temperatures can be a strong determinant of hay yields in the spring [31]. In the lowlands, the growing season often spans from May to September and is generally cool and wet.

The first settlers brought with them the traditional ways of Scandinavian farming, cereal production, and animal husbandry. However, the country was for the most part poorly suited for arable production, so farm animals became the main support for the first settlers' livelihoods. Iceland's farming system was primarily based on sheep, cattle, goats, and horses, and in the earliest centuries of settlement, pigs [2]. Products from sheep and cattle provided the bulk of the food supply during the Middle Ages. Sheep husbandry was the main farming activity in Iceland, but productivity without external inputs was very low. The natural resources of the island, although at a high environmental cost, were able to support ~300,000 sheep, by grazing year-round and with the hay obtained from wetlands [25]. However, not much is known about the total number of sheep until 1703 when record keeping began [32].

Historically, a small proportion of Iceland was cultivated for annual crops, including grass for hay [30]. The low efficiency in haymaking prior to mechanization in the late nineteenth century set a limit to the number of animals that could survive the winter: reliance on winter grazing could drastically reduce the number of animals in hard winters [33]. Historical winter grazing contributed significantly to accelerated land degradation close to some farms [34], as livestock offtake of vegetation would almost always exceed the limited vegetation productivity in winter. It has been proposed that during this time grazing pressure on mountain rangelands was relatively low, as part of the animals, the ewes, were kept close to the farms or in shielings, as they had to be milked twice a day [35]. The number of sheep increased considerably after 1850 with increased export of live lambs to Britain, when sheep farming became the main agricultural activity in Iceland [32].

By the end of the nineteenth century, grasslands and cultivated hay meadows expanded, with the introduction of mechanization that allowed ploughing and harrowing [35]. As well, the increased use of artificial fertilizers in the early twentieth century allowed increased agricultural production [31]. A law passed in 1923 aimed at increasing agricultural production by encouraging the drainage of wetlands to allow cultivation of hay fields [36]. Large-scale mechanized wetland draining began after World War II and was promoted by the Icelandic government as a way of ensuring food security in Iceland by creating new areas for grazing and agricultural production. Draining the water-logged wetland soils results in a rapid short-term increase in biomass production because of the oxidation of organic matter and release of nutrients. Mechanized draining of Icelandic lowland wetlands allowed a rapid increase in winter fodder production and, consequently, in the numbers of animals [25, 30]. It has been estimated that >4,000 km^2 of wetland areas have been drained in Iceland, and about 70% of the wetlands in lowland areas have been disturbed [36]. Such disturbance has negative effects on hydrology as well as on aquatic and aboveground wildlife and biodiversity, including important portions of world populations of several breeding bird species [14]. Drained wetlands are prime agricultural lands and approximately half of the cultivated land in Iceland is on drained wetland soils, but currently less than 15% of the drained areas are used for agricultural purposes. The sheep and dairy industries still rely on haymaking for winter feed, with hayfields usually 30–100 ha on each farm (Fig. 2).

The number of sheep reached a historical maximum in 1977, of around 900,000 winterfed ewes [32]. Overproduction in the sheep sector called for a revision of the legal framework for agricultural policies in the early 1980s [30, 37]. The introduction of a quota system to promote structural adjustment of production to domestic demand reduced the number of sheep by half. Since then, sheep numbers have remained relatively stable but are still high compared to historical numbers before the mid-nineteenth century [38]. Despite the large reductions in sheep numbers, grazing pressure on the highland ranges has not declined at the same rate, as successful breeding programmes with higher average number of lambs per ewe and increased carcass weights have maintained the amount of meat produced despite the lower number of ewes in the recent decades [30].

Nowadays, Icelandic agriculture relies mainly on grassland-based livestock production (sheep farming and dairy production), cultivation (haymaking and to some extent cereal production), and horticulture. Animal production covers 90% of the domestic demand of meat, 96% of eggs, and 99% of milk products [39]. The sheep

Fig. 2 Farm in Northwest Iceland showing drained wetlands used for haymaking and grazing of dairy cattle in summer. Hayfields are ploughed every 7–15 years for regeneration. Sheep are grazed away from the farm in summer, often in extensive communal grazing areas in the highlands. (Photograph by Ólafur Arnalds [46])

industry produces about 9,000 tons of carcasses per year, and beef production is about 4,600 tons [40]. Around 35% of sheep meat is exported due to surplus production [39]. As of 2020, there were in Iceland around 26,000 dairy cows and about 400,000 sheep [40]. The number of horses, mainly used for recreational purposes, has been estimated at around 77,000 [30]. Other domestic animals include around 1,500 goats and about 3,000 pigs [40].

Compared to other countries in Europe, agricultural production in Iceland is very small [30]. Still, the majority of the land area is used for different agricultural uses, but this agriculture is very different from the cultivation-based European agriculture – sheep production in Iceland is an open range grazing culture of extensive rangelands, often of limited productivity. The area of cultivated land in Iceland is ~1,300 km^2, about 7% of the area below 200 m a.s.l. which is the area suitable for agriculture, and consists mainly of permanent hay fields [30]. Large patches of natural or seminatural habitats surround the hay- and arable fields, forming a heterogeneous mosaic of areas with different management intensity [15].

The number of farms in Iceland declined rapidly after World War II [30], and in 2010 there were around 2,600 farms [40]. Farms in Iceland are on average over 600 ha, much larger size than in other European countries [30]. Some farms have extensive joint grazing areas used for grazing by sheep and cattle and often farmers' communities have access to communal summer grazing areas, usually located in the highlands or in mountain areas. The highland communal grazing areas are typically public lands owned by the state where the grazing rights belong to the farmers of the community. These rangelands provide a substantial amount of the forage for sheep in the summer [33]. Horses are mainly grazed in the lowlands while sheep are grazed both in highland and lowland areas. It has been recently estimated that about 61% of the surface of Iceland is used for grazing (71% when glaciers, water, and urban areas are not considered) [41]. Many of the communal grazing areas have undergone severe land degradation which still continues in many areas, as discussed below.

Sheep and dairy production in Iceland are heavily subsidized [42]. Government support for dairy and sheep production is ranked among the highest in the world [43] and subsidies represent about half of the income of the sheep production industry [44]. Originally the aim was to subsidize prices of agricultural products to consumers and to ensure steady incomes to farmers, but later other aims became important, such as promoting rural development, product quality, and environmental aspects [44]. For instance, in 2003, a new legislation, the Quality Management in Sheep Farming (QMS) tied part of the subsidies to responsible land use, linking subsidy payment to criteria for quality and land condition of the grazing areas [45]. However, the aim of developing a green subsidy system for sheep grazing in Iceland failed [42], with most of the commons that required a land improvement plan or protection from grazing still being grazed at high grazing pressures.

5 A Framework Model for Land Degradation in Iceland (Ice-LaCoRe)

Centuries of agricultural use in Iceland under marginal natural conditions have caused severe and large-scale land degradation, which is a main environmental concern still today. However, land degradation needs to be placed into a broader context. A model for land degradation in Iceland was recently proposed by Arnalds [46]. The model was based on the 'pressure-state-response' model presented by FAO [47], and was similar to other frameworks of grassland degradation [48] that recognize three distinct elements: drivers, processes, and consequences. Such models identify pressures (drivers) as external forces or changes that cause degradation. These drivers include, for example, increases in grazing pressure or any other change in land use that can influence soil quality. States or processes are measurable changes in land condition, including vegetation, soils, nutrients, and water, that can be evaluated using indicators of land condition [47]. Finally, consequences and responses characterize the results of the process of land degradation and include efforts by land users and governments to remedy any degradational change.

We build on the model of Arnalds [46] to propose a framework model for land condition response in Iceland (Ice-LaCoRe; Fig. 3). This model distinguishes the underlying reasons (A) for particular land uses (B), including their type and pressure, that lead to changes in ecosystem processes (C) that determine the present condition of the land (D); the societal consequences of these changes (E) can in turn influence the drivers of land use. We separate here the drivers (A) from the land use type and pressure (B), as the former represent societal factors like subsidies, traditions, land tenure systems, interest groups, and legal frameworks, which drive or maintain certain type of land use or pressure, and need to be addressed as such. Changes in ecosystem processes (C) determine the shifts from one ecosystem state to another, for example, from woodland to heathlands, barren areas, or vice versa [1]. These changes are largely determined by environmental conditions that affect ecosystem

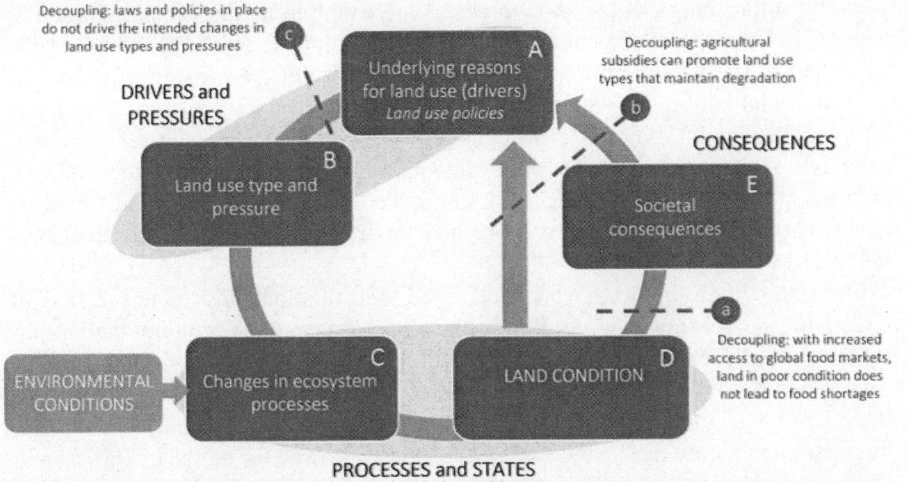

Fig. 3 A framework model for land condition response in Iceland (Ice-LaCoRe), indicating pressures and drivers (**A**, **B**), processes and states (**C**, **D**), and consequences (**E**). This land condition response cycle can be decoupled (dashed lines), (**a**) when land condition does not lead to societal consequences: for example, with globalization of food markets and other means of securing food availability, land in poor condition does not lead to food shortages that will trigger societal responses; (**b**) when links between land condition or its societal consequences and land use policies are broken, for example, through agricultural subsidies irrespective of land condition, degradation may continue because land condition is not perceived as a problem; or (**c**) when laws and policies are in place but are not effective in driving the intended changes in land use types and pressures

resilience and the rate of these processes. For instance, degradation is typically more rapid at higher elevations (less resilience) and recovery of vegetation cover following grazing exclusion takes much longer time under harsher environmental conditions of the highlands than in the lowlands [49]. Changes in land use type and pressure can lead to ecosystem degradation or ecosystem recovery, depending on their strength and direction. In the present model, changes in ecosystem processes (C) are similarly separated from the current land condition (D). Land condition as such is often omitted from land degradation frameworks [47], but is a key element of the Ice-LaCoRe model. Land condition is the state of the land at a given point in time and can be assessed using various indicators of ecosystem functions and properties, such as vegetation type, ground cover, and soil factors [50]. Land condition reflects thus the cumulative changes (C) that have occurred over millennia in Iceland as a result of long-term land use (B) driven by various underlying reasons (A).

To illustrate how this framework model works, we may consider land use policies aimed at increased food production (A) leading to implementation of more intensive land uses (B). Sustained intensive use will induce changes in ecosystem processes (C), especially when coupled with unfavourable environmental conditions, including low resilience, that eventually lead to ecosystem degradation and land in poor condition (D). Ecosystems in poor condition are less productive and can lead to negative consequences for societies (E), which may in turn respond to address these issues by influencing policy development. One example of this situation was the beginning of the twentieth century, when recognition of extensive rangeland degradation led to the first organized societal responses, with the passing of a law on vegetation and soil conservation and the establishment of the Icelandic Soil Conservation Service in 1907 [51]. Sustainable rangeland management and ecosystem restoration started to gain further policy attention in the early 1960s, with an Act on Land Reclamation passed in 1965 [37]. Improved legal status of rangeland management was attained with the approval of a new environmental Act on Land Reclamation in 2018 [52].

Elements within the framework model and their relationships are not static and have changed over time (Fig. 3). For example, the drivers of agricultural land use in Iceland radically changed from subsistence agriculture to the heavily subsidized, mechanized agriculture of the modern western society. In the past, people were dependent on the land and food scarcity determined human survival. Mechanization of agriculture, mainly during the twentieth century, allowed for more haymaking and increased winter fodder production, which in turn allowed for an increased number of domestic animals with the subsequent increase in grazing pressures on sheep grazing areas. Environmental conditions have also changed over time. It has been argued that the onset of human activity in Iceland had particularly strong consequences because it was combined with a colder period, the Little Ice Age (1450–1850; [53]), which implied more rapid declines in vegetation cover than before settlement [54].

In addition, some of the relationships in the land condition response cycle might be disrupted by different decoupling mechanisms (Fig. 3). For instance, globalization of food markets can decouple the societal consequences from land condition (a in Fig. 3), as the reduced productivity of land in poor condition has no longer implications to societies because food can be acquired as needed from elsewhere. The link between land condition and the drivers of land degradation can be decoupled as well (b in Fig. 3), for example, through governmental subsidies to agriculture. In Iceland, agricultural subsidies were introduced in the twentieth century, mainly after World War II, in an effort to stimulate domestic production. In 2019, sheep subsidies amounted to a total of around 5.2 billion Icelandic króna (ca. 36 million euros; [42]), largely exceeding the income generated by the sheep products [44]. Even though there has been an attempt to link these incentives to sustainable land uses through the cross-compliance schemes, these subsidies have been qualified as 'perverse incentives' [42], as in the long run they can adversely affect the economy and the environment [55]. In this sense, maintaining the sheep industry comes at the cost of maintaining the poor state of many areas without societal consequences (decoupling b in Fig. 3).

Finally, a particular form of decoupling occurs when soil conservation and nature protection laws and regulations fail to change land use and land use pressures as intended (c in Fig. 3). For example, since the early 1990s, the Icelandic government has tried to promote sustainable grazing management by introducing new laws and regulations to help achieve targets for improved ecological conditions [37]. One such mechanisms was the inclusion of improved rangeland management as a compulsory pillar to the Quality Management in Sheep Farming (QMS). The QMS is a cross-compliance scheme that came into force in 2003, with the aim of securing sustainable rangeland grazing. Participation in the scheme is voluntary and sheep farmers that participate and comply with the requirements of sustainable land use receive ~30% higher subsidy payments than non-participating farmers [45]. However, recent evaluations indicate that the QMS scheme and other interventions have failed to facilitate long-term system transition towards sustainable rangeland management and have not increased the understanding of what sustainable rangeland management involves among relevant stakeholder groups, including the agricultural sector and local authorities [42, 45, 52, 56]. The current administrative structure has not facilitated the intended changes aimed at improved rangeland management [52]. The institutional setting has not adopted adaptive governance approaches most likely because of the lack of robust vertical and horizontal integration of the governance system. For example, there is a lack of formal platforms for active participation and information sharing between and within stakeholder groups and with the administration entities [45, 52].

6 Present Condition of the Land

The current condition of Icelandic ecosystems reflects a long history of land use. The impacts of land use depend on multiple factors that influence ecosystem resilience, including elevation, slope, drainage (e.g. wetland vs dryland), proximity to the active volcanic zone, the presence of active unstable sand sources, and land use intensity [46]. Land degradation has repeatedly been highlighted as a main environmental concern in Iceland [57]. Initial conservation efforts focused on halting rapid erosion and sand encroachment, with increased public awareness in the latter part of the twentieth century [51]. Over the last 30 years, several initiatives have attempted to determine land condition on a national scale, based on measurable indicators, including soil erosion, energy, water, and nutrient fluxes and vegetation composition (e.g. [58]).

Soil erosion has been and continues to be a main environmental problem in Iceland [3]. Soil erosion processes can be initiated by the formation of small, isolated erosion spots that break up the vegetation cover. These erosion spots might be formed by grazing and trampling of livestock, or by other processes like drifting sand and volcanic tephra that can suffocate vegetation and accelerate subsequent erosion [59]. Once the soil is exposed it is easily eroded, especially thick soils within the volcanic active zone and at higher elevation in the highlands [3]. The process accelerates as the erosion spots become larger and more numerous, and subsequently coalesce to form large bare ground patches and eventually lead to the complete removal of vegetation and soil cover [60]. The barren surfaces are subjected to intense erosion by wind and water, sometimes resulting in severe dust storms [3]. As soil frost prevents water infiltration in winter on barren surfaces, surface runoff events are common during rainstorm and snow-thaw events [61].

Fig. 4 A distinctive erosion form in Iceland are erosion escarpments called 'rofabard' in Icelandic. These erosion forms can be dominant in communal grazing areas, where fully vegetated ecosystems are gradually replaced by barren surfaces (Photograph by Ólafur Arnalds [46])

A distinctive form of active soil erosion in Icelandic landscapes are erosion escarpments called 'rofabards' (Fig. 4). Rofabards are present in about 20,000 km^2 of Icelandic landscapes and form in thick but non-cohesive Andosols that overlay more cohesive materials, as a result of severe wind erosion, frost-heaving, and water erosion [3]. Rofabards represent the advancing front of the erosion process and their retreat leaves behind unstable desert surfaces. Active erosion processes, together with frequent freeze–thaw cycles and water erosion in winter, retard natural vegetation re-establishment on the barren surfaces. Many of these areas continue to be grazed in spite of limited vegetation cover, further reducing the possibility of natural regeneration. However, some extensive desertified sandy areas, especially in South Iceland have been protected from grazing over the last century and vegetation has successfully re-established [57].

An assessment of soil erosion in Iceland based on recognizable erosion forms, such as rofabards, encroaching sand, or erosion spots, was conducted in the early 1990s [62]. This mapping was the first effort to assess soil erosion at a national scale and confirmed the poor state of the land, with about 40% of Iceland consisting of poorly vegetated surfaces, and severe soil erosion still affecting vegetated ecosystems in many parts of the country [62]. Subsequent assessments of land condition in Iceland have been largely based on the conceptual model developed by Aradóttir et al. [60], which included six land condition classes or states, from pristine conditions to collapsed systems. Specific schemes have been developed for assessment of land condition of horse and sheep pastures for field agents and farmers [63, 64]. More recently, Barrio et al. [1] developed a conceptual state-and-transition model for Iceland with the aim of understanding landscape scale vegetation changes and the drivers of these changes.

Fig. 5 The most recent assessment of land condition in Iceland was conducted under the framework of the monitoring programme GróLind. Terrestrial ecosystems in Iceland are classified into five categories of land condition from poorest (GL values 5–12) to best land condition (GL values 26–30). Only the two best categories (GL values 22 or higher) are considered well suited for grazing [58]. Map reproduced from GróLind/Icelandic Soil Conservation Service with permission

The most recent assessment of land condition in Iceland was conducted under the framework of the GróLind monitoring programme [41, 65]. The assessment built on previous efforts in mapping soil erosion in Iceland [62], the land condition classes developed by Aradóttir et al. [60], and the recent classification of Icelandic terrestrial ecosystems into habitat types [66]. The results of this assessment indicate that 45% of terrestrial ecosystems in Iceland are in poor condition, a large part of these being relatively barren surfaces, and only 26% of the land cover can be considered in good condition (GróLind land condition categories 4 and 5) [58](Fig. 5).

7 Agriculture, Land Degradation, and Climate: The Future?

In addition to losses of productivity, agricultural land degradation can lead to pronounced losses of carbon from ecosystems, with consequences for ongoing climate change. Carbon losses related to agricultural land degradation are due to three main processes: direct loss of carbon by soil erosion, reduced carbon storage in soils and vegetation, and direct CO_2 emissions from drained wetlands.

The erosion of soils rich in organic matter can cause the release of considerable amounts of carbon dioxide. Pristine Icelandic Andosols are rich in carbon, typically ranging between 25 and 50 kg C/m^2 in dryland soils and 100–300 kg/m^2 in wetlands [3]. Estimates of current carbon losses due to soil erosion are between 0.05 and 0.1 million tons of carbon per year. It has been estimated that in Iceland, 120–500 million tons of soil organic carbon have eroded during the past millennium, and approximately half of this amount has been oxidized and released to the atmosphere [67].

Degradation processes in Iceland also involve reduction of carbon stocks in soils. Undisturbed Icelandic Andosols usually contain 6–12% carbon in surface horizons while degraded heathlands commonly have 2–5% C [3]. The total losses of CO_2 from drylands due to land degradation has been estimated to be 2–8 million tons CO_2 per year with a total of >1,000 million tons CO_2 emitted to the atmosphere since the time of settlement [68]. Even though these values are subject to a high level of uncertainty, they provide the first estimates of the carbon footprint of land degradation in Iceland.

Finally, drained wetlands are a large-scale source of CO_2 emissions, or about 7.5 million tons per year higher than natural emissions if draining had not occurred [68]. These emissions greatly exceed the near five million tons CO_2 of total emissions caused by other human activities like transportation and heavy industry [69]. Wetland draining nearly ceased in the 1990s, but emissions of the drained areas will continue for a long time after drainage [69].

The recognition of increasing greenhouse gas concentrations in the atmosphere has added another aspect to the severity of agricultural disturbance of Icelandic ecosystems. Agricultural land use contributes to greenhouse gas emissions that exceed the total emissions from other sources [69]. The poor state of the land in many areas means that there is potential for sequestering carbon back into the ecosystems in association with land restoration projects. Socioeconomic drivers such as the creation of new grazing lands, motivated re-vegetation efforts over most of the twentieth century, but more recently climate change mitigation and nature conservation have become important drivers of ecosystem restoration in Iceland [51]. Many of the severely degraded areas contain limited amounts of soil carbon [67] but the volcanic nature of the soils and the steady dust deposition allow for relatively high carbon sequestration rates in soils. Reclamation of Icelandic desertified areas may result in an average carbon sequestration rate of 0.6 t C per hectare and year in soils [3] and 0.01–0.5 t C per hectare and year in above and belowground biomass [70]. Ecosystem restoration and soil conservation are indeed one of Iceland's key strategies in mitigating climate change and were included as part of Iceland's action plan within the Kyoto protocol [37].

The necessary changes in agricultural production can be facilitated by changes in the structure of the agricultural subsidy system. Icelandic consumers are calling for environmentally-friendly food production, and lamb meat consumption has halved since the early 1980s [40]. The subsidy system should focus more on meeting these changes while phasing out lamb production in areas of collapsed ecosystems. Future attempts to reduce the impact of agriculture will need to take land condition into account and find ways to sequester carbon back into soils and vegetation. Moderate grazing pressures on areas that are now in reasonably good condition can lead to carbon sequestration in these systems [68]. By excluding grazing from degraded areas and restoring systems such as the birch forests that once covered large areas, Icelandic agriculture could potentially achieve carbon neutrality [68]. Such changes can also facilitate other means of food production and rural employment opportunities, especially within the rapidly growing tourist industry and by subsidizing ecosystem restoration efforts on a large scale. Carbon sequestration efforts in relation to afforestation and ecosystem restoration can also represent a new mainstay of rural subsidies.

A new government report on the future of agriculture in Iceland emphasizes sustainable land uses, reductions in greenhouse gas emissions, and innovation [71]. To the extent possible, it is necessary to diversify agricultural production in Iceland towards methods and products that reduce overall environmental impact, including carbon emissions, such as increased production of cereals, vegetables, and more diverse horticultural products. Changes should also consider food security in the future [39]. Recognition of the poor state of the land is a key element for the road forward and calls for the identification of barriers and decoupling mechanisms that prevent improved land use. Enhanced education and improved function of governance systems are also important priorities. Lastly, we emphasize the need of establishing large-scale ecosystem restoration projects, in line with current global emphasis, including the United Nations' Decade of Ecosystem Restoration (https://www.decadeonrestoration.org/).

8 Final Considerations

The relatively poor condition of many Icelandic ecosystems can be mostly attributed to agricultural use over the past millennium, especially grazing and wood harvesting. In the literature, Iceland has been used as an example to illustrate a case of severe land degradation and desertification (e.g. [72]). Agricultural land uses, in combination with the harsh environmental conditions and frequent volcanic activity have been repeatedly pointed out as the drivers of extensive land degradation and extensive soil erosion in Iceland. The framework model for land condition response in Iceland (Ice-LaCoRe) proposed here helps in separating the underlying reasons, drivers, processes, states, and consequences of land degradation, where decoupling mechanisms favour inaction in dealing with the poor state of the land.

In the light of the poor condition of many Icelandic ecosystems land restoration has emerged as a key priority. Large-scale ecosystem restoration programmes restore ecological values for diverse land use in the future and sequester carbon in both soils and vegetation, thus linking climate and nature restoration targets. Combining possible carbon sequestration in soils and vegetation with enhanced ecological restoration efforts, millions of tons of CO_2 could be added to ecosystems from the atmosphere annually, thus making a carbon-neutral Iceland a possibility within decades. Simultaneously, further land degradation should be prevented. A thorough revision of agricultural policies in Iceland is thus needed, giving priority to environmental issues in rural and agricultural subsidy policies. Such regulations should carefully consider land condition and sheep grazing should be limited to areas where ecosystems are in good condition and where sheep production is vital for the viability of rural communities.

References

1. Barrio IC, Hik DS, Thórsson J, Svavarsdóttir K, Marteinsdóttir B, Jónsdóttir IS (2018) The sheep in wolf's clothing? Recognizing threats for land degradation in Iceland using state-and-transition models. Land Degrad Dev 29(6):1714–1725
2. McGovern TH, Vésteinsson O, Friðriksson A, Church M, Lawson I, Simpson IA et al (2007) Landscapes of settlement in Northern Iceland: historical ecology of human impact and climate fluctuation on the millennial scale. Am Anthropol 109(1):27–51
3. Arnalds Ó (2015) The soils of Iceland. World soil. Springer, Dordrecht, p 180
4. Palter JB (2015) The role of the Gulf stream in European climate. Annu Rev Mar Sci 7:113–137
5. Bjornsson H, Jonsson T, Gylfadottir SS, Olason EO (2007) Mapping the annual cycle of temperature in Iceland. Meteorol Zeitschrift 16(1):45–56
6. Crochet P, Jóhannesson T, Jónsson T, Sigurdsson O, Björnsson H, Pálsson F et al (2007) Estimating the spatial distribution of precipitation in Iceland using a linear model of orographic precipitation. J Hydrometeorol 8(6):1285–1306
7. Ólafsson H, Furger M, Brümmer B (2007) The weather and climate of Iceland. Meteorol Zeitschrift 16(1):5–8
8. Thórhallsdóttir TE (1997) Tundra ecosystems of Iceland. In: Wielgolaski FE (ed) Ecosystems of the world. Elsevier, Amsterdam, pp 85–96
9. Buckland PC, Dugmore A (1991) If this is a refugium, why are my feet so bloody cold? The origins of the Icelandic biota in the light of recent research. In: Maizels JK, Caseldine C (eds) Environmental change in Iceland: past and present [Internet]. Kluwer Academic Publishers, pp 107–125. http://link.springer.com/10.1007/978-94-011-3150-6
10. Thórisson S (1984) The history of reindeer in Iceland and reindeer study 1979-1981. Rangifer 4(2):22
11. Hersteinsson P, Angerbjörn A, Frafjord K, Kaikusalo A (1989) The arctic fox in Fennoscandia and Iceland: management problems. Biol Conserv 49(1):67–81
12. Bonesi L, Palazon S (2007) The American mink in Europe: status, impacts, and control. Biol Conserv 134(4):470–483
13. Bengtson S, Nilsson A, Rundgren S (1989) Population structure and dynamics of wood mouse Apodemus sylvaticus in Iceland. Ecography 12(4):351–368
14. Gunnarsson TG, Gill JA, Appleton GF, Gíslason H, Gardarsson A, Watkinson AR et al (2006) Large-scale habitat associations of birds in lowland Iceland: implications for conservation. Biol Conserv 128(2):265–275
15. Jóhannesdóttir L, Alves JA, Gill JA, Gunnarsson TG (2018) Use of agricultural land by breeding waders in low-intensity farming landscapes. Anim Conserv 21(4):291–301
16. Jóhannesdóttir L, Gill JA, Alves JA, Brink SH, Arnalds Ó, Méndez V et al (2019) Interacting effects of agriculture and landscape on breeding wader populations. Agric Ecosyst Environ 272: 246–253. https://doi.org/10.1016/j.agee.2018.11.024
17. Haraldsson HV, Ólafsdóttir R (2006) A novel modelling approach for evaluating the preindustrial natural carrying capacity of human population in Iceland. Sci Total Environ 372(1):109–119

18. European Commission (2021 [cited 2003 Jan 20]) Eurostat [Internet]. Eurostat. https://ec.europa.eu/eurostat/
19. Lawson I, Gathorne-Hardy F, Church M, Newton A, Edwards K, Dugmore A et al (2007) Environmental impacts of the Norse settlement: palaeoenvironmental data from Mývatnssveit, northern Iceland. Boreas 36(1):1–19
20. Hallsdóttir M (1995) On the pre-settlement history of Icelandic vegetation. Búvísindi 9:17–29
21. Vickers K, Erlendsson E, Church MJ, Edwards KJ, Bending J (2011) 1000 years of environmental change and human impact at Stora-Mork, southern Iceland: a multiproxy study of a dynamic and vulnerable landscape. Holocene 21(6):979–995. http://hol.sagepub.com/cgi/doi/10.1177/0959683611400201
22. Eddudóttir SD, Erlendsson E, Tinganelli L, Gísladóttir G (2016) Climate change and human impact in a sensitive ecosystem: the Holocene environment of the Northwest Icelandic highland margin. Boreas 45:715–728
23. Erlendsson E, Edwards KJ (2009) The timing and causes of the final pre-settlement expansion of Betula pubescens in Iceland. Holocene 19(7):1083–1091. http://hol.sagepub.com/cgi/doi/10.1177/0959683609341001
24. Ólafsdóttir R, Schlyter P, Haraldsson HV (2001) Simulating Icelandic vegetation cover during the Holocene. Implications for long-term land degradation. Geogr Ann 83(4):203–215
25. Fridriksson S (1972) Grass and grass utilization in Iceland. Ecology 53(5):785–796
26. Bjarnason H (1971) Um friðun lands og frjósemi jarðvegs. Ársrit Skógræktarfélags Íslands 1971:4–19
27. Wöll C (2008) Treeline of mountain birch (Betula pubescens Ehrh) in Iceland and its relationship to temperature [Internet]. Technical University Dresden. http://skemman.is/stream/get/1946/7559/20046/1/11855_Birch_treeline_in_Iceland_Christoph.pdf
28. Snorrason A, Traustason B, Kjartansson BÞ, Heiðarsson L, Ísleifsson R, Eggertsson Ó (2016) Náttúrulegt birki á Íslandi. Náttúrufræðingurinn 86(3–4):97–111
29. Arnalds O, Gisladottir FO, Sigurjonsson H (2001) Sandy deserts of Iceland: an overview. J Arid Environ 47(3):359–371
30. Helgadóttir Á, Eythórsdóttir E, Jóhannesson T (2013) Agriculture in Iceland – a grassland based production. Grassl Sci Eur 18:30–43
31. Bergthorsson P (1985) Sensitivity of Icelandic agriculture to climatic variations. Clim Chang 7: 111–127
32. Guðbergsson G (1996) Í norðlenskri vist. Um gróður, jarðveg, búskaparlög og sögu [The influence of human habitation on soil and vegetation in three counties in North-Iceland]. Búvísindi Icel Agric Sci 10:31–89
33. Thorsteinsson I (1986) The effect of grazing on stability and development of northern rangelands: a case study of Iceland. In: Guðmundsson Ó (ed) Grazing research at northern latitudes. NATO ASI series, series A: life sciences, vol 108, pp 37–43
34. Simpson IA, Guðmundsson G, Thomson AM, Cluett J (2004) Assessing the role of winter grazing in historic land degradation, Mývatnssveit, Northeast Iceland. Geoarchaeology 19(5): 471–502
35. Þórhallsdóttir AG, Júlíusson AD, Ögmundardóttir H (2013) The sheep, the market and the soil: environmental destruction in the Icelandic highlands 1880–1910. In: Jørgensen D, Sorlin S (eds) Northscapes: history, technology and the making of northern environments. University of British Columbia Press, Vancouver, pp 153–173
36. Arnalds O, Gudmundsson J, Oskarsson H, Brink SH, Gisladottir FO (2016) Icelandic inland wetlands: characteristics and extent of draining. Wetlands 36(4):759–769. https://doi.org/10.1007/s13157-016-0784-1
37. Fannarsson BS, Barkarson BH, Pétursson JG, Helgadóttir SS (2018) Reform of the Icelandic soil conservation law. In: Ginzky H, Dooley E, Heuser IL, Kasimbazi E, Markus T, Qin T (eds) International yearbook of soil law and policy 2017. Springer, Cham, pp 223–243
38. Marteinsdóttir B, Barrio IC, Jónsdóttir IS (2017) Assessing the ecological impacts of extensive sheep grazing in Iceland. Icel Agric Sci 30:55–72

39. Sturludóttir E, Þorvaldsson G, Helgadóttir G, Guðnason I, Sveinbjörnsson J, Sigurgeirsson ÓI et al (2021) Fæðuöryggi á Íslandi [Food security in Iceland]. Rit LbhÍ 139:56
40. Statistics Iceland (2021) Statistics Iceland [Internet]. https://statice.is/
41. Stefánsson JH, Þorvaldsdóttir S, Hauksdóttir I, Þórarinsdóttir EF, Marteinsdóttir B, Brink SH (2020) Kortlagning beitarlanda sauðfjár á Íslandi. Soil Conservation Service of Iceland, Reykjavík
42. Arnalds O (2019) Development of perverse environmental subsides for sheep production in Iceland. Agric Sci 10:1135–1151
43. OECD (2020) Chapter 12. Iceland. In: Agricultural policy monitoring and evaluation 2020. OECD Publishing, Paris
44. Sveinbjörnsson J, Kristófersson DM (2021) Afkoma sauðfjárbænda á Íslandi og leiðir til að bæta hana [Performance of Icelandic sheep farms]. Rit LbhÍ 142:48
45. Þorláksdóttir JS (2015) Connecting sustainable land use and Quality Management in Sheep Farming: effective stakeholder participation or unwelcome obligation? University of Iceland, Reykjavík
46. Arnalds Ó (2020) Ástand lands og hrun íslenskra vistkerfa [Land condition and the collapse of Icelandic ecosystems]. Rit LbhÍ 130:1–79
47. Benites JR, Shaxson TF, Vieira M (1997) Land condition change indicators for sustainable land resource management [Internet]. http://www.fao.org/3/W4745E/w4745e09.htm
48. Tiscornia G, Jaurena M, Baethgen W (2019) Drivers, process, and consequences of native grassland degradation: insights from a literature review and a survey in Río de la Plata grasslands. Agronomy 9(5):8–12
49. Magnússon SH, Svavarsdóttir K (2007) Áhrif beitarfriðunar á framvindu gróðurs og jarðvegs á lítt grónu landi. Fjölrit Náttúrufræðistofnunar 49:1–67
50. Herrick JE, Schuman GE, Rango A (2006) Monitoring ecological processes for restoration projects. J Nat Conserv 14:161–171
51. Aradóttir ÁL, Petursdottir T, Halldorsson G, Svavarsdottir K, Arnalds O (2013) Drivers of ecological restoration: lessons from a century of restoration in Iceland. Ecol Soc 18(4):33
52. Petursdottir T, Baker S, Aradottir AL (2020) Functional silos and other governance challenges of rangeland management in Iceland. Environ Sci Pol 105:37–46. https://doi.org/10.1016/j.envsci.2019.12.006
53. Geirsdóttir Á, Miller GH, Thordarson T, Ólafsdóttir KB (2009) A 2000 year record of climate variations reconstructed from Haukadalsvatn, West Iceland. J Paleolimnol 41(1):95–115
54. Haraldsson HV, Ólafsdóttir R (2003) Simulating vegetation cover dynamics with regards to long-term climatic variations in sub-arctic landscapes. Glob Planet Change 38(3–4):313–325
55. Myers N (1998) Lifting the veil on perverse subsidies. Nature 392(6674):327–328
56. Stefansson JH (2018) Of sheep and man. Analysis of the Agri-environmental cross-compliance policies in the Icelandic sheep grazing regime. University of Iceland, Reykjavík
57. Crofts R (2011) Healing the land [Internet]. Soil Conservation Service of Iceland, Reykjavík, p 212. https://www.moldin.net/uploads/3/9/3/3/39332633/ad_lesa_og_lækna_landi.pdf
58. Marteinsdóttir B, Þórarinsdóttir EF, Halldórsson G, Stefánsson JH, Þórsson J, Svavarsdóttir K et al (2020) Stöðumat á ástandi gróður- og jarðvegsauðlinda Íslands. Soil Conservation Service of Iceland, Gunnarsholt
59. Vilmundardóttir OK, Magnússon B, Gisladóttir G, Magnússon SH (2009) Áhrif sandfoks á mólendisgróður við Blöndulón. Náttúrufræðingurinn 78:125–138
60. Aradóttir ÁL, Arnalds Ó, Archer S (1992) Hnignun gróðurs og jarðvegs. In: Arnalds A (ed) Græðum Ísland. Reykjavik, Landgræðsla Ríkisins, pp 73–82
61. Orradottir B, Archer SR, Arnalds O, Wilding LP, Thurow TL (2008) Infiltration in Icelandic andisols: the role of vegetation and soil frost. Arct Antarct Alp Res 40(2):412–421
62. Arnalds Ó, Thorarinsdottir EF, Metusalemsson S, Jónsson Á, Arnason A (2001) Soil erosion in Iceland. Agricultural Research Institute, Soil Conservation Service
63. Magnússon B, Elmarsdóttir Á, Barkarsson BH (1997) Hrossahagar. Aðferðir til að meta ástand lands, Reykjavík

64. Jónsdóttir S (2010) Sauðfjárhagar. Landgræðsla Ríkisins, Gunnarsholt
65. Marteinsdóttir B, Þórarinsdóttir EF, Halldórsson G, Þórsson J, Svavarsdóttir K, Einarsson MÞ et al (2017) GróLind - Mat og vöktun á gróður- og jarðvegsauðlindum Íslands. Ársskýrsla, Gunnarsholt
66. Ottósson JG, Sveisdóttir A, Harðadóttir M (2016) Vistgerðir á Íslandi. Fjölrit Náttúrufræðistofnunar 54:1–299
67. Óskarsson H, Arnalds Ó, Gudmundsson J, Gudbergsson G (2004) Organic carbon in Icelandic Andosols: geographical variation and impact of erosion. Catena 56(1–3):225–238
68. Arnalds Ó, Guðmundsson J (2020) Loftslag, kolefni og mold [Climate, carbon and soil]. Rit LbhÍ 133:1–51
69. Keller N, Stefani M, Einarsdóttir SR, Helgadóttir ÁK, Guðmundsson J, Snorrason A et al (2020) National inventory report. Emissions of greenhouse gases in Iceland from 1990 to 2018. Submitted under the United Nations Framework Convention on Climate Change and the Kyoto Protocol. Reykjavík
70. Aradóttir ÁL, Svavarsdóttir K, Jónsson TH, Gudbergsson G (2000) Carbon accumulation in vegetation and soils by reclamation of degraded areas. Icel Agric Sci 13:99–113
71. Bjarnason B, Sveinsdóttir HH (2021) Ræktum Ísland! Landbúnaður á 21. öld. Umræðuskjal [Internet]. Reykjavík, Iceland; https://www.stjornarradid.is/library/01%2D%2DFrettatengt%2D%2D-myndir-og-skrar/ANR/Landbunadur/210414_ANR_RaektumIsland_V6.pdf
72. Imeson A (2012) Desertification, land degradation and sustainability. Paradigms, processes, principles and policies. Wiley, Chichester

Agricultural Land Degradation in Italy

Demetrio Antonio Zema, Giuseppe Bombino, and Santo Marcello Zimbone

Contents

1 Introduction 180
2 Main Characteristics of Climate and Soils 181
  2.1 Climate 181
  2.2 Soils 184
3 Overview of the Agricultural Sector 186
4 Soil Degradation in Italy 188
5 Land Degradation in the Italian Agriculture 191
  5.1 Desertification 193
  5.2 Erosion 197
  5.3 Salinization 201
  5.4 Soil Sealing and Compaction 203
  5.5 Organic Matter Reduction 205
  5.6 Soil Contamination 208
  5.7 Others 209
6 Future Perspectives and Conclusions 213
References 216

Abstract We present an updated state of the art about the land degradation in the Italian agriculture. After preliminary outlines about the main characteristics of the Italian climate and soils, we analyse the risks of desertification, erosion, salinization, sealing/compaction, contamination and organic matter reduction of soils. Minor factors of soil degradation (overgrazing, forest fires, pollution due to microplastics and agrochemicals) are also discussed. This review shows that, in the Italian agricultural soils, the rates of sealing, compaction and organic matter decline are close to the values recorded in several European countries. Soil erosion and landslides are the major degradation factors in steeper agricultural areas of the mid mountains and hills, and particularly in the internal areas of Southern Italy. High

D. A. Zema (✉), G. Bombino, and S. M. Zimbone
"Mediterranean" University of Reggio Calabria, Department "AGRARIA", Reggio Calabria, Italy
e-mail: dzema@unirc.it; giuseppe.bombino@unirc.it; smzimbone@unirc.it

Paulo Pereira, Miriam Muñoz-Rojas, Igor Bogunovic, and Wenwu Zhao (eds.), *Impact of Agriculture on Soil Degradation II: A European Perspective*, Hdb Env Chem (2023) 121: 179–222, DOI 10.1007/698_2022_925,

Published online: 14 December 2022

erosion rates are the main reason of reductions in soil organic matter, which worsens soil fertility and induces biodiversity loss. Unsuitable soil management and unsustainable agricultural exploitation aggravate these land degradation factors. In contrast, natural areas have been subjected to severe environmental regulations (e.g. the national and regional parks) for many years, and these environments are not particularly vulnerable. The final considerations provide insight regarding the possible countermeasures to limit the land degradation rates in the affected areas and ensure soil conservation in the other zones.

Keywords Contamination, Desertification, Erosion, Salinization, Sealing/ compaction, Soil health

1 Introduction

In 2019, the last year before economy was affected by the impacts of the COVID-19 pandemic, the domestic product of the Italian agriculture was 34.2 billion € at market prices, a share of 2.1% of the global domestic product (GDP, equal to 1.789 billion €, corresponding to a GDP per capita of 29.23 €, current values). Another share of 1.9% of the Italian GDP comes from agro-food industry, whose domestic product is 30.6 billion € [1]. The incidence of the Italian agro-food sector on the national GDP has been decreasing from the beginning of the last century due to the development of industry and services. In spite of this decreasing trend, the Italian agro-food products make up the so-called Made in Italy brand (together with clothing-fashion, furniture and automation-mechanics), which represents the excellence of Italy worldwide [2, 3].

Agriculture is practiced in rural areas, covering 25.5% of the Italian territory (302,073 km^2). The country displays a wide North–South extension (1,180 km of latitude) and a narrower West–East extension (longitude width of 530 km). This noticeable latitude extension, together with the intrinsic geological and morphological characteristics – large variability of soils, as well as hilly landscape, 41.6% of the total area, followed by mountains (35.2%) and lowlands (23.2%) – determines a large variability of conditions for agricultural activities. This variability has enhanced diversified cultivation systems with highly specialized agro-food products [1].

In spite of the peculiarities of Italian agricultural and agro-food systems, croplands and forestlands are affected by several threats that have often determined continuous and intense, with heavy impacts to some areas for several decades. The on-going land degradation rates affecting croplands, pasturelands and forestlands of Italy might be aggravated because of the expected climate change scenarios, which forecast temperature increases and precipitation amount decreases with consequent extreme drought and rainstorm events [4].

To face land degradation in Italy, several laws have been enacted in the last 20 years from national and regional authorities, often addressing the relevant EU's directives and regulations. To summarize, the milestone national law is the Legislative Decree n. 152 issued on 3rd April 2006 ("Environment Overall Text"). This Text reports the national regulations about soil conservation and desertification risk, water protection from pollution and water resources management. The Italian regions are allowed to issue local regulations in these fields, following the Legislative Decree n. 152.

An updated state of the art about the land degradation rates and trends in the Italian agriculture is needed to give insight to land managers and authorities about the possible countermeasures to limit these rates in the degraded areas and ensure soil conservation in the less affected zones. This chapter provides an overview of the land degradation status in the Italian agricultural lands. To this aim, we revise national studies focused on the triggering factors and rates of land degradation in rural areas. First, we outline the main characteristics of the Italian climate and soils (Sect. 2), followed by sections focused on desertification, erosion, salinization, sealing/compaction, contamination and organic matter reduction of soils. Other minor risks are also analysed (overgrazing, forest fires, pollution due to microplastics and agrochemicals). Another specific aim of this overview is the identification of the most critical drivers of land degradation in rural areas, on which research paths and authorities' efforts should be directed.

2 Main Characteristics of Climate and Soils

2.1 Climate

Italy is in the temperate zone of the boreal hemisphere and geographically lies inside the Mediterranean Basin. Its shape is elongated along a latitude gradient (7–19 East parallel) and quite narrow along the longitude gradient (36–47 North meridian). Due to this elongation, which determines different supply of solar radiation to its area, and to the effects of sea (which is a reservoir of heat and vapour) and orography (with the barrier of Alps and Apennines Mountain chains against cold streams from Northern Europe and North Atlantic), the Italian climate shows a large variability in precipitation and temperature.

Figure 1 reports the distribution of the mean annual temperature and precipitation across the national territory. The mean air temperature of Italy is 12.6°C, this value being variable from -11°C on the Alps to around 20°C in some inland parts of Sicily [5]. The mean annual precipitation is 932 mm with the lowest and highest values of 434 (South Sardinia) and 2,254 (Friuli-Venezia Giulia, North-East) mm, respectively. Snow covers the Alps and Apennines in winter and the flat areas of Northern and Central Italy in extremely cold years. Precipitation has a marked seasonality (which is typical of the semi-arid Mediterranean climate), as shown by the ratio between the difference between long-term rainfall of the wettest and driest months

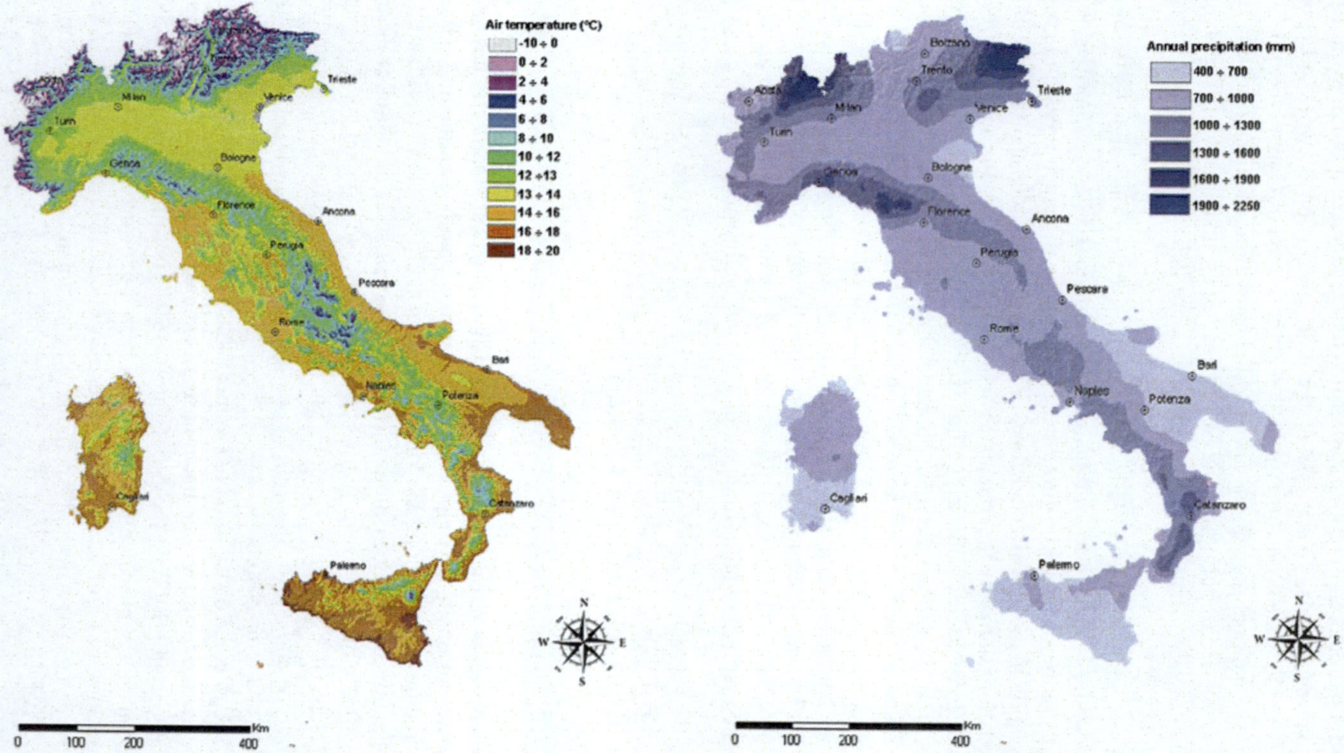

Fig. 1 Mean annual value of temperature (left) and precipitation (right) from 1960 to 2008 in Italy (source: [5])

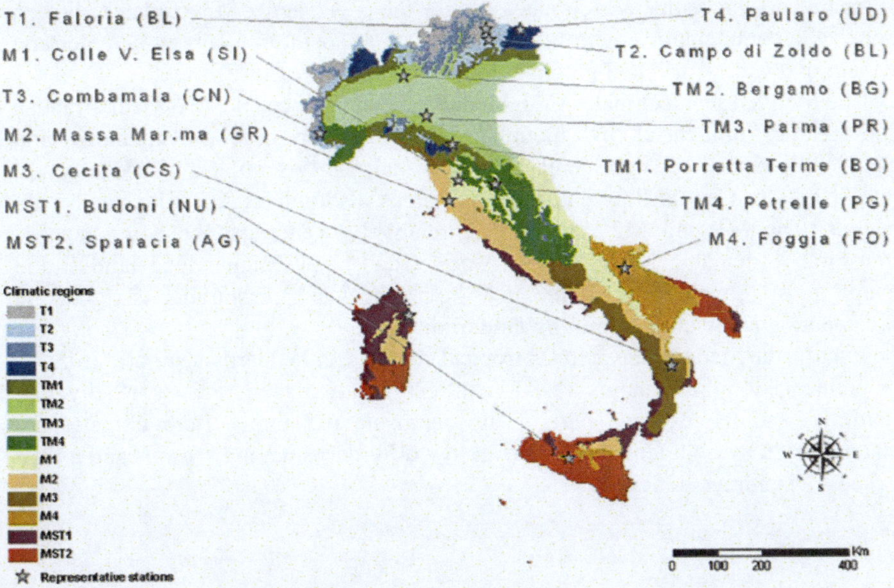

Fig. 2 Map of the climatic regions of Italy with representative weather stations (source: [5])

and the total annual rainfall that is equal to 11%. The potential evapo-transpiration is on average 1,000 mm/year, while the lowest and highest values are recorded in the Alps and the Northern Apennines (600 mm/year) and some parts of Apulia, Sicily and Sardinia (1,300 mm/year), respectively. The mean annual evapo-transpiration is close to precipitation, but, while the latter exceeds evapo-transpiration in the Alps, the Northern Po Valley and parts of the Apennines, large areas of Southern Italy (Sicily, coastal Calabria, Apulia, Sardinia and Basilicata) suffer from a precipitation deficit, being dry for large periods throughout the year.

According to the climatic classification of the soil database for Europe [6, 7], the Italian territory comprises 14 climatic regions (Fig. 2).

The Alps lay on the temperate climates (T1-temperate continental and T2-temperate sub-continental climate influenced by mountains, T3-temperate to warm temperate sub-continental, partly arid, and T4-temperate mountainous). The colder T1 climate characterizes the highest mountains with permanent snow cover, while the T2 and T3 climates are typical of lower mountains; T3 climatic region receives less precipitation than T2. The T4 climate is warmer and wetter compared to the other three temperate climates.

The Apennines are dominated by Mediterranean intermediate to TM1-sub-oceanic and TM4-mountainous warm temperate climates, while low hills of Northern and Central Italy as well as the Po Valley are in the TM2-Mediterranean sub-oceanic to sub-continental climate, influenced by mountains, and TM3-Mediterranean sub--oceanic to sub-continental climate. The TM1 or TM4 climates are much warmer compared to the temperate climates, while the TM2 and TM3 climates show a noticeable continentality, with high contrasts in temperature between summer and winter, and well-distributed precipitation throughout the year.

The Mediterranean climates (M1-Mediterranean sub-oceanic, M2-Mediterranean sub-oceanic, influenced by mountains, M3-Mediterranean mountainous and M4-Mediterranean sub-continental to continental, partly semi-arid to arid) are typical of hills of Central Italy and mountain areas of Southern Apennines and main islands. The M1 and M2 climates are relatively wetter and have less evapo-transpirative demand, M3 climate is warmer and shows higher total rainfall and evapo-transpirative demand, while, in M4 climate, evapo-transpiration is higher and precipitation is lower, determining land aridity.

Finally, the Mediterranean to subtropical climates, either partly semi-arid (MST2) or influenced by mountains (MST1), affect the two main islands (Sicily and Sardinia) and the extreme part of the peninsula (Calabria). Both climates are characterized by high temperatures with low differences among seasons and a high evapo-transpirative deficit [5].

2.2 Soils

Italian soils are classified into *regions*, *sub-regions* and *systems* [8]. The soil regions database is the highest informative level of the Italian soil map of Italy and is linked to the European soil database [9]. The characteristics of these soil regions are related to typical climate and parent material associations. Soil regions are classified according to geographical area: (1) alluvial and coastal plans and associated low hills; (2) hills; (3) Alps; (4) Apennines; (5) Etna and Sardinian mountains. Hilly soils show the largest pedodiversity according to the distribution in regions. The soil system distribution of Italy is also characterized by a large variability (Fig. 3). The most common systems at the national level are Haplic Cambisols (Calcaric), followed by Haplic Regosols (Calcaric), Haplic Cambisols (Eutric), Haplic Calcisols, Vertic Cambisols, Cutanic Luvisols, Leptic Phaeozems, Haplic Luvisols (Chromic), Haplic Cambisols (Dystric) and Fluvic Cambisols (Fig. 4) [8].

According to the common soil classifications by WRB-IUSS [10] and Soil Taxonomy by Soil Survey Staff [11], the soil map of national territory consists of 25 prevalent Reference Soil Groups (RSGs) on the 32 RSGs of WRB, and 10 out of 12 soil orders provided by the Soil Taxonomy, which further shows the large pedodiversity of Italy. Cambisols and Luvisols are the most common RSGs of Italy, while Inceptisols are the most abundant soil order (Soil Taxonomy) (Fig. 4). However, many soil profiles of the Italian database cannot be classified in a given soil system unit, due to the very small area covered.

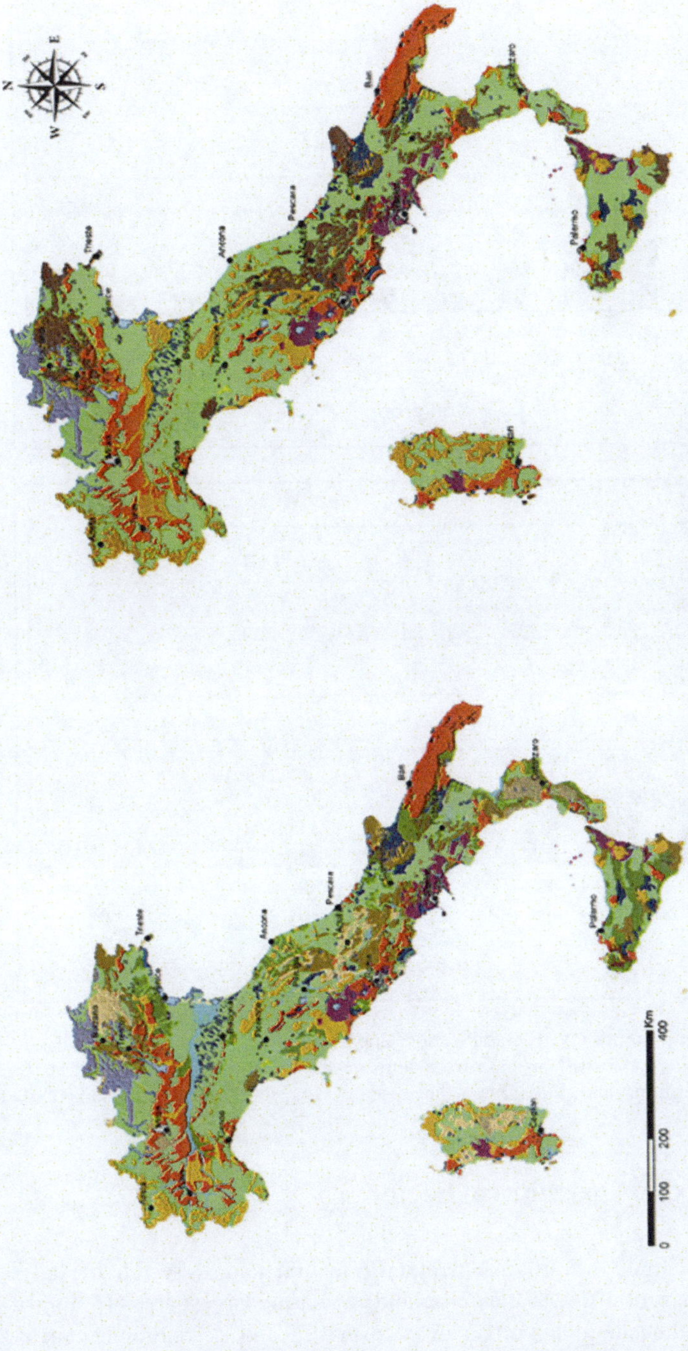

Fig. 3 Map of the dominant reference soil groups (WRB classification, left, and Soil Taxonomy classification, right) of Italy (source: [8])

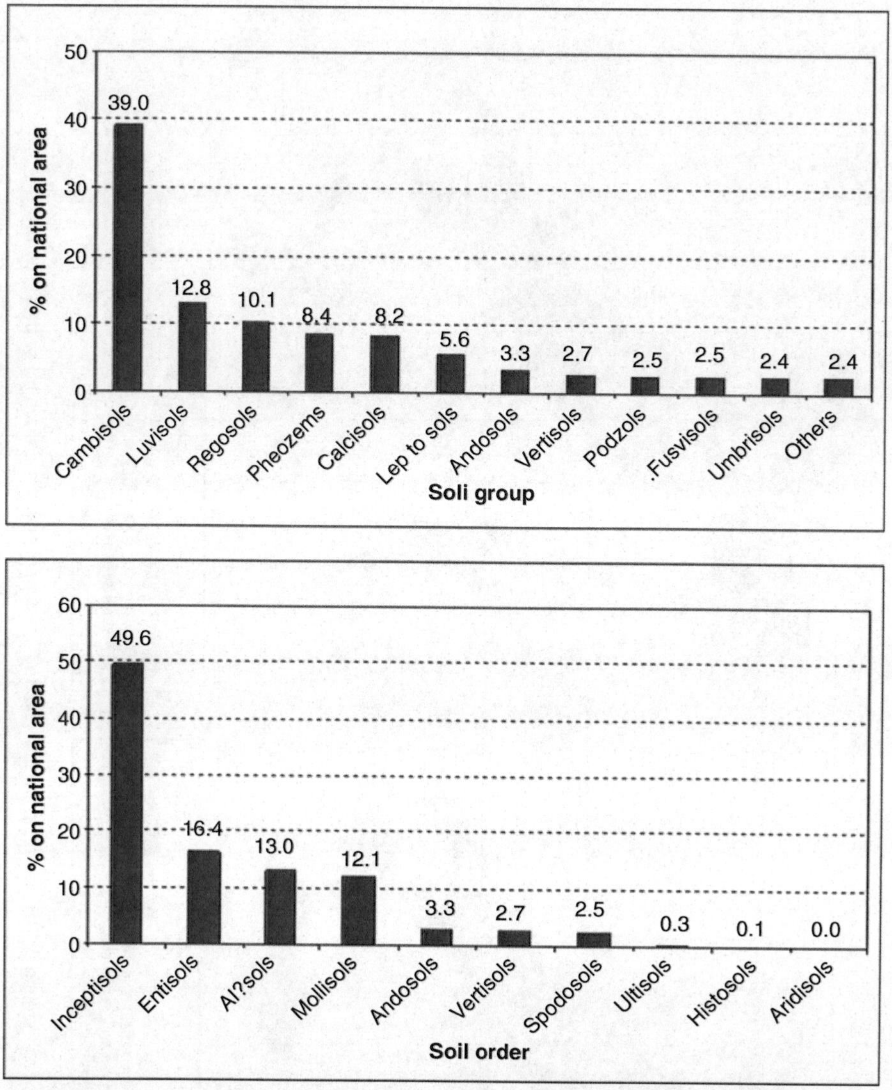

Fig. 4 Distribution of soil groups (WRB classification, upper) and orders (Soil Taxonomy classification, lower) in Italy. (source: [8] on data from https://esdac.jrc.ec.europa.eu/Library/Data/250000/Italy/MapRegions.htm, and https://www.nrcs.usda.gov/wps/portal/nrcs/detail/soils/use/?cid=nrcs142p2_054013)

3 Overview of the Agricultural Sector

According to the Institute for Environmental Protection and Research of Italy (ISPRA) data and maps of 2019, in Italy croplands cover the largest area (15.5 million hectares), while the natural and urban areas cover 12.9 and 1.7 million hectares, respectively. However, agricultural areas have decreased (and, to a lower extent, also woodlands) and artificial areas have generally increased over the last 5 years.

Forest areas cover about 35% of the Italian territory and are mainly distributed on Alps, Apennines and mountain chains of the main islands (Fig. 5). These areas have

Fig. 5 Dominant soil cover of Italy (source: [12] based on data from Corine Land Cover, 2012—first level)

been progressively increasing in the last decades due to both the abandonment of marginal croplands and the large reforestation works. Reforestation is currently less common than in the past, with a focus on small degraded areas. The main forest types are conifers and broadleaves, while coppices, previously covering large areas on mountains, have been progressively been converted into high stands. Unfortunately, large areas of the Italian forestland, including national and regional national parks, are affected by wildfire risk and erosion.

Agriculture is mainly practiced in mixed croplands with annual and permanent crops (Fig. 6). Annual crops are cultivated in rotation in the same field, or also intercropping on different fields each year, while permanent crops are cultivated separately. Regarding farm size, open fields are increasing, while closed fields are reducing. This has increased soil susceptibility to erosion, since the common natural and human works separating field, such as ditches, tree rows, stone walls, tracks, are disappearing, and decreased the organic matter content of the Italian soils [13, 14]. The effects of *set aside* policy have determined the conversion of cropland to tree orchards, mainly olive, citrus, grapevine and other fruits, in order to increase the farm profit. Olive and citrus groves are mainly cultivated in southern regions, due to the more favourable weather conditions for these species. In Northern and Central Italy, industrial crops, e.g., cereals, sugar beets and vegetables, are more common. Cattle's breeding is currently concentrated in large farms, with a progressive pasture and meadow replacement with annual crops and forage, which led to soil and water pollution due to uncontrolled disposal of animal manure and wastewater. Moreover, overgrazing in mountain prairies has been recorded with consequent soil compaction and erosion, and marginal areas have been abandoned with uncontrolled development of shrubs and weeds. The reduction in terraced fields and cropland on contour lines in mountain agriculture are other practices that are increasing the land degradation risks. Moreover, due to landslides, erosion and cultivation practices, a general surface levelling (that is, soil removal from reliefs and filling up of valleys) has been observed with the removal or burial of natural and old drainage networks [13, 14].

In some areas, large open fields have been covered by greenhouses for production of vegetables and flowers, which are often cultivated with large amounts of organic and mineral fertilizers and soil conditioners. Another undesired effect of greenhouse cover is the heavy reduction of salt leaching due to precipitation, and this determines soil salinization. Besides this agro-forestry evolution recorded in the last decades, new agricultural management strategies are developing, such as the organic and integrated agriculture. These management techniques are able to contrast organic matter decline of Italian soils with consequent improvements of hydrological properties of soil, e.g., higher water infiltration and retention [13].

4 Soil Degradation in Italy

Italy is geographically divided into three main areas, North, Centre and South based on differences in climate, landscape, soil characteristics and natural resource availability as well as socio-economic factors. This determines gradients in economic

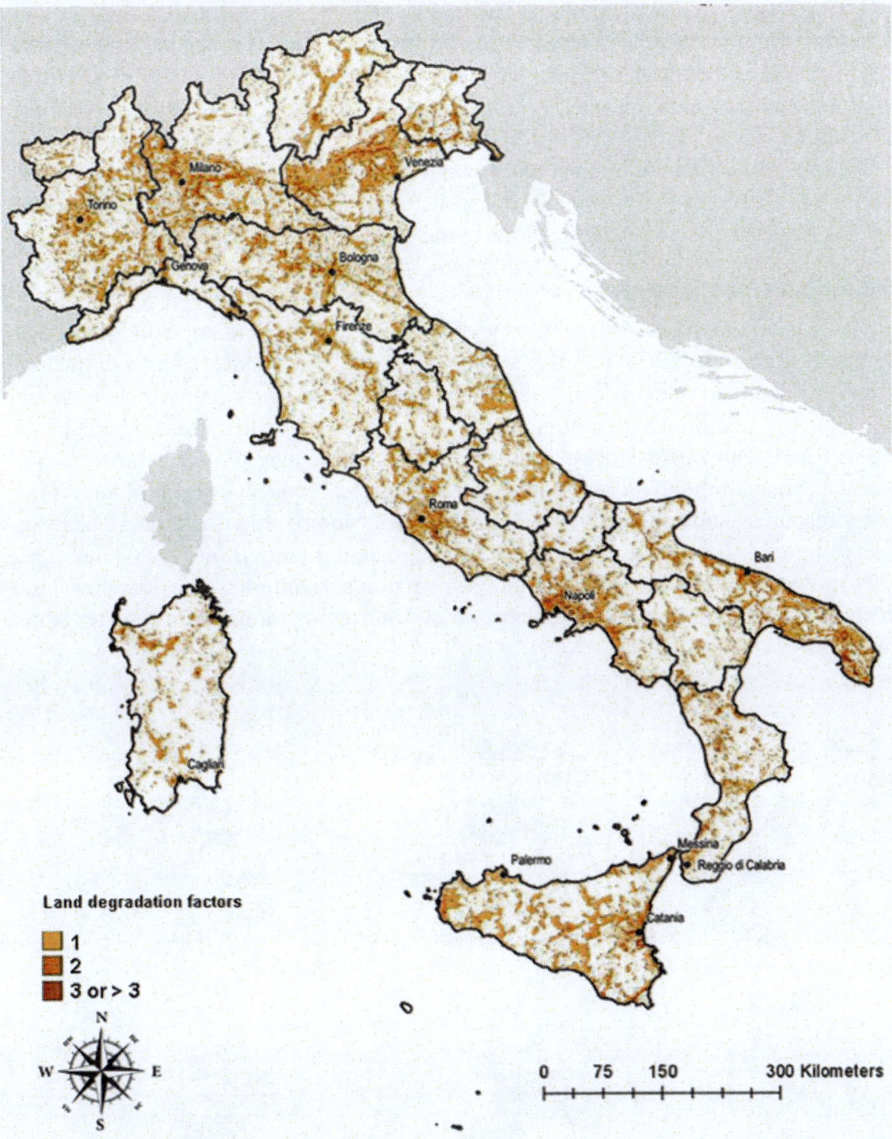

Fig. 6 Map of degraded areas in Italy (source: [12])

structure, population density, urbanization level and income distribution, and the differences among these areas heavily weigh on the spatial distribution of land degradation risk [15, 16]. More specifically, while Northern Italy, characterized by wet and continental climate, is one of the most developed regions of Europe, Central Italy is a patchy land with urban areas mainly located on the coast and traditional rural landscapes in the internal mountain zones. Southern Italy, Sicily and Sardinia, whose climate is dry and semi-arid, suffer from backwards in economy and welfare [15, 17].

It is generally acknowledged that large areas of the Italian territory, as many other countries of the Mediterranean Basin, are vulnerable to land degradation [18, 19]. These hotspots are primarily the coastal and upland areas of Southern Italy and the two main islands, but also some lowlands of Northern Italy (where population density is higher) are at high risk [17, 20]. ISPRA has reported in a map of Italy the areas affected by one, two or three and more degradation factors (Fig. 6). According to this map, more than 30% of the territory of Sicily, Lazio, Veneto and Apulia is affected by soil degradation. About 11,000 km^2 (more than 3% of the national territory) have been exposed to more than one degradation factor, requiring control and protection [12].

Several authors have studied factors, effects and rates of land degradation in Italy in the last two decades, but the results of these studies are often contrasting [18, 20]. identified the following factors of the reduced land quality in Italy semi-arid climatic conditions (including the climate change trend): soil's tendency to erosion and salinization, forest fires, intensification of agricultural activities, progressive abandonment of traditional agricultural landscapes, over-exploitation of water resources and urbanization of coastal areas. Moreover, a quantitative analysis weighting each process in relation to land degradation carried out by Salvati et al. [20] showed that climate, erosion and compaction/agricultural intensification were the processes with a potentially higher role in determining vulnerability to the Italian soils (Fig. 7).

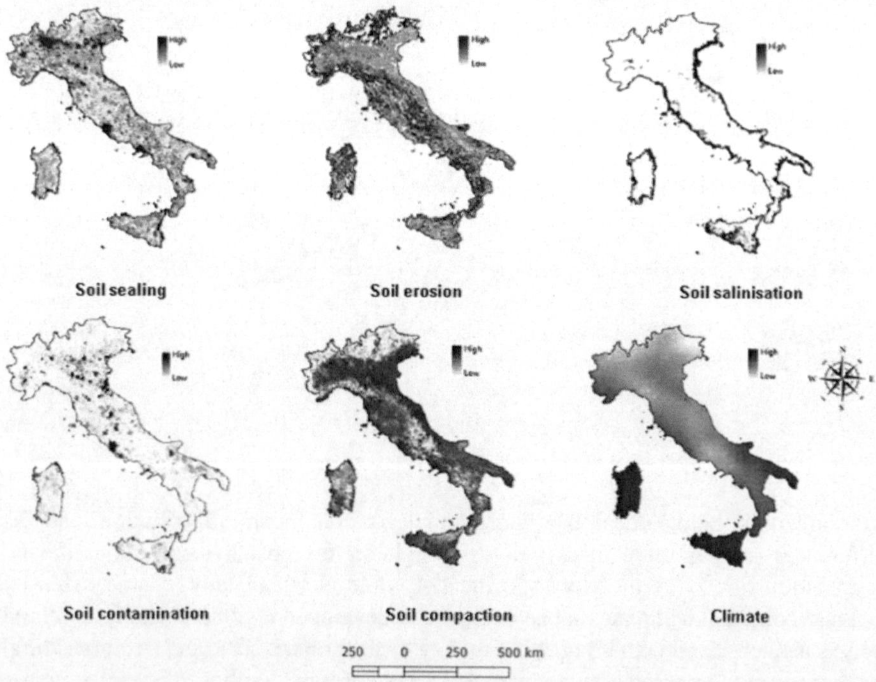

Fig. 7 Geographical distribution of the thematic indicators of vulnerability to each degradation process considered in [20], same source

Dazzi et al. [14] identified as the main threats for soil degradation in Italy the following processes: erosion, sealing/consumption, pollution/contamination, salinization/alkalization, decline in organic matter and diversity loss. According to these authors, other soil threats, such as forest fires and landslides, are also drivers for land degradation in Italy.

Some of these processes (sealing, salinization, erosion, contamination and compaction/agricultural intensification) [20–22] have been also recognized as the most representative causes for oil degradation in Mediterranean areas [17, 20, 23]. In Italy, land sensitivity to degradation has increased throughout the last century, particularly in areas with specific socio-economic characteristics [24, 25]. An analysis of the changes in the Italian landscape throughout the last 50 years was carried out by [18, 24–26]. These authors showed that: (1) land vulnerability has been noticeably increasing in Northern and Central Italy, while is stable in Southern Italy; (2) marginal areas with low *pro-capite* income, agricultural specialization and population ageing show a reduction in ecological conditions; (3) areas providing highly-specialized services in economy and tourism are characterized by relatively good environmental conditions and moderate-to-low vulnerability to land degradation [27]; (4) socio-economic contexts with intermediate population density, environmentally and economically sustainable agricultural systems and dynamic tertiary sector are characterized by moderate land vulnerability; and (5) local districts with high-quality, traditional productions (such as wine, olive oil, fruits and rain-fed cereals), intermediate and stable population density, and diversified economic structure) are areas with the highest long-term reduction in the sensitivity to land degradation. This means that, in Italy, local systems with a dynamic economic structure do not necessarily show higher pressure on the environment and are compatible with low and stable levels of land vulnerability compared to lower-income areas. These findings somewhat contrast to the widespread belief that in Italy economic performance and environmental quality follow an opposite North–South gradient [28, 29].

Unsustainable management is among the heaviest factors of land degradation in Italy, and, in agriculture, its effects may be aggravating the land degradation risks. Crop production results in over-exploitation of the land [25], while intensive agriculture usually determines heavy changes in physical, chemical and biological properties of soils, with possible degradation of its quality [30, 31]. On the other hand, the abandonment of previously cultivated land as well as marginal and mountainous lands may seriously alter soils, and deforestation may expose soils to negative hydrological impacts with increased runoff and erosion rates (e.g. [32]).

5 Land Degradation in the Italian Agriculture

Generally, land degradation risk in agricultural areas in Italy is presumably a result of the different development levels between Northern and Southern areas over the last years and depends on unbalances in the territorial distribution of land resources, economic polarization and socio-economic disparities [16, 33].

As earlier outlined, agriculture is presumably the economic sector with the highest positive or, at the same time, negative impacts on land degradation, since the agricultural activities may decrease or increase soil vulnerability to degradation [33]. The Italian agriculture has generally undergone two opposite temporal trends in the last decades: (1) agricultural intensification and specialization in those croplands with large availability of natural resources and easy access to technologies; and (2) cropland abandonment in those marginal areas that have been affected by population and economic decreases [33]. In more detail, the first trend (crop intensification) has been recorded since the 1990s in Northern Italy, but favourable conditions (e.g. climate, soil, water availability) for agriculture have determined land degradation only in small areas, mainly due to local anthropogenic factors [16, 33]. Moreover, some rural districts in these areas have adopted less intensive but profitable agricultural models, which are characterized by innovation, technology, value-added and quality products, and close relationships to territory. By contrast, in Southern Italy crop intensification has been lower and the weight of agriculture on the regional gross domestic product (GDP) and employment level for workers are higher compared to North; here, the sensitivity to land degradation has increased with crop productivity [33]. Moreover, due to the Common Agricultural Policy (CAP), which has brought increased competitivity of markets, higher mechanization and industrialization levels, and over-exploitation of water resources, land degradation has locally increased [16, 33].

The potential impacts of land degradation on agricultural profitability in Italy were quantified by Salvati and Carlucci [33], who used selected quantitative indicators at the national and district scales. More specifically, these authors estimated the potential costs of land degradation using a standard index of land sensitivity to degradation (the so-called ESAI) and calculated a "depletion factor" in relation to the agricultural value added in about 800 local districts across the entire country. The procedure processed the changes in the ESAI between 1990 and 2000, and other economic variables, such as the per capita value added, share of agriculture in the total production, and agricultural profitability. At the national level, the land degradation risk in agriculture represents about 0.5% on the agricultural value added with a cost that has been estimated between 5 and 15 €/ha. At the regional level, the maximum cost is even 457 €/ha in some districts of Southern Italy, representing 21.5% of the agricultural value added. Sicily (Southern Italy) suffers from low agricultural profitability and high sensitivity to land degradation, but also some regions of Northern (Lombardy, Veneto, and Emilia-Romagna) and Central-Southern Italy (Molise, Apulia, and Basilicata), with intermediate to high agricultural profitability, show moderate but increasing levels of sensitivity to land degradation. According to the evaluations of Salvati and Carlucci [33], in spite of these local contexts, more than 65% of the Italian territory is not prone to land degradation risks, 25% is exposed to moderate risks, while only small areas (10% of the country) are affected by on-going land degradation. Land elevation, agriculture income and position with reference to cities are the factors that mostly influence land degradation, and this influence heavily weighs on the environmental costs. Land degradation in urban and peri-urban areas has experienced high land degradation risks for long time [33].

In addition to this assessment, water availability for agriculture should be considered as an important factor for land degradation in croplands. Due to the climatic conditions (large semi-arid areas in the country), water is scarce in summer, when irrigation requirement is high, due to insufficient precipitation (with large interannual variability) and high solar radiation, which result in high evapo-transpiration [20, 34]. In addition to these climatic factors all economic sectors (e.g. industry, energy, agriculture) in Italy compete for water and additionally, often the low water quality for irrigation limits crop growth and yields.

Moreover, the scarce vegetation cover due to low precipitation input exposes the agricultural soil to water erosion in steep cultivated hillslopes [20, 35, 36]. In this regard, the rural abandonment of people migrating to cities is an aggravating factor of land degradation of hills and mountains, since maintenance protection of land and rural landscapes is stopped. This leads to the need of land management practices targeted to soil conservation (e.g. crop cover, terracing, reforestation, water harvesting) [35–37], since these practices promote water infiltration and reduce surface runoff.

5.1 Desertification

Literature defines "desertified area" an unproductive and sterile area for agricultural or forestry use, due to soil degradation processes [17, 38]. Generally speaking, in Italy desertification affects small areas in southern regions (Basilicata, Apulia, Calabria, Sardinia and Sicily), reducing the potential productivity of agricultural soils (Fig. 8). This reduction is mainly due to the combination of unsustainable anthropogenic impacts on the natural environment and semi-arid or even arid conditions [36, 37]. In other areas, the potential risk desertification is increasing, since the climatic conditions are progressively worsening (due to decrease in precipitation amount and increase in rainfall intensity) and crop systems and urbanization have been intensified [36, 39]. Desertification is expected to affect not only arid and marginal areas, but also the most productive agro-ecosystems [36, 40].

The analysis of the current rates and potential risks of desertification in Italy has been carried out at two levels. First, at the government level, the Ministry for Environment and Territory has produced the Italian Atlas of areas with desertification risk, based on the spatial assessments of "sensitivity and vulnerability indexes of desertification risks, organised into a system of soil degradation processes (deposition, erosion, urbanization, salinization, aridity)" [38] (Fig. 9). This Atlas has been considered as a tool to prepare the national action plan to face off drought and desertification, following the guidelines of the United Nations Convention to Combat Desertification (UNCCD). To assess this risk, the Soil Aridity Index (SAI) – "the mean annual number of days when the moisture control section is dry in soils with a Mediterranean type of climate" – was used. The SAI was calculated for 13,000

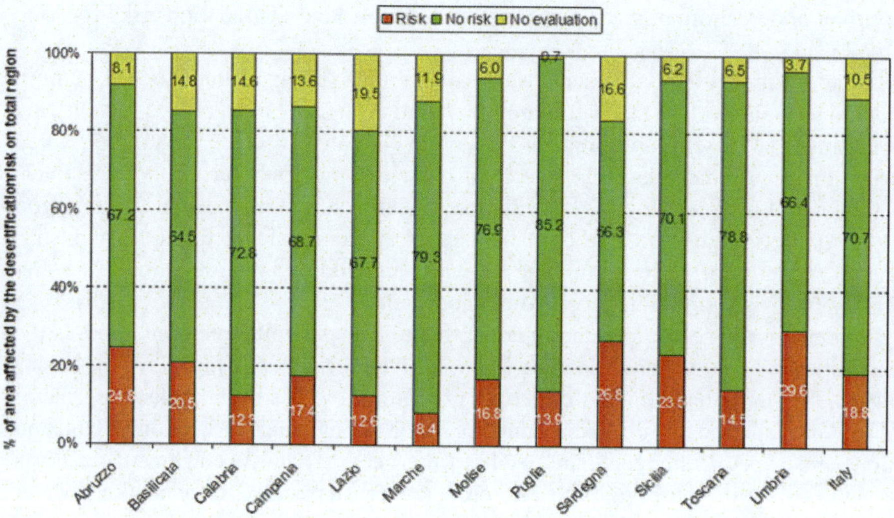

Fig. 8 Distribution of the desertification risk among the Italian regions (source: [38])

samples of Italian soils and the related map, produced at 1:250,000 scale, identified three SAI classes according to the influence of increasing water scarcity on agriculture and forestry. To calculate the SAI, soil regions considered at the highest potential risk of desertificaton were prioritized, that is the coastal areas of Central Italy and all the Southern regions [41]. The results of this study, whose geographical distribution is reported in the Atlas, showed that 52% of the national area was affected by a potential risk (the whole area of Sicily, Sardinia, Apulia, Calabria, Basilicata and Campania regions in the Southern Italy, and a large part of Lazio, Abruzzo, Molise, Tuscany, Marche and Umbria regions in Central Italy), due to both climatic and pedoclimatic factors (Fig. 9). At a higher level of land degradation risk, the Atlas reported that 21% of Italy is subject to land degradation, of which 4% already showed functional sterility – and thus is already unproductive – 5% is sensitive to desertification and another 12% can be considered vulnerable to this process. The protected areas (regional and national parks) are about 10% of the territory at potential risk of desertification, but these areas cover only 20% of unproductive, vulnerable or sensitive lands [38, 41].

Regarding other factors of desertification reported in the Atlas, erosion is the most important process, while soil aridity is the second main driver. Areas affected by aridity are 19% of the national territory, while salt-affected soils are more than 3% (many coastal areas of Sicily, Sardinia, Apulia and Tuscany). According to Costantini and Abate [41], irrigation using sustainable water management may be able to limit soil aridity and salinization, but only 4% of the area at potential risk of desertification is currently equipped with irrigation networks. In some areas (generally irrigated alluvial plains of internal land), brackish irrigation waters are pumped from wells bored in highly saline aquifers. Moreover, farmers use private wells, and

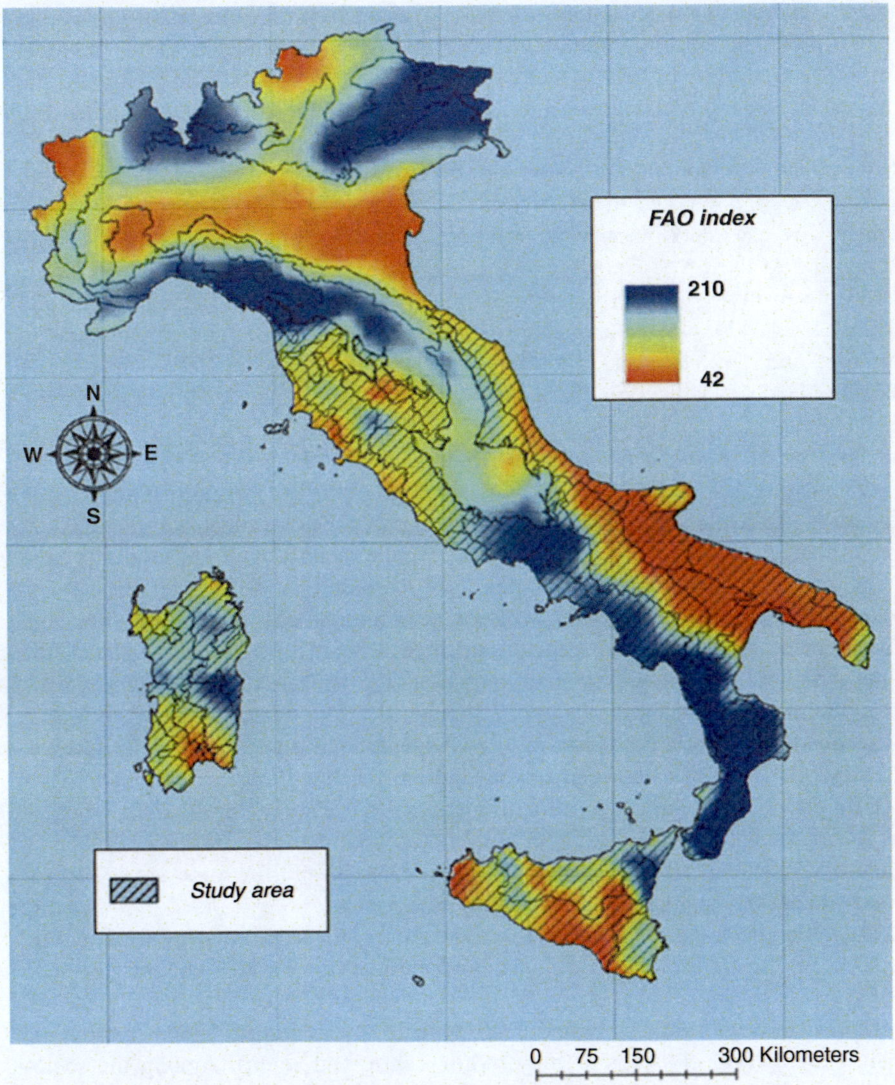

Fig. 9 Map of climatic aggressivity in Italy expressed by the FAO index (source: [38])

this shows a basically poor management of the collective irrigation systems, especially in Southern Italy [42–44]. Overall, sustainable irrigation management with water-saving techniques only covers about 10% of the sensitive and vulnerable lands in Southern Italy. The agro-environmental measures for row crops and pastures are not sufficient to reduce the area at potential risk of desertification [41].

A second analysis of desertification at the national level was carried out in the last decade by Salvati et al. [36]. This research group has studied the evolution of the

Italian vulnerable land due to desertification over the past five decades (from 1960 to 2010), producing multi-temporal maps of desertification indexes at the national scale. The sensitivity of Italy to land desertification has been assessed by monitoring changes in some indicators over time. These indicators were related to climate, soil, vegetation and land management, which were considered as the most important driving factors of land degradation and, especially, desertification. The individual variables were combined in an Environmental Sensitive Area Index (ESAI, [45]), which was calculated in several temporal windows between 1960 and 2010. According to the ESAI framework, the national territory was classified into four levels of sensitivity to desertification (non-affected, potentially affected, fragile and critical level) [24, 46]. The results of this study showed a general increase in "fragile" and "critical" land areas between 1960 and 1990 in Italy, while, more recently, land desertification distribution was spatially limited [24]. In more detail, the number and extent of areas that are sensitive to desertification increased throughout the last 50 years in the southern regions. Currently, more than one-third of Southern Italy fall in the "critical" class, and thus this area is particularly fragile regarding desertification, due to natural characteristics of this land coupled to anthropogenic actions. As a matter of fact, Southern Italy is prone to aridity, and the climate influence on desertification has increased in the most recent years. Moreover, the agricultural systems of southern regions are affected by poor land management, slow economic growth and high human impacts on croplands (the latter mainly due to the pressures of tourism and urbanization). The anthropogenic factors sum up to the natural sensitivity to desertification, which has been worsening over time [46]. Also in Northern Italy, where land conditions are more sustainable, the desertification risk has been increased from the last decades. This increase is mainly due to the reduction in rainfall amounts, but also to other factors, such as increasing population density, agricultural intensification, urbanization (with consequent soils sealing), land fragmentation urbanization, and, in some areas, to progressive abandonment of agricultural land [46].

Salvati and Bajocco [46] and Wilson and Juntti [47] stated that socio-economic factors are also important causes for land desertification besides climatic features. For this reason, Salvati [15] analysed the relationships between the spatial distribution of socio-economic factors and land vulnerability to desertification in the three main geographical divisions of Italy (north, centre and south). Among the socio-economic factors, the authors evaluated more than one hundred indicators of demography, human settlements, labour market and human capital, rural development, income and wealth, to discriminate vulnerable from non-vulnerable areas. The results of this evaluation highlighted that, at the national level, land vulnerability to desertification is due to four human-driven impacts, such as increase in population density, crop intensification, unsustainable agricultural practices (including irrigation management) and fragile economy. Intensive-farming and urbanization are more important in discriminating vulnerable from non-vulnerable areas compared to poor economic conditions and depopulation [15]. The latter author reports also the distribution of vulnerable lands to desertification in Italy in relation to the selected socio-economic indicators. This distribution basically follows four spatial gradients:

(1) a gradient in economic development between North and South, quantified by differences in *pro-capite* income and unemployment rate; (2) a socio-economic gradient between coastal and inland areas, shown by the spatial distribution of urban areas; (3) a morphological gradient, due to land elevation and quantified by the incidence of irrigated land on the total agricultural area; (4) a land management gradient between rural and urban areas, indicated by the share of workers in the agricultural sector. As resulting from these four gradients, in Italy the vulnerable land is concentrated in flat areas with specialized agriculture (mainly in the Po River valley) in Northern Italy, along the coasts of both Adriatic and Tyrrhenian Sea in central Italy, as well as in the inland areas of Basilicata, Sicily and Sardinia in Southern Italy [15]. The importance of the demographic processes (e.g. population structure and dynamics, ageing, family size) and economic factors (e.g. available income, revenues from taxes, infrastructures) is more important to discriminate vulnerable and non-vulnerable areas in southern regions than in Northern Italy in determining sensitivity to desertification. Regarding the main economic drivers, scarcity of capitals, lack of investments (particularly in marginal rural areas), shortage of labour, poor public services, inadequate incentives for sustainable practices and restricted institutional support may have increased this sensitivity. In Southern Italy, unsustainable agricultural practices heavily impact on soil, water and, more in general, crop system quality, noticeably increasing the desertification risk [48]. By contrast, human settlement weighs more in Northern Italy, where peri-urbanization may be one of the most important factors increasing land vulnerability [15, 40]. In non-vulnerable areas, some natural areas and excellent landscapes, such as the national and regional parks, are under strict environmental regulations.

5.2 *Erosion*

The European Environment Agency (EEA) has estimated that more than 110 and 42 million hectares in Europe (about 12% and 4% of its total area) is affected by water and wind erosion, respectively [49]. In Italy, over 20% of the agricultural land is exposed to moderate to severe risks of water erosion, corresponding to an annual rate higher than 10 tons/ha. Erosion goes beyond the topsoil layer, as demonstrated by ISPRA, which has mapped 485,000 landslides in Italy, covering an area of more than 2 million hectares (7% of the country area). Large areas on intensively cultivated hillslopes with row crops (in Central and Southern Italy) are exposed to critical erosion risks. Moreover, woodland with steep slopes in the Alps and the Apennines of Northern Italy are also prone to landslides [50, 51].

Other estimations by the Organisation for Economic Co-operation and Development (OECD) [14, 52] estimated that 30% of the Italian land is exposed to a soil erosion risk with annual rates higher than 10 tons/ha. The study by OECD showed that some anthropogenic actions (cultivations on fragile soils, overgrazing in hilly and mountainous areas, scarce adoption of soil conservation practices in agriculture) and natural factors (e.g. increasing trends of drought and high-intensity rainfall events) have left almost constant or even have been aggravating these risks. In contrast, several countries in the world have adopted soil conservation practices, improving soil quality, and converted agricultural lands into pastures or forests, increasing vegetation cover, with subsequent reductions in annual erosion rates [14].

The high sensitivity of Italian soils to erosion has prompted further research to assess soil loss at different spatial and temporal scales. Investigations at the finer scales have mainly focused on soil degradation due to erosion and landslides in catchments affected by severe erosion and geological instability. Despite the importance of mapping the erosion risk also at the coarse spatial scale, only in the last two decades Italy has adopted plans to provide soil erosion maps, in order to support rural development or adoption of soil conservation measures in agriculture and forestry funded by CAP of European Union. These studies have been carried out both at national and regional scales. Italian regions under the convergence objective of the EU regional policy have produced 1:250,000 maps of potential risk of soil erosion applying the USLE-family models [14]. Several other studies have evaluated soil erosion in specific areas or catchments of Italy, using different hydrological models (see the studies by [13, 14] for the comprehensive list of these applications).

At the national scale, Knijff et al. [53] used the USLE model to assess the actual and potential soil erosion, in order to identify those regions that are prone to this risk. This study was preliminary to the production of the Soil Map of Italy at 1:250,000-scale and 250-m spatial resolution. However, due to the limitations of the USLE models, only rill and interrill erosion was estimated in this map, while gully, wind, tillage or mass erosion was not considered; moreover, USLE assumed lacks of anti-erosive practices (P factor equal to one) [54]. According to this map, the annual soil loss over 10 tons/ha, which is over the tolerance range proposed by USDA (2 and 10 tons/ha, depending on soil type) is exceeded almost everywhere in Italy, except in flat areas. The actual soil erosion risk is very high in the Alps (Northern Italy), Apennines (Central Italy) as well as coastal areas of Calabria and Central Sicily (Southern Italy), all these areas showing annual erosion rates over 40 tons/ha [14] (Fig. 10).

However, Knijff et al. [53] did not validate their data, due to the lack of erosion observations for Italy, and therefore the USLE-estimations of soil erosion could not be accurate. Some years later, an indirect validation of the erosion estimates proposed by [53] for Italy was proposed by Van Rompaey et al. [55]. These authors modelled the sediment volumes exported by 20 Italian reservoirs, using the spatially distributed WaTEM/SEDEM model [56, 57]. The differences between estimations and observations were over 60%. Erosion was noticeably over-predicted in mountain forest catchments of the Alpine environment, while the estimations of sediment yields in hilly cropped areas were more accurate [54].

Another validation attempt of the erosion estimations by Knijff et al. [53] in Italy was made by Grimm et al. [58], applying a revised version of the USLE model. These authors improved the estimation of the R-factor (rainfall erosivity), using the data from a large number of meteorological stations, adopted new interpolation procedures, and calculated the K-factor (soil erodibility), considering soil crusting and sealing processes. The errors in annual erosion found in the original model were

Fig. 10 Map of the soil erosion risk in Italy (source: [14])

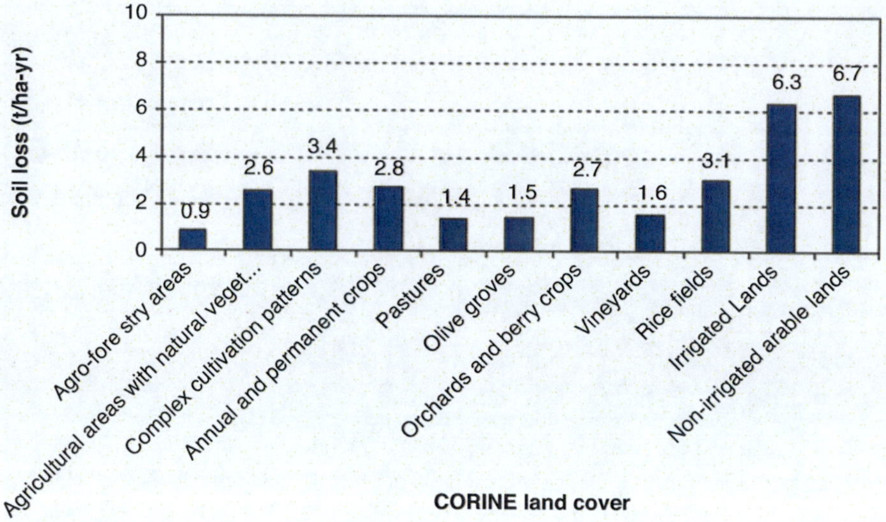

Fig. 11 Mean value of soil erosion risk for agricultural areas of CORINE land cover map (data source: [14])

high in the mountain areas (Alps and Apennines) and in the central part of Sicily, where the estimations by Knijff et al. [53] were affected by large over- and under-prediction. Conversely, the errors were lower than 5 tons/ha in most part of the country (Fig. 11) [14].

Overall, these applications of the USLE-family erosion models in Italy, in spite of the validations, showed low accuracy in both absolute values and spatial patterns of erosion predictions. This is not due to the intrinsic model issues, but rather to the low resolution and poor quality of input data [59, 60]. However, the soil erosion risk map of Italy produced applying USLE was found to be a useful support to identify the agricultural areas that are prone to soil erosion and to adopt soil conservation practices [14].

More recently, spatial estimations of soil erosion in Italy were part of European maps, based on the use of:

- the MESALES model (Modèle d'Evaluation Spatiale de l'ALéa Erosion des Sols – Regional Modelling of Soil Erosion Risk, [61]) at the 1:1,000,000 scale, 1 km × 1 km resolution and five categories (very low, low, medium, high and very high) of soil erosion risk;
- the PESERA (Pan-European Soil Erosion Risk Assessment) project, adopting a process-based and spatially distributed model, at 1:1,000,000 scale and 1 km × 1 km resolution [62].

According to the PESERA-map, potential soil loss exceeds the USDA limits in 30% of the Italy area. Moreover, the PESERA model showed that both permanently and not irrigated arable lands can be exposed to mean annual soil erosion over 6 tons/ha, while agro-forestry areas are affected by the lowest mean soil loss (lower than 1 tons/ha). Some flat regions (i.e. the Po Valley) are at erosion risk, even if limited, while, in some mountainous and hilly areas (i.e. Alps and Apennines), the risk of erosion is very low, thanks to the forest cover. Overall, this study has confirmed that erosion in Italy is not only influenced by land slope (Fig. 10).

To take into account the large influence of climatic conditions on soil erosion, [63] studied the spatial and temporal distribution of rainfall erosivity in Italy and developed a national-scale grid-based map. At the spatial scale, this study showed that the highest average annual R-factor occurs in the north-western part of Italy (Friuli-Venezia Giulia), while Trentino-Alto Adige, Valle d'Aosta and Po Valley (Northern Italy), the eastern part of Apennines (Central Italy) as well as Basilicata, Sicily and Sardinia regions (Southern Italy) show a rainfall erosivity below the national average. However, in the semi-arid part of Italy the average rainfall erosivity is slightly higher compared to the alpine and continental areas of this country. From a temporal approach, the seasonal distribution of R-factor throughout the year is not significantly different between the alpine and continental areas. By contrast, the semi-arid zones are affected by a remarkable seasonality of rainfall erosivity with the highest values recorded in the coldest months [63].

Finally, according to Piccarreta et al. [64] and Costantini and Lorenzetti [50], soil erosion in some parts of Southern Italy is due to changes in land use as well as some effects of the agricultural policy. For example, a 20-year period of set aside for remodelled areas that have been applied following the Reg. CE 2078/92 has presumably generated increases in soil erosion and land degradation in Southern Italy [50, 64].

5.3 *Salinization*

A too high content or an accumulation of salt in soil compared to normal conditions determines salinization with degradation and reduction of its quality. In this case, seed germination and thus crop growth become often impossible, and the so-called saline or salt-affected soils suffer from reduction in fertility and damage to microbial communities. Measured data about salt content or electrical conductivity in soils over the Italian territory are quite scarce and limited to few case studies [65]. Therefore, a reliable map of distribution of saline soils based on direct and widespread measurements does not exist for Italy [13, 14]. Indirect indications about this soil degradation process have been drawn using modelling approaches [17] or proxies [66], in order to identify the location of saline soils on a national scale [65]. Only an exploratory estimation at this large scale has been carried out by Dazzi [67] and Dazzi C and Lo Papa [14].

In Italy, soil salinization affects over 3 million hectares [14], about 10% of the country area. In spite of this large incidence, some scientists state that soil salinity plays a marginal effect on land degradation in Italy or is confined to some small areas

[50, 67, 68]. However, the latter authors report that the extension of saline soil across the national area is higher than currently estimated. As a matter of fact, saline soils are distributed across all the regions, but with different extent. This distribution is confirmed by the data of the National Soil Database, which shows that soils in almost all the Italian regions are affected by salinization [14].

In more detail, two gradients (north to south and internal areas to coasts) characterize the spatial distribution of soil salinity in Italy. The southern part of the country (especially Sicily) as well as the Tyrrhenian and Adriatic coastal areas (mainly Apulia, Basilicata and Sardinia) shows larger areas with saline soils compared to the northern regions (Fig. 12) [20]. Moreover, also flat areas, such as the Po River valley, are affected by soil salinity [14, 65].

Regarding the salinization risk of agricultural soils, a detailed investigation has been carried out only in the southern regions, when the national database of the electrical conductivity of irrigated and non-irrigated soils of Southern Italy has been

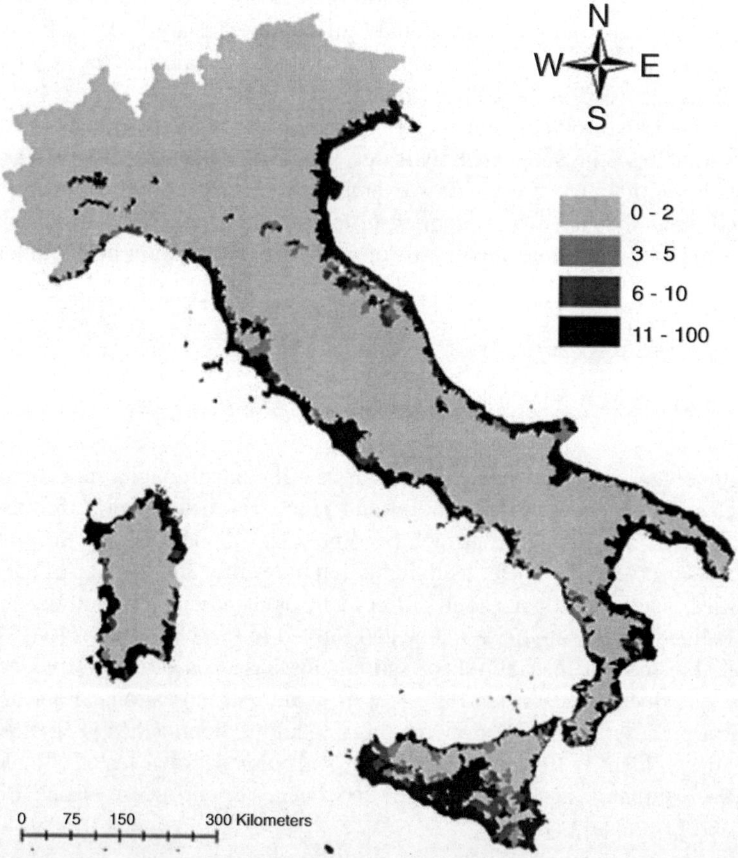

Fig. 12 Map of distribution of soil affected by the salinization risk in Italy (source: [65])—values are the percentages of area exposed to the risk

compiled. This part of the country is considered as being the most affected by the salinization risk, also because Southern Italy relies on the use of water of poor quality for crop irrigation [14, 69]. This is due to the limited availability of resources for irrigation in surface water bodies (lakes and rivers), which forces the use of wastewater (both urban wastewater and agro-industrial wastewater, such as olive oil mill wastewater and citrus processing water [70, 71] or groundwater). The use of irrigation resources that contain noticeable salt concentrations, since water is pumped by wells in coastal areas (affected by seawater intrusion), increases soil salinization. In contrast, in other parts of Italy, the diversification of the water resources for crop irrigation and the limited use of groundwater in agriculture limit the soil salinization risk [20]. Despite the irrigation with poor-quality water, the mean electrical conductivity of agricultural soils in Southern Italy is far from indicating an actual salinization risk. The soils cultivated and irrigated for vegetable production or the meadows are exceptions, since their electrical conductivity is quite high, indicating significant increase in salinity. In contrast, the other irrigated croplands are not noticeably affected by soil salinity, since salt leaching due to precipitation in the wet seasons is able to dilute the excessive saline content, particularly in agricultural soils with prevalent loamy texture and good drainage characteristics [14, 50].

5.4 Soil Sealing and Compaction

In Italy, soil sealing and compaction are leading to the reduction in water and air infiltration capacity, due to covering with impervious surface (such as building roofs, roads, railways or other civil infrastructures) and deterioration of its structure because of the traffic of heavy machinery on agricultural or forest soils. Generally, compaction affects soils with poor organic matter content. Regarding soil sealing due to anthropogenic activities, Italy has undergone a large process of urbanization throughout the last five decades, when very large natural areas were sealed at a rate (the so-called soil consumption) of about 100 ha/day [72]. The percentage of artificial coverage is 7.1% of the national territory (in the 1950s it was 2.7%, +160%) with an area larger than 21,000 km^2 [1]. Soil consumption in Italy is on average 14 ha/day, a value that is in any case far from the zero net consumption's objective established for 2050. Soil consumption in the coastal strip is higher compared to other areas in the country. A percentage of 23.4% of the strip within 300 m from sea, 19.7% between 300 m and 1,000 m and 9.3% between 1 km and 10 km is urbanized, compared to national mean of 7% [12].

Moreover, urbanization, since the earlier 1980s, has not followed the pattern of population growth, because soil consumption and subsequent sealing has been caused by extensions of commercial and industrial activities and building of productive and residential infrastructures in dispersed areas across the country [73]. Soil consumption has been higher in Northern Italy compared to southern regions (Fig. 13), and particularly in the neighbouring of metropolitan areas or in coastal

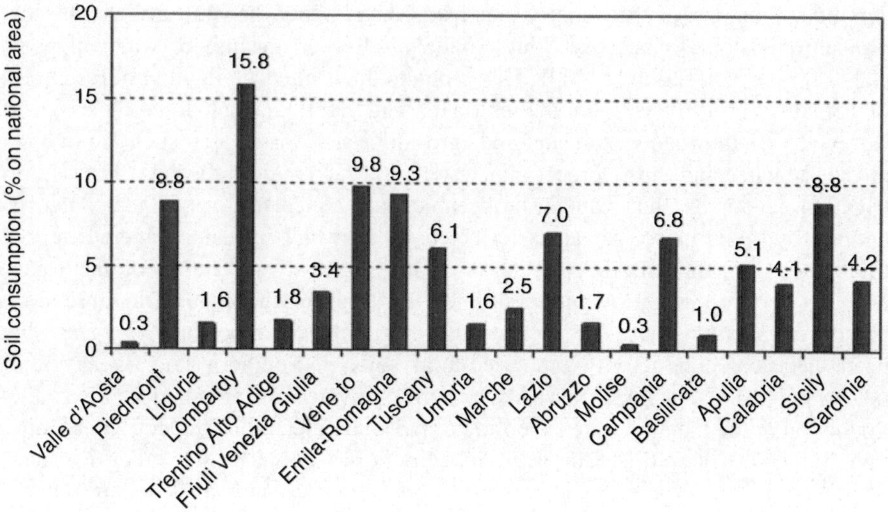

Fig. 13 Distribution of soil consumption among the Italian regions (data source: [72])

zones [14, 50]. Of course, these urbanization trends have reduced the areas devoted to agriculture, forestry and animal breeding; therefore, these processes have brought threats to vegetation and fauna with evident ecological damage. This reduction has also been due to the installation of thousand photovoltaic plants (largely supported by national subsidies to renewable energy production), which covered large natural areas, that were previously marginal or even cultivated.

However, some authors argue that land vulnerability to soil sealing is relatively low-throughout Italy, this process being concentrated around cities, in lowland and coastal areas [20]. In this regard, recent studies have shown that the increase in population density has not been linked to land degradation, since population has decreased in highly vulnerable land, but not in moderately vulnerable areas from the 1980s [74].

Regarding soil compaction, Italy shows large areas prone to this degradation risk, about 30% of the national territory being affected by a noticeable reduction in soil bulk density [75]. According to Salvati et al. [20], spatial patterns of soil compaction follow a latitude gradient from north to south. Compacted soils have covered previously cultivated areas of both hilly and flat zones of Italy, with more severe effects in soils with fine texture and low content of organic carbon; soil compaction has been surveyed also in rice cultivations and other wetlands, degraded natural areas and non-irrigated vegetable crops. In contrast, mountain grasslands, pasture-lands and woodlands as well as irrigated crops have been less affected by the compaction risk, since supporting soils are porous [50].

5.5 *Organic Matter Reduction*

It is well known that a stable content of organic matter (OM) in agricultural soils is essential for crop growth and production. The optimal OM amount for soil is debatable, since it depends on soil, plant and climate, among other characteristics [76]. A minimum threshold of 1.8% can be assumed value for an adequate supply of elements to plants and a limited infiltration of pollutants from the soil surface to the groundwater, in line with [50, 75, 77–79]. An evaluation of OM content in soil can be affected by a high uncertainty, due to the large number and mutual interrelations of factors that influence this parameter. Despite such uncertainty, mean data about OM in Italian soils can be derived by the European map of organic carbon content in the 30-cm topsoil. This map, having a spatial resolution 1 km × 1 km, has been produced by the Joint Research Centre (JRC) and funded by European Union [80–82]. Compared to the reference value above (2% of OM content for healthy soils), it has been estimated that over 30% of the Italian soils shows a lower value (mainly in southern regions, such as Apulia, and in the islands, Sicily and Sardinia, but also in large areas of the Po River valley (Fig. 14) [13, 14, 49, 50, 80–83]; 80% of these soils is well below the target (3.5%) established in the EU Resource Efficiency Roadmap [50]. One-third of Italian soils showing decreases in OM content is subjected to agriculture or is natural land. In more detail, the difference in OM content of soils between cropland and marginal or woodland areas is high [50]. On average, a minimum OM content has been surveyed in rice cultivations (2.0%) and vineyards (2.1%), while grasslands and rangelands contain over 3% of OM, and forest soils show percentages even over 6% (Fig. 15); however, these mean values are affected by a large spatial and temporal variability, as shown by the large standard deviation. This wide range in OM content is due to several factors, such as the natural variability of soils, climatic conditions and land management, and the latter is considered as the main influencing factor [84]. In areas from Central and Southern Italy in particular, which rely on irrigation for agricultural production, the OM content of irrigated soils (particularly for production of vegetables and fruits) is significantly lower compared to rain-fed croplands (Fig. 15). Besides this variability, Italian soils are also affected by a noticeable temporal variability in the OM content. According to the estimations by Fantappiè et al. [85], the OM content in soils of Italy has decreased on average from 2.5 to 2.0% throughout the last five decades, and this decrease was mainly attributed to land management and secondarily to the climate trends [50, 86].

Fantappiè et al. [85] produced national maps of organic carbon in the 50-cm upper layer of Italian soils. According to these maps, the carbon stock decreased by 3.3 Pg (mean of the period 1979–1988) to 2.7 Pg (1989–1998), when agricultural intensification was recorded; then this stock increased to 2.9 Pg in 1999–2008, presumably because of the effects of the European Agricultural Policy that enhanced carbon sequestration by supporting the adoption of agro-environmental practices for environmental protection of rural areas. Forestlands in Alps, Apennines and Sardinia show the highest organic stocks, while intensive croplands of hilly and flat areas of

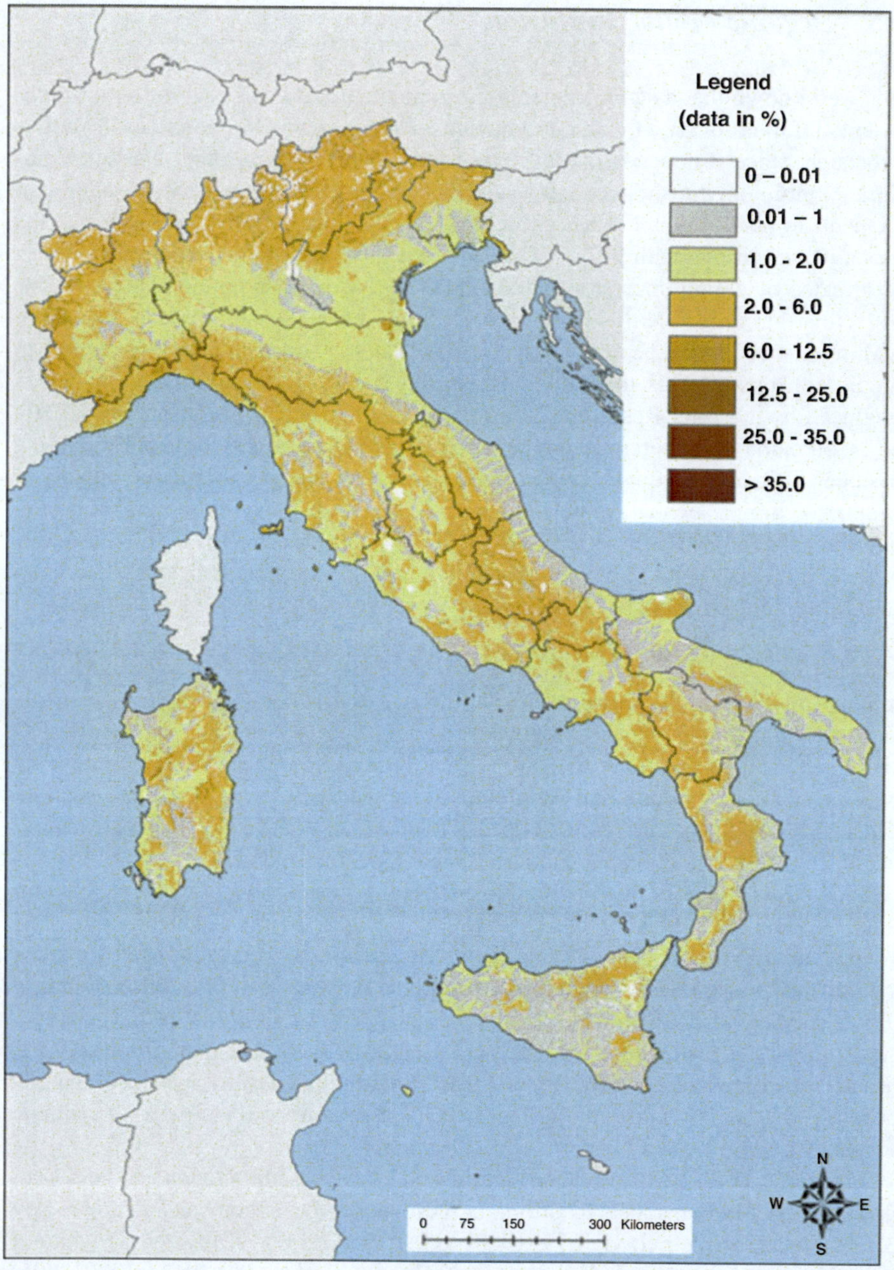

Fig. 14 Map of the carbon content of soils (0–30 cm) in Italy (source: [14, 80])

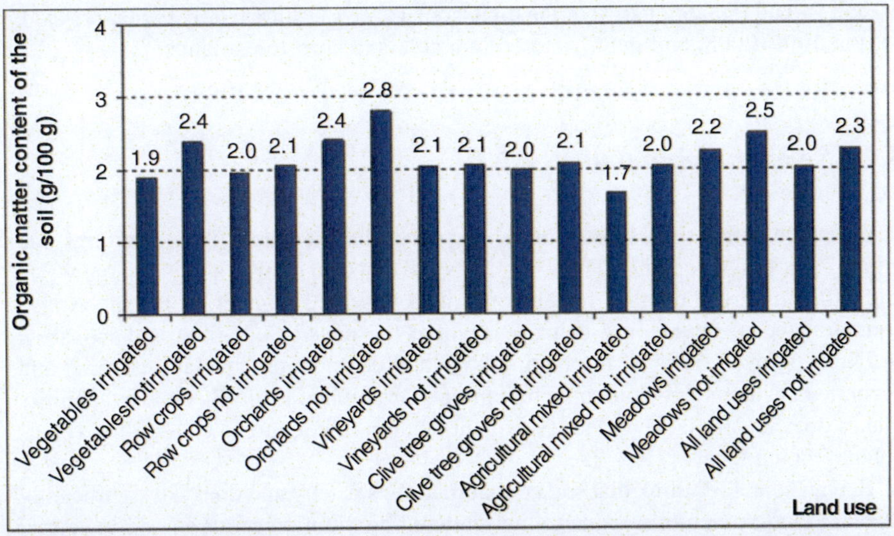

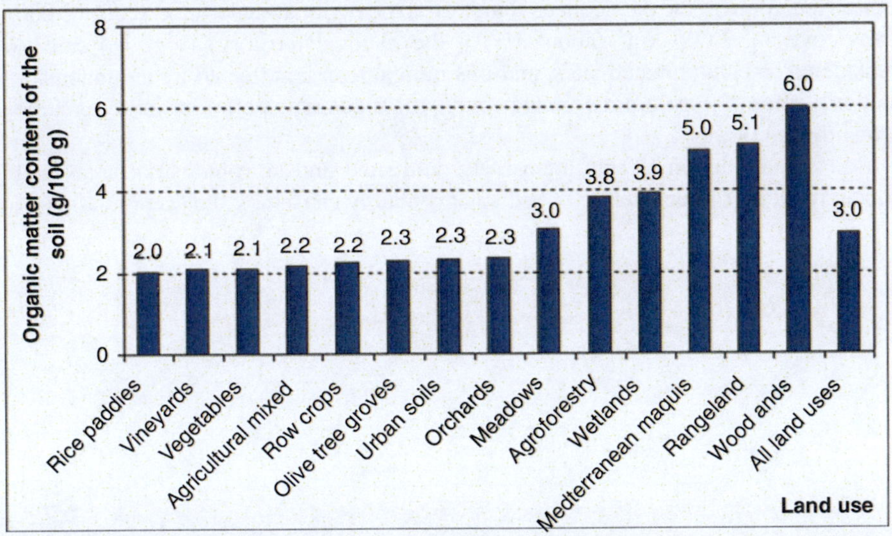

Fig. 15 Mean organic matter content of the Italian soils classified for land uses (not irrigated, upper, and irrigated, lower) (source: [50] based on data from the national soil database, available from: http://www.soilmaps.it)

Southern Italy have the minimum stocks. Moreover, the most noticeable changes over time have been recorded in Po River valley (positive variations) and in hilly and plain areas of Central and Southern Italy (negative variations) [13, 85].

Other estimations of organic carbon stock in Italian soils have focused on the top 30-cm of agricultural soils [87] since these soils are considered as the most sensitive to OM variations [88]. This assessment has revealed that the whole stock of organic carbon is 0.49 Pg, about 20% of the total stock (2.9 Pg) estimated by Fantappiè et al. [85] in 1999–2008, and mainly stored in arable and agro-forest lands.

5.6 Soil Contamination

Soil contamination is a human-driven process observed both in agricultural and urbanized areas, where the use of chemicals and disposal of civil and industrial waste and effluents are the main factors of soil pollution. However, soil can be also contaminated in developing countries, where the density of urban and industrial activities is lower. Despite the high risk of land degradation, spatially-covered and accurate data about soil contamination in Italy are limited [20]. This low availability makes difficult a reliable and precise quantification of contaminated areas as well as contamination levels [13, 14].

Salvati et al. [20] think that soil contamination is patchy and does not significantly increase land vulnerability, due to the scattered distribution of urban and industrial areas. According to the data collected in Italy at the regional scale since 2000, 12,000 areas covering 2,600 km^2 (about 1% of the national territory) (Fig. 16) can be considered as contaminated sites, and this number can be affected by an underestimation, since these figures relate only to industrial pollution and exclude brownfields [14].

Soil contamination affects intensively cultivated and agro-industrial areas and contamination of intensively cropped soils is mainly caused by the accumulation of

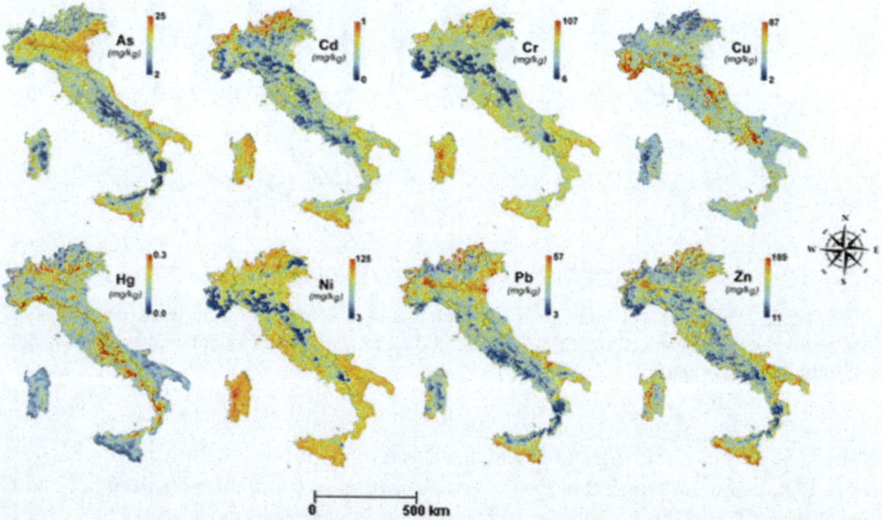

Fig. 16 Maps of estimated concentrations of heavy metals in topsoil in Italy (source: [14] based on EU maps and geodatabase of Land Management and Natural Hazard Unit of Joint Research Centre, [89])

nitrates used as fertilizers under animal slurry form. These nitrates can leach into groundwater or contaminate surface water, when disposed into rivers or lakes. Soil contamination by nitrates has been observed in cultivated areas of the Po River valley and intensive croplands of Northern Italy. However, several regions have carefully applied the L.D. n. 152 of 2006, which followed the EU's Directive issued in 1991; the agricultural areas in the regional territory that are more vulnerable to nitrate contamination have been identified, and the tolerance limit against an excessive land spreading of nitrates on soils has been set up. Thanks to these rules as well as to careful monitoring procedures of vulnerable water and soils, decreases in soil and water contamination due to nitrate use have been recently observed [14].

Another source of contamination for soil and water is the excessive land application of pesticides and other inorganic fertilizers to crops and soils, and this abuse also affects the croplands of both Northern and Southern Italy. Unfortunately, the data about soil pollution due to these contaminants are scarce and scattered, which makes a systematic diagnosis of soil degradation practically impossible; in some regions, soil contamination due to extensive use of xenobiotic compounds seems to be severe [14].

5.7 *Others*

5.7.1 Overgrazing

Overgrazing, caused by the excessive exploitation of pasture, is a residual but not negligible form of land degradation in rural areas. Unfortunately, in Italy, as in many other Mediterranean countries, spatial data about overgrazing in marginal areas or in rural communities lack or are very limited [90]. To partially fill this gap, these authors investigated the relations between the grazing pressure and some socio-economic variables related to population, labour, life quality, rural development and environment in Italy. This study highlighted that the rural municipalities exposed to unsustainable grazing and high erosion rates are concentrated in Central and Southern Italy and that these land degradation factors sum up to desertification risk (Fig. 17). The overgrazing pressure also depends on farming diffusion and land availability versus competing land uses. Moreover, the impacts of overgrazing on overall land degradation rates are mainly associated with the lower economic performance of agricultural areas of Central and Southern Italy compared to the North [28, 90].

5.7.2 Forest Fires

Wildfires completely remove the vegetal cover of forests and determine heavy changes in soil properties [91–93]. These effects increase the runoff and erosion rates by some orders of magnitude, with removal of OM and nutrients from burned

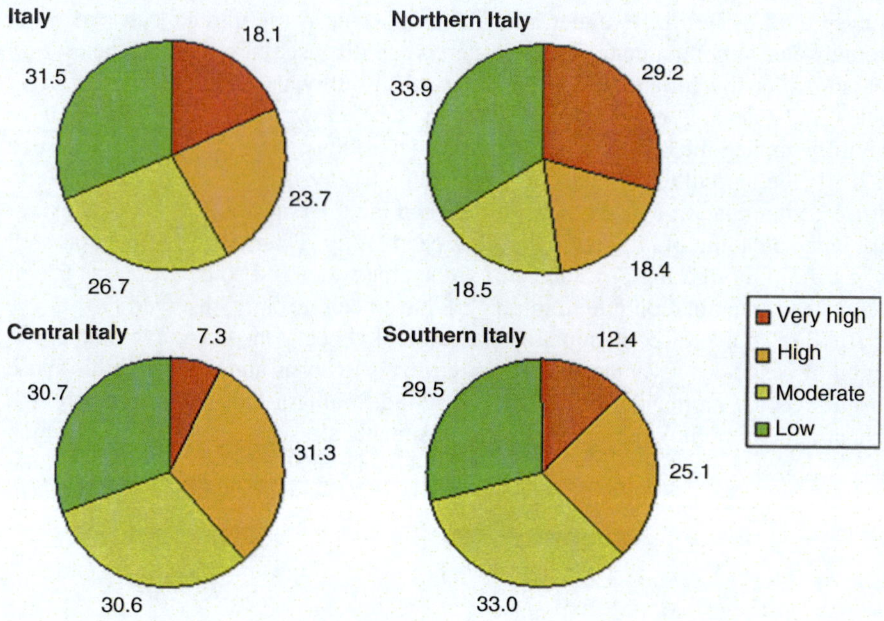

Fig. 17 Map of the grazing pressure in Italy (source: [90])—values are expressed as Grazing Pressure Indicator, suggested by the same authors

forests and transport downstream [94–96]. Each year wildfires destroy large forest areas in Italy and often affect sub-urban zones surrounding forestlands. Fire is mainly due to fraudulent or unintentional actions of men, including burning of weed and pruning residues, pasture renewal and recreational activities. These anthropogenic actions sum up to the climatic vulnerability to fire (because of long and hot summers with dry and strong wind) of large areas across the national territory [13, 14, 97].

According to data from the Italian Commando of Forest, Environment and Agro-food Police (or "Carabinieri Forestali d'Italia"), in 2019 the number of fires was 7,526 with a fire-affected area of 45,719 ha. More than 50% of the burned area and 65% of the number of fires affected three regions (Sicily, Sardinia and Calabria), as in the previous years. The area burned by fire was on average 6.10 ha, and Piedmont was the region with the highest mean area (29.24 ha), followed by Sicily (14.70 ha) and Apulia (9.63 ha) (Fig. 18). The investigations of the Carabinieri Forestali d'Italia reveal that the main cause of the fires is malicious and that the most frequent reason is the renewal of the pasture.

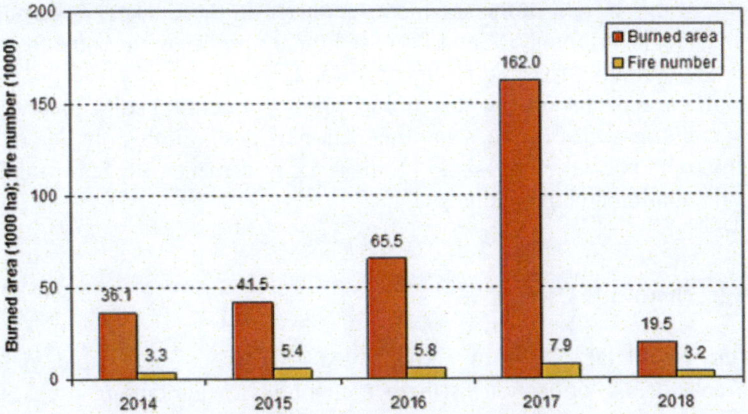

Fig. 18 Burned area and fire number in Italy throughout the period 2014–2018 (data source: Italian National Corp of Forest Guards, Carabinieri Forestali, 2019)

5.7.3 Microplastics

In agricultural lands, microplastics (MPs) – tiny particles of plastic polymer smaller than 5 mm [98, 99] – derive from mulching materials, soil conditioner (e.g. composting sludge) and distribution of sludge from treatment plants [99, 100]. Particles of MPs are transported by runoff from croplands through the river channels and are then poured in water bodies (sea, reservoirs and lakes) with evident pollution risks.

Italy is the second leading country in Europe (after Germany) in plastic demand [101]. Since the Italian territory is facing on extremely long coasts (more than 8,000 km) and its hydrographic network is well developed, it is evident that this country is exposed to MP pollution of the surrounding lands. To give a rough indication, it has been estimated that the Po River, which catchment covers a large part of Northern Italy and hosts about 25% of the Italian population with 55% of national cattle load and 35% of the agricultural products [102], pours more than 120 tons of MP per year to the Adriatic Sea [101]. This waste flow generates MP concentration from 2.92 ± 4.86 to 23.30 ± 45.43 particles per kg of beach sediments of its Delta [102].

According to Campanale et al. [103] and Guerranti et al. [104], in Italy comprehensive data on MPs in sea and freshwater environments are lacking, and this lack is common to almost all countries facing the Mediterranean Sea. However, several studies have investigated the MP concentration in surface waters and sediments in the Mediterranean Basin, and especially the Italian coastline [105], although the related database is still limited to some case studies. MP particles have been detected in 74% of water samples collected in lakes of Alpine area [103, 106]. A study by Fischer et al. [107] reported up to 3.36 (Chiusi Lake, Tuscany) and 4.42 (Bolsena Lake, Lazio) particles/m^3 in surface waters, as well as 112 (Bolsena) to 234 (Chiusi) particles/kg of dry weight in the sediments collected in these lakes. Regarding the Italian coastline, [108] found PE and PP in sediment samples in the same frequency as the most commonly produced polymers [105, 109].

Although these data show the presence of MP in agricultural soils, Italy was one of the first countries of Europe to issue strict national rules limiting pollution of water ecosystems (with special emphasis to plastics), in agreement with European directives and international conventions [103].

5.7.4 Agrochemicals Use

Despite the beneficial effects of agrochemicals for crop growth and yield, an intensive and diffuse utilization of these compounds in agriculture poses severe concerns for environmental protection, and especially for soil quality and human health concerns [110, 111]. Agrochemicals are not only able to contaminate soils, crops and deep waters (with possible inclusion in food chains), but runoff in surface waters and sediments, generating severe off-site impacts [110].

Data of 2018 from EUROSTAT indicate that Italy is not among the European countries with the higher application rates of agrochemicals in agriculture, although the agrochemical use poses this country at the highest rank (together with Spain and France) because of its large agricultural area. Regarding the pesticide use in Italy, their presence is significantly lower than in the other EU's countries, with 53% of the soils containing pesticide residues [110].

In 2018, about 114,000 tons of agrochemicals for plant protection was sold in the Italian market, corresponding to a content of active principles of about 54,000 tons. Of these products, 47% are fungicides, 18% insecticides and acaricides, 17.7% herbicides and 17.3% classified as others. These data evidence a significant decline (−23.7%) in the use of agrochemicals compared to the last 10 years (2008–2018), and particularly for the use of active principles (fungicides −38.7%, insecticides and acaricides −36.7%, herbicides −18.3% and others −18.8%, with a mean of −32.9%), except organic compounds that show a countertrend (+130%) [12]. This decline is a consequence of three factors: first, the long-term effect of the adoption of the National Action Plan for the sustainable use of plant protection products (PAN) in January 2014 (decrease of 12% in total amount and 8.9% in active principles in 2018 compared to 2014); second, the noticeable reduction in farmers' investments in agriculture; and third, a more rational use of technical means that avoid excessive release of agrochemicals in the environment [12].

At the national level, the application rate of active principles distributed per unit of surface is equal to 6.0 kg/ha. The distribution shows that a share of 54.3% is applied in the northern regions, while 11.5% and 34.2% of the total amount is distributed in central and southern regions. Veneto, Emilia-Romagna, Apulia, Sicily, Piedmont, Lombardy and Campania cover over 75% of national consumption. A similar distribution is surveyed for the active principles, with 53.2% consumed in the northern regions and 35.6% and 11.2%, respectively, in the southern and central regions. The grape-cultivated lands show the largest absolute use and application rates of agrochemicals. The active principle that has been surveyed in larger concentrations in freshwater and groundwater is the glyphosate and its metabolite, the aminomethylphosphonic acid (AMPA). It resulted that 28.5% (for glyphosate) and 58.4% (for AMPA) of the surveyed soils in the monitored sites were over the environmental quality limits; this percentage was 6% in groundwater measurements [12].

Overall, this decreasing trend is encouraging towards the protection of quality of natural resources, although not constant in terms of the used amount of agrochemicals [12].

Regarding the use of fertilizers, in 2019, about 65% of the total amount was applied in the northern regions, 15% in central areas, and the remaining 20% in Southern Italy and in the main islands. The total consumption amounted to less than 2.7 million tons. In more detail, the use of organic fertilizers was reduced by 1.2%, the organic-minerals by 11.5% and mineral fertilizers by 4.4% compared to the previous years (2008–2018) [1].

Finally, from 1990s to present, Italian organic farming has grown significantly, both in terms of covered area and number of farms. According to the latest EUROSTAT review, in 2017 Italy was at the fourth ranked in Europe (28 monitored countries) in percentage of agricultural area covered by organic farming. Overall, organic crops covered 15.5% of the national agricultural area and 6.1% of total farms. Apulia and Calabria showed the largest areas with organic farming, while pasture meadows, forage and cereals are the main crops oriented to organic farming. Also, olive and vine started the conversion to organic farming, and in the cattle breeding sector, poultry and sheep are the most common types of organic livestock [12].

According to a survey carried out in Greece, Italy and United Kingdom, and targeted to stakeholders associated with pesticide exposure in agriculture, agricultural workers are open to follow recommendations about safety practices, while inhabitants of agricultural areas are quite reluctant in adopting protective measures, since the latter presumably do not percept the effects of pesticides on their own health [112].

6 Future Perspectives and Conclusions

This review has shown that Italian agricultural soils are threatened by several land degradation processes. To summarize, soil sealing mainly affects the flat areas close to urban coastal zones with high density of population and economic activities and these areas are also subject to the high flooding and hydrogeological risks. While the rates of sealing, compaction and organic matter decline of Italian soils are in the same order of magnitude than those in several European countries, landslides, soil erosion and salinization, which are thought to increase over time and enlarge in space, are more dangerous compared to other countries. Moreover, the mitigation of these risks is more challenging that elsewhere because of the large variability of climatic and geomorphological conditions, intrinsically unfavourable, that challenges the implementation of homogenous soil conservation techniques [50, 113]. Unsuitable environmental management and land planning contribute to aggravate soil degradation risks [50, 114].

Soil erosion and landslides are the major degradation factors in steeper agricultural areas of the mid mountains and hills, and particularly in the internal areas of Southern Italy. High erosion rates are the main reasons for the reductions in soil organic matter, which worsens soil fertility – with subsequent reductions in quality and quantity of crops – and induces biodiversity loss. Unsuitable soil management and agricultural exploitation aggravate these land degradation factors, such as the use of heavy machinery with consequent soil compaction. Unsuitable soil management (for instance, the use of heavy machinery which excessively compacts soils) and agricultural intensification (with excessive uptake of organic matter and nutrients or supply of agrochemicals) aggravate land degradation. In contrast, natural areas have been subjected to severe environmental regulations (e.g. the national and regional parks) for many years, and these environments are not particularly vulnerable. This highlights, on the one hand, the responsibility of authorities and public administrations that failed to control soil quality levels and, on the other hand, the need for effective policies to fight against soil degradation [15, 115].

Although the Italian farmers have been always quite reluctant in adopting sustainable soil conservation practices [116, 117], soil quality of croplands is not far from the equilibrium over time with some exceptions recorded in small areas [118–120]; in contrast, pasturelands are prone to soil degradation due to climate change trends [13, 86].

Moreover, we agree with Dazzi and Lo Papa [14], who stated that many other soil degradation processes, which still are less visible than the most common factors, affect the soil resources in Italy. These include the reduction in the cultural value of soil, degradation of the traditional landscapes and decrease in pedodiversity. These authors have ascribed these degradation processes to crop intensification, unsuitable soil management practices and unsustainable cultivation techniques, which furthermore aggravate the hydraulic unsafety and hydrogeological risk of the country. According to Costantini and Lorenzetti [50], soil degradation in agriculture is currently underestimated, and this is particularly felt in Italy, which is known to be one of the richest places in the world in terms of soil and landscape diversity.

The analysis of the land degradation conditions in the Italian agriculture carried out so far has highlighted the needs for actions specifically dedicated to remove the residual vulnerability of territories to degradation factors. These actions cannot disregard the need to improve a deeper knowledge of the current degradation risks that has to consider the large variability of climatic and geomorphological conditions of the country. As a matter of fact, each degradation risk has been investigated by several actors (national authorities, local administrations, academic institutions and so on), often without a supervisor that organizes a systematic and synergistic approach. Moreover, the degradation risks affecting the Italian agriculture have been often studied in combination with other production sectors. The information related to the soil degradation risks relies on data at different temporal and spatial scales, often lacking homogenous indicators of land degradation or vulnerability. This makes the analysis and comparisons with other countries quite hard and time-consuming. For instance, the maps and databases of degradation risks analysed in this study have been produced by more than three authorities (ISPRA, Ministry, National Research Council and some universities). Even the erosion map was prepared by eminent but foreign scientists, who cannot obviously be fully aware about the specific conditions of the Italian agriculture. Moreover, a well-known erosion model has been used, but the input parameters have been drawn from databases sourced for other environmental contexts. This is a proof that, despite the intrinsic climate and pedological diversity of Italy, locally-validated and thus reliable estimation models are still not available for the Italian territory [14]. Furthermore, the surveys needed to collect field data have been carried out only in few sites and often not for a specific diagnosis at the national level (e.g. for scientific investigations based on local case studies). In this regard, national monitoring campaigns of degradation factors (not only in agriculture, but also in all civil and productive sectors and at the national scale) are needed, and the use of the most recent remote sensing techniques (based on satellite data) is required to cover the entire Italian territory with money and human resource savings. The availability of national-scale databases for land degradation in Italy would also allow the application of global models for climate change analysis, which could make forecasting the effects of expected global warming in agriculture simpler.

The studies about land degradation in Italy have been often carried out using a technical approach (except the papers of Salvati et al. [e.g. 16, 18] and some few other examples), and the socio-economic dimensions of the Italian agriculture have been in general neglected. In other words, the analysis of implications among the land degradation trends, social conditions of farmers and economical structure of agricultural producers is still insufficient. In contrast, the co-operation between policy makers and farmers, as the main stakeholders in agriculture and forestry, could provide an opportunity to promote an economically and environmentally sustainable agro-forestry. For instance, the farmers' needs for adopting soil conservation practices and, at the same time, leaving unaffected crop productivity require a careful attention by the authorities devoted to the environment protection. Also education activities that are targeted to improve the technical skills of farmers can also play an important role in stimulating the conservation of land resources at the farm level [33, 121]. It is well known that the scientific literature proposes effective and low-cost practices for land management and soil conservation (e.g. cover crops, crop rotation, reduced tillage, water-saving irrigation methods and prescribed fire) for a sustainable agricultural production and natural resource protection. However, the knowledge of their benefits and practical applicability in the different local contexts is often insufficient and even lacking. Therefore, the effectiveness of these techniques has not been yet evaluated in large parts of the Italian croplands and forested areas. These practices have been sometimes adopted in high-performance agricultural districts, but these experiences have still remained geographically limited; conversely, a "towing" action from the most advanced local contexts for low-performance districts may allow reversing some land degradation trends (especially soil compaction and salinization) in agriculture and forestry [33, 122]. Obviously, the large-scale adoption of the most sustainable land management practices must be supported by national and local subsidies for the actual involvement of farmers (particularly in marginal areas), while penalties to discourage unsustainable practices of land management should be adopted to reduce the environmental sensitivity to degradation [33]. However, the large money requirement collides with the ongoing reductions of financial supports at the European level, which makes fund shortage alarming for agriculture and forestry.

Finally, we are in close agreement with Costantini and Lorenzetti [50], who state that mitigation of land vulnerability in agriculture requires a "major effort to adopt a nation-wide campaign dedicated to the implementation of specific, locally tailored agro-techniques across all agricultural land uses". This effort is the prerequisite of both profitable agro-forestry activities and sustainable land conservation issues, targeted to a quick and simple achievement of United Nations Sustainable Development Goals in the 2030 Agenda for Sustainable Development, adopted by all Member States in 2015.

References

1. CREA Research Centre for Agricultural Policies and Bioeconomy (2020) L'agricoltura italiana conta 2019
2. Fortis M, Carminati M (2009) Sectors of excellence in the Italian industrial districts. Edward Elgar Publishing, Cheltenham
3. Aiello G, Donvito R, Grazzini L, Halliburton C, Wagner B, Wilson J, Godey B, Pederzoli D, Shokola I (2015) An international comparison of "Made in Italy" in the fashion, furniture and food sectors: an observational research study in France, Russia and The United Kingdom. J Glob Fash Market 6:136–149. https://doi.org/10.1080/20932685.2015.984822
4. Collins M, Knutti R, Arblaster J, Dufresne J-L, Fichefet T, Friedlingstein P, Gao X, Gutowski WJ, Johns T, Krinner G (2013) Long-term climate change: projections, commitments and irreversibility. In: Climate change 2013-the physical science basis: contribution of Working Group I to the Fifth Assessment Report of the Intergovernmental Panel on Climate Change. Cambridge University Press, pp 1029–1136
5. Costantini EA, Fantappié M, L'Abate G (2013) Climate and pedoclimate of Italy. In: The soils of Italy. Springer, pp 19–37
6. Finke P, Hartwich R, Dudal R, Ibanez J, Jamagne M, King D, Montanarella L, Yassoglou N (1998) Georeferenced soil database for Europe. Manual of procedures Version 1:184
7. Hartwich R, Baritz R, Fuchs M, Krug D, Thiele S (2005) Erläuterungen zur Bodenregionenkarte der Europäischen Union and ihrer Nachbarstaaten 1: 5,000,000 (version 2.0). , Hannover. Bundesanstalt für Geowissenschaften und Rohstoffe (BGR)
8. Costantini EA, Barbetti R, Fantappie M, L'Abate G, Lorenzetti R, Magini S (2013) Pedodiversity. In: The soils of Italy. Springer, pp 105–178
9. European Soil Bureau (1999) The European Soil Database. ISPRA, Italy
10. WRB-IUSS (2015) World reference base for soil resources. World Soil Resources Reports 106
11. Soil Survey Staff (2010) Keys to soil taxonomy, 11th edn
12. ISPRA (Istituto Superiore per la Protezione e la Ricerca Ambientale (2021) Stato dell'Ambiente
13. Costantini EA, Dazzi C (2013) The soils of Italy. Springer
14. Dazzi C, Lo Papa G (2013) Soil threats. In: The soils of Italy. Springer, pp 205–245

15. Salvati L (2014) A socioeconomic profile of vulnerable land to desertification in Italy. Sci Total Environ 466–467:287–299. https://doi.org/10.1016/j.scitotenv.2013.06.091
16. Salvati L, Zitti M (2008) Regional convergence of environmental variables: empirical evidences from land degradation. Ecol Econ 68:162–168
17. Costantini EAC, Urbano F, Aramini G, Barbetti R, Bellino F, Bocci M, Bonati G, Fais A, L'Abate G, Loj G (2009) Rationale and methods for compiling an atlas of desertification in Italy. Land Degrad Dev 20:261–276
18. Salvati L, Tombolini I, Perini L, Ferrara A (2013) Landscape changes and environmental quality: the evolution of land vulnerability and potential resilience to degradation in Italy. Reg Environ Chang 13:1223–1233. https://doi.org/10.1007/s10113-013-0437-3
19. UNCCD (1994) United Nations Convention to Combat Desertification. UNEP, Paris. http://www.unccd.int/convention/text/convention.php. Accessed Apr 2021
20. Salvati L, Bajocco S, Ceccarelli T, Zitti M, Perini L (2011) Towards a process-based evaluation of land vulnerability to soil degradation in Italy. Ecol Indic 11:1216–1227. https://doi.org/10.1016/j.ecolind.2010.12.024
21. Eckelmann W, Baritz R, Bialousz S, Bielek P, Carré F, Hrušková B, Jones RJ, Kibblewhite M, Kozak J, Le Bas C (2006) Common criteria for risk area identification according to soil threats. Office for Official Publications of the European Communities, Luxembourg
22. Kibblewhite M, Rubio JL, Kosmas C, Jones R, Arrouays D, Huber S, Verheijen F (2007) Environmental assessment of soil for monitoring desertification in Europe. In: 8. Conference of the parties of the United Nations convention to combat desertification. Cranfield University, p 62
23. Montanarella L (2007) Trends in land degradation in Europe. In: Climate and land degradation. Springer, pp 83–104
24. Salvati L, Zitti M, Perini L (2016) Fifty years on: long-term patterns of land sensitivity to desertification in Italy. Land Degrad Dev 27:97–107. https://doi.org/10.1002/ldr.2226
25. Salvati L, Carlucci M (2014) Zero net land degradation in Italy: the role of socioeconomic and agro-forest factors. J Environ Manag 145:299–306. https://doi.org/10.1016/j.jenvman.2014.07.006
26. Salvati L, Carlucci M (2011) The economic and environmental performances of rural districts in Italy: are competitiveness and sustainability compatible targets? Ecol Econ 70:2446–2453. https://doi.org/10.1016/j.ecolecon.2011.07.030
27. Niedertscheider M, Erb K (2014) Land system change in Italy from 1884 to 2007: analysing the North–South divergence on the basis of an integrated indicator framework. Land Use Policy 39:366–375
28. Floridi M, Pagni S, Falorni S, Luzzati T (2011) An exercise in composite indicators construction: assessing the sustainability of Italian regions. Ecol Econ 70:1440–1447
29. Salvati L, Zitti M, Carlucci M (2014) Territorial systems, regional disparities and sustainability: economic structure and soil degradation in Italy. Sustainability (Switzerland) 6:3086–3104. https://doi.org/10.3390/su6053086
30. Lucas-Borja ME, Zema DA, Plaza-Álvarez PA, Zupanc V, Baartman J, Sagra J, González-Romero J, Moya D, de las Heras J (2019) Effects of different land uses (abandoned farmland, intensive agriculture and forest) on soil hydrological properties in southern Spain. Water 11: 503
31. Shabanpour M, Daneshyar M, Parhizkar M, Lucas-Borja ME, Zema DA (2020) Influence of crops on soil properties in agricultural lands of northern Iran. Sci Total Environ 711:134694
32. Parhizkar M, Shabanpour M, Khaledian M, Cerdà A, Rose CW, Asadi H, Lucas-Borja ME, Zema DA (2020) Assessing and modeling soil detachment capacity by overland flow in forest and woodland of Northern Iran. Forests 11:65. https://doi.org/10.3390/f11010065
33. Salvati L, Carlucci M (2010) Estimating land degradation risk for agriculture in Italy using an indirect approach. Ecol Econ 69:511–518. https://doi.org/10.1016/j.ecolecon.2009.08.025

34. Salvati L, Zitti M, Ceccarelli T (2008) Integrating economic and environmental indicators in the assessment of desertification risk: a case study. Appl Ecol Environ Res 6:129–138. https://doi.org/10.15666/aeer/0601_129138
35. Salvati L, Scarascia MEV, Zitti M, Ferrara A, Urbano V, Sciortino M, Giupponi C (2009) The integrated assessment of land degradation. Ital J Agron 4:77–90. https://doi.org/10.4081/ija.2009.3.77
36. Salvati L, Zitti M, Ceccarelli T, Perini L (2009) Developing a synthetic index of land vulnerability to drought and desertification. Geogr Res 47:280–291
37. Salvati L, Zitti M (2007) Long term demographic dynamics along an urban-rural gradient: implications for land degradation. Biota 8:61–69
38. Costantini EA, Urbano F, Bonati G, Nino P (2007) Atlante nazionale delle aree a rischio di desertificazione
39. Salvati L, Zitti M (2005) Land degradation in the Mediterranean basin: linking bio-physical and economic factors into an ecological perspective. Biota 5:67–77
40. Salvati L, Zitti M (2009) Assessing the impact of ecological and economic factors on land degradation vulnerability through multiway analysis. Ecol Indic 9:357–363
41. Costantini EAC, L'Abate G (2009) A soil aridity index to assess desertification risk for Italy. Land degradation and rehabilitation: dryland ecosystems. In: Papers presented at the fourth international conference on land degradation, Cartagena, Murcia, Spain, 12–17 September 2004 40, pp 231–242
42. Zema DA, Nicotra A, Zimbone SM (2018) Diagnosis and improvement of the collective irrigation and drainage services in Water Users' Associations of Calabria (Southern Italy). Irrig Drain 67:629–644
43. Zema DA, Nicotra A, Mateos L, Zimbone SM (2018) Improvement of the irrigation performance in Water Users Associations integrating data envelopment analysis and multi-regression models. Agric Water Manag 205:38–49
44. Zema DA, Nicotra A, Tamburino V, Zimbone SM (2015) Performance assessment of collective irrigation in Water Users' Associations of Calabria (Southern Italy). Irrig Drain 64:314–325
45. Basso F, Bove E, Dumontet S, Ferrara A, Pisante M, Quaranta G, Taberner M (2000) Evaluating environmental sensitivity at the basin scale through the use of geographic information systems and remotely sensed data: an example covering the Agri basin (Southern Italy). Catena 40:19–35
46. Salvati L, Bajocco S (2011) Land sensitivity to desertification across Italy: past, present, and future. Appl Geogr 31:223–231. https://doi.org/10.1016/j.apgeog.2010.04.006
47. Wilson GA, Juntti M (2005) Unravelling desertification: policies and actor networks in Southern Europe. Wageningen Academic Publishers
48. Salvati L, Mancini A, Bajocco S, Gemmiti R, Carlucci M (2011) Socioeconomic development and vulnerability to land degradation in Italy. Reg Environ Chang 11:767–777
49. JRC (Joint Research Centre) (2011) The state of soil in Europe
50. Costantini EAC, Lorenzetti R (2013) Soil degradation processes in the Italian agricultural and forest ecosystems. Ital J Agron 8:233–243. https://doi.org/10.4081/ija.2013.e28
51. ISPRA (Istituto Superiore per la Protezione e la Ricerca Ambientale) (2013) Linee guida per la valutazione del dissesto idrogeologico e la sua mitigazione attraverso misure e interventi in campo agricolo e forestale
52. OECD (2008) Environmental performance of agriculture in OECD countries since 1990
53. Knijff JVD, Jones RJA, Montanarella L, Van der Knijff JM (1999) Soil erosion risk assessment in Italy. Office for Official Publications of the European Communities EUR 19022, Luxembourg, 32 pp.
54. Wischmeier WH, Smith DD (1978) Predicting rainfall erosion losses: a guide to conservation planning. Department of Agriculture, Science and Education Administration
55. Van Rompaey AJ, Bazzoffi P, Jones RJ, Montanarella L, Govers G (2003) Validation of soil erosion risk assessments in Italy. European Commission, Joint Research Centre, Luxembourg

56. Van Oost K, Govers G, Desmet P (2000) Evaluating the effects of changes in landscape structure on soil erosion by water and tillage. Landsc Ecol 15:577–589
57. Van Rompaey AJ, Verstraeten G, Van Oost K, Govers G, Poesen J (2001) Modelling mean annual sediment yield using a distributed approach. Earth Surf Process Landf 26:1221–1236
58. Grimm M, Jones RJ, Rusco E, Montanarella L (2003) Soil erosion risk in Italy: a revised USLE approach. European Soil Bureau Research Report 11:23
59. Bezak N, Mikoš M, Borrelli P, Alewell C, Alvarez P, Anache JAA, Baartman J, Ballabio C, Biddoccu M, Cerdà A (2021) Soil erosion modelling: a bibliometric analysis. Environ Res 197:111087
60. Borrelli P, Alewell C, Alvarez P, Anache JAA, Baartman J, Ballabio C, Bezak N, Biddoccu M, Cerdà A, Chalise D (2021) Soil erosion modelling: a global review and statistical analysis. Sci Total Environ 780:146494
61. Le Bissonnais Y, Montier C, Jamagne M, Daroussin J, King D (2002) Mapping erosion risk for cultivated soil in France. Catena 46:207–220
62. Kirkby MJ, Irvine BJ, Jones RJ, Govers G, PESERA Team (2008) The PESERA coarse scale erosion model for Europe. I.–Model rationale and implementation. Eur J Soil Sci 59:1293–1306
63. Borrelli P, Diodato N, Panagos P (2016) Rainfall erosivity in Italy: a national scale spatio-temporal assessment. Int J Digit Earth 9:835–850. https://doi.org/10.1080/17538947.2016.1148203
64. Piccarreta M, Capolongo D, Boenzi F, Bentivenga M (2006) Implications of decadal changes in precipitation and land use policy to soil erosion in Basilicata, Italy. Catena 65:138–151
65. Canfora L, Salvati L, Lo G (2017) Saline soils in in Italy: distribution, ecological processes and socioeconomic issues. Riv Econ Agrar 72:63–77. https://doi.org/10.13128/REA-21964
66. Perini L, Ceccarelli T, Sorrenti S, Salvati L, Zitti M (2008) La desertificazione in Italia: processi, indicatori, vulnerabilità del territorio. Bonanno
67. Dazzi C (2006) Saline waters and soil quality. Ital J Agron 1:467–474
68. Tóth G, Adhikari K, Várallyay G, Tóth T, Bódis K, Stolbovoy V (2008) Updated map of salt affected soils in the European Union. Threats to soil quality in Europe, pp 61–74
69. Crescimanno G, Marcum KB, Reina C, Versaci A (2009) Investigating soil–plant relationships for sustainable management of irrigation with saline water in a Sicilian vineyard. WIT Trans Ecol Environ 125:525–535
70. Zema DA, Calabro PS, Folino A, Tamburino V, Zappia G, Zimbone SM (2019) Wastewater management in citrus processing industries: an overview of advantages and limits. Water 11: 2481. https://doi.org/10.3390/w11122481
71. Zema DA, Esteban Lucas-Borja M, Andiloro S, Tamburino V, Zimbone SM (2019) Short-term effects of olive mill wastewater application on the hydrological and physico-chemical properties of a loamy soil. Agric Water Manag 221:312–321. https://doi.org/10.1016/j.agwat.2019.04.011
72. Legambiente (2010) Un'altra casa?
73. Munafò M, Salvati L, Zitti M (2013) Estimating soil sealing rate at national level—Italy as a case study. Ecol Indic 26:137–140
74. Salvati L (2012) The spatial nexus between population growth and land degradation in a dry Mediterranean region: a rapidly changing pattern? Int J Sustain Dev World Ecol 19:81–88
75. APAT (Agenzia per la protezione dell'ambiente e per i servizi tecnici) (2007) Il suolo la radice della vita
76. Loveland P, Webb J (2003) Is there a critical level of organic matter in the agricultural soils of temperate regions: a review. Soil Tillage Res 70:1–18
77. Huber S, Prokop G, Arrouays D, Banko G, Bispo A, Jones RJA, Kibblewhite MG, Lexer W, Moller A, Rickson RJ (2008) Environmental assessment of soil for monitoring. Volume I: indicators & criteria. European Communities

78. Johnston AE (1991) Soil fertility and soil organic matter. Advances in soil organic matter research: the impact on agriculture and the environment the. Royal Society of Chemistry, Melksham, Wiltshire, pp 299–314
79. Körschens M, Weigel A, Schulz E (1998) Turnover of soil organic matter (SOM) and long-term balances—tools for evaluating sustainable productivity of soils. Z Pflanzenernähr Bodenkd 161:409–424
80. Jones, R. J., Hiederer, R., Rusco, E., Loveland, P. J., & Montanarella, L. (2004). The map of organic carbon in topsoils in Europe. European Commission, Directorate General Joint Research Centre
81. Panagos P, Borrelli P, Meusburger K, Alewell C, Lugato E, Montanarella L (2015) Estimating the soil erosion cover-management factor at the European scale. Land Use Policy 48:38–50
82. Panagos P, Van Liedekerke M, Montanarella L, Jones RJ (2008) Soil organic carbon content indicators and web mapping applications. Environ Model Softw 23:1207–1209
83. Schils R, Kuikman P, Liski J, Van Oijen M, Smith P, Webb J, Alm J, Somogyi Z, Van den Akker J, Billett M (2008) Review of existing information on the interrelations between soil and climate change.(ClimSoil). Final report
84. Smith P, Davies CA, Ogle S, Zanchi G, Bellarby J, Bird N, Boddey RM, McNamara NP, Powlson D, Cowie A (2012) Towards an integrated global framework to assess the impacts of land use and management change on soil carbon: current capability and future vision. Glob Chang Biol 18:2089–2101
85. Fantappiè M, L'Abate G, Costantini EAC (2010) Factors influencing soil organic carbon stock variations in Italy during the last three decades. In: Land degradation and desertification: assessment, mitigation and remediation. Springer, pp 435–465
86. Fantappiè M, L'Abate G, Costantini EAC (2011) The influence of climate change on the soil organic carbon content in Italy from 1961 to 2008. Geomorphology 135:343–352
87. Chiti T, Gardin L, Perugini L, Quaratino R, Vaccari FP, Miglietta F, Valentini R (2012) Soil organic carbon stock assessment for the different cropland land uses in Italy. Biol Fertil Soils 48:9–17
88. Janssens I, Freibauer A, Ciais P, Smith P, Nabuurs G-J, Folberth G, Schlamadinger B, Hutjes R, Ceulemans R, Ernst Detlef S, Valentini R, Dolman AJ (2003) Europe's terrestrial biosphere absorbs 7 to 12% of European anthropogenic CO2 emissions. Science 300:1538–1542. https://doi.org/10.1126/science.1083592
89. Lado LR, Hengl T, Reuter HI (2008) Heavy metals in European soils: a geostatistical analysis of the FOREGS Geochemical database. Geoderma 148(2):189–199. https://doi.org/10.1016/j.geoderma.2008.09.020
90. Salvati L, Carlucci M (2015) Towards sustainability in agro-forest systems? Grazing intensity, soil degradation and the socioeconomic profile of rural communities in Italy. Ecol Econ 112:1–13. https://doi.org/10.1016/j.ecolecon.2015.02.001
91. Certini G (2005) Effects of fire on properties of forest soils: a review. Oecologia 143:1–10. https://doi.org/10.1007/s00442-004-1788-8
92. Pereira P, Francos M, Brevik EC, Ubeda X, Bogunovic I (2018) Post-fire soil management. Curr Opin Environ Sci Health 5:26–32. https://doi.org/10.1016/j.coesh.2018.04.002
93. Zema DA (2021) Postfire management impacts on soil hydrology. Curr Opin Environ Sci Health 21:100252. https://doi.org/10.1016/j.coesh.2021.100252
94. Moody JA, Shakesby RA, Robichaud PR, Cannon SH, Martin DA (2013) Current research issues related to post-wildfire runoff and erosion processes. Earth Sci Rev 122:10–37
95. Shakesby RA (2011) Post-wildfire soil erosion in the Mediterranean: review and future research directions. Earth Sci Rev 105:71–100
96. Wittenberg L, Pereira P (2021) Fire and soils: measurements, modelling, management and challenges. Sci Total Environ 776:145964. https://doi.org/10.1016/j.scitotenv.2021.145964
97. EEA (European Environment Agency) (2010) Mapping the impacts of natural hazards and technological accidents in Europe. An overview of the last decade

98. Peng G, Xu P, Zhu B, Bai M, Li D (2018) Microplastics in freshwater river sediments in Shanghai, China: a case study of risk assessment in mega-cities. Environ Pollut 234:448–456
99. Zhang Y, Pu S, Lv X, Gao Y, Ge L (2020) Global trends and prospects in microplastics research: a bibliometric analysis. J Hazard Mater 400:123110
100. Steinmetz Z, Wollmann C, Schaefer M, Buchmann C, David J, Tröger J, Muñoz K, Frör O, Schaumann GE (2016) Plastic mulching in agriculture. Trading short-term agronomic benefits for long-term soil degradation? Sci Total Environ 550:690–705
101. Bellasi A, Binda G, Pozzi A, Galafassi S, Volta P, Bettinetti R (2020) Microplastic contamination in freshwater environments: a review, focusing on interactions with sediments and benthic organisms. Environments 7:30
102. Piehl S, Mitterwallner V, Atwood EC, Bochow M, Laforsch C (2019) Abundance and distribution of large microplastics (1–5 mm) within beach sediments at the Po River Delta, Northeast Italy. Mar Pollut Bull 149:110515
103. Campanale C, Bagnuolo G, Dierkes G, Massarelli C, Uricchio VF (2019) Qualitative and quantitative screening of organic pollutants associated on microplastics from Ofanto River (South Italy). In: International conference on microplastic pollution in the Mediterranean Sea. Springer, pp 175–182
104. Guerranti C, Perra G, Martellini T, Giari L, Cincinelli A (2020) Knowledge about microplastic in Mediterranean tributary river ecosystems: lack of data and research needs on such a crucial marine pollution source. J Mar Sci Eng 8:216
105. Akdogan Z, Guven B (2019) Microplastics in the environment: a critical review of current understanding and identification of future research needs. Environ Pollut 254:113011
106. Sighicelli M, Pietrelli L, Lecce F, Iannilli V, Falconieri M, Coscia L, Di Vito S, Nuglio S, Zampetti G (2018) Microplastic pollution in the surface waters of Italian Subalpine Lakes. Environ Pollut 236:645–651
107. Fischer EK, Paglialonga L, Czech E, Tamminga M (2016) Microplastic pollution in lakes and Lake shoreline sediments–a case study on Lake Bolsena and Lake Chiusi (Central Italy). Environ Pollut 213:648–657
108. Vianello A, Boldrin A, Guerriero P, Moschino V, Rella R, Sturaro A, Da Ros L (2013) Microplastic particles in sediments of Lagoon of Venice, Italy: first observations on occurrence, spatial patterns and identification. Estuar Coast Shelf Sci 130:54–61
109. Frias JP, Gago J, Otero V, Sobral P (2016) Microplastics in coastal sediments from Southern Portuguese shelf waters. Mar Environ Res 114:24–30
110. Silva FC, Vieira DCS, van der Spek E, Keizer JJ (2019) Effect of moss crusts on mitigation of post-fire soil erosion. Ecol Eng 128:9–17. https://doi.org/10.1016/j.ecoleng.2018.12.024
111. Stolte J, Tesfai M, Oygarden L, Kvaerno S, Keizer J, Verheijen F, Panagos P, Ballabio C, Hessel R (2016) Soil threats in Europe: status, methods, drivers and effects on ecosystem services: deliverable 2.1 RECARE project. European Commission DG Joint Research Centre
112. Remoundou K, Brennan M, Sacchettini G, Panzone L, Butler-Ellis MC, Capri E, Charistou A, Chaideftou E, Gerritsen-Ebben MG, Machera K (2015) Perceptions of pesticides exposure risks by operators, workers, residents and bystanders in Greece, Italy and the UK. Sci Total Environ 505:1082–1092
113. Corti G, Cocco S, Brecciaroli G, Agnelli A, Seddaiu G (2013) Italian soil management from antiquity to nowadays. In: The soils of Italy. Springer, pp 247–293
114. Terribile F, Basile A, Bonfante A, Carbone A, Colombo C, Langella G, Iamarino M, Manna P, Minieri L, Vingiani S (2013) Future soil issues. In: The soils of Italy. Springer, pp 303–348
115. Sirami C, Nespoulous A, Cheylan J-P, Marty P, Hvenegaard GT, Geniez P, Schatz B, Martin J-L (2010) Long-term anthropogenic and ecological dynamics of a Mediterranean landscape: impacts on multiple taxa. Landsc Urban Plan 96:214–223
116. Bombino G, Denisi P, Gómez JA, Zema DA (2021) Mulching as best management practice to reduce surface runoff and erosion in steep clayey olive groves. Int Soil Water Conserv Res 9: 26–36. https://doi.org/10.1016/j.iswcr.2020.10.002

117. Bombino G, Denisi P, Gómez J, Zema D (2019) Water infiltration and surface runoff in steep clayey soils of olive groves under different management practices. Water 11:240. https://doi.org/10.3390/w11020240
118. Gardi C, Sconosciuto F (2007) Evaluation of carbon stock variation in Northern Italian soils over the last 70 years. Sustain Sci 2:237–243
119. Lugato E, Zuliani M, Alberti G, Delle Vedove G, Gioli B, Miglietta F, Peressotti A (2010) Application of DNDC biogeochemistry model to estimate greenhouse gas emissions from Italian agricultural areas at high spatial resolution. Agric Ecosyst Environ 139:546–556
120. Morari F, Lugato E, Berti A, Giardini L (2006) Long-term effects of recommended management practices on soil carbon changes and sequestration in North-Eastern Italy. Soil Use Manag 22:71–81
121. Tanrivermis H (2003) Agricultural land use change and sustainable use of land resources in the Mediterranean region of Turkey. J Arid Environ 54:553–564
122. Pérez-Sirvent C, Martınez-Sanchez MJ, Vidal J, Sánchez A (2003) The role of low-quality irrigation water in the desertification of semi-arid zones in Murcia, SE Spain. Geoderma 113: 109–125

Agricultural Land Degradation in Slovenia

Matija Zorn

Contents

1 Introduction 224
2 Land-Use Change, Organic Matter, and Organic Carbon 225
3 Legislation on Agricultural Land Degradation 231
4 Soil Sealing 235
5 Soil Contamination 242
6 Soil Erosion 249
7 Conclusion 254
References 257

Abstract Slovenia is among the European countries with poorer natural farming conditions, which are further challenged by several degradation processes. Among them agricultural land is primarily threatened by soil sealing, contamination, and erosion. The disregard for agricultural land and soil as an important natural resource keeps coming to the fore in siting major infrastructure and commercial structures. Most soil on agricultural land is not polluted, but nonetheless, some areas are contaminated with certain inorganic pollutants (e.g., cadmium, lead, arsenic, and copper) and organic pollutants (e.g., pesticides). Agricultural land is also at risk of soil erosion by water and wind.

Keywords Land degradation, Slovenia, Soil contamination, Soil erosion, Soil sealing

---

M. Zorn (✉)
ZRC SAZU Anton Melik Geographical Institute, Ljubljana, Slovenia
e-mail: matija.zorn@zrc-sazu.si

Paulo Pereira, Miriam Muñoz-Rojas, Igor Bogunovic, and Wenwu Zhao (eds.),
Impact of Agriculture on Soil Degradation II: A European Perspective,
Hdb Env Chem (2023) 121: 223–262, DOI 10.1007/698_2022_927,

Published online: 20 December 2022

1 Introduction

Even though Slovenia is small (20,271 km^2), it is characterized by great landscape diversity because it lies at the intersection of various European macroregions – the Alps, the Pannonian Basin, the Dinaric Alps, and the Mediterranean [1]. Despite its wealth of landscapes, a range of geographical indicators ranks it among European countries with poor natural farming conditions [2]. This is shown by some terrain indicators that are strongly connected with nearly all other natural factors. 91.3% of Slovenia's surface is at an elevation of over 200 m, and 34.9% is over 600 m. 83.7% of its surface has an inclination of over 2°, 70.5% over 6°, and 50.7% over 12°. Plains cover only 15.3% of Slovenia and, because these are flood-prone areas, just under a quarter of all agricultural land can be found on them. Another important factor is the lithological composition, due to which over 40% of the country is karstified and thus less suitable for agriculture [3]. The countryside is also characterized by intensive farmland abandonment and afforestation (Sect. 2) [4], and the agricultural land is often threatened by unsustainable use, which affects soil fertility [5].

What is especially alarming in Slovenia is the low share of agricultural land per capita, which is significantly below the EU average. In 2007, Slovenia recorded 2,447 m^2 of utilized agricultural land per capita (compared to the EU average of 3,469 m^2), corresponding to 24% of all land in Slovenia (compared to the EU average of 40%). The share of fields and gardens in Slovenia was only 866 m^2 per capita (compared to the EU average of 2,094 m^2), corresponding to 9% of all land in Slovenia (compared to the EU average of over 24%) [6]. These figures did not change significantly between 2000 and 2016 [7]. The situation is especially alarming from the perspective of Slovenia's food security [8].

This paper presents the major agricultural land degradation processes in Slovenia, especially soil sealing (Sect. 4), contamination (Sect. 5), and erosion (Sect. 6). It also tackles some related issues such as land-use change and organic matter (Sect. 2). Acidification, salinization, and desertification are not discussed because they are not an issue in Slovenia [5, 9]. More acidic soils can only be found in areas with predominantly noncarbonate bedrock but, because carbonate rock predominates in Slovenia, the soils have a high buffering capacity [10]. To a smaller extent, acidification may result from excessive nutrient runoff or large quantities of acidic mineral fertilizers [11]. Similarly, soil salinization is not a major soil degradation processes in Slovenia [12, 13]. It may occur in areas affected by sea salt during high tide, such as at the mouths of rivers (e.g., the Dragonja, the Rižana, and Strunjan Creek), or along roads (e.g., the tree-lined avenue north of Logatec) that are excessively salted in winter [10, 12, 14].

Despite the awareness of agricultural land degradation processes, also evident from the legislation (Sect. 3), the Slovenian Court of Auditors wrote the following in 2010 [6]: "In 2010, the Slovenian Government, the Ministry of Agriculture, Forestry, and Food, the Ministry of the Environment and Spatial Planning, and the Slovenian Farmland and Forest Fund failed to safeguard agricultural land," establishing for 2010 alone that at least 2,321 ha of agricultural land (0.35% of all agricultural land), including 970 ha of the best agricultural land, was recategorized to other uses as part of municipal spatial plans. In addition, a further 297 ha was recategorized to other uses as part of national spatial plans (178 ha was allocated to highway construction); added to this should also be illegal changes to agricultural land [6]. The situation has not improved in the last decade, as the Slovenian Court of Auditors concluded in 2021 that the utilized agricultural area declined by 1.8% or 11,120 ha in the period 2015–2019 and that by the end of 2020 no Slovenian municipality has defined in its spatial plans the areas of permanently protected agricultural land [15].

Data on agricultural land degradation in Slovenia can be found in numerous publications (see reference list), but those that have attempted to summarize all the important degradation processes are not many. Among the most informative are a conference volume entitled Soil Protection Strategy in Slovenia [16], which covers soil erosion, urbanization of agricultural land, soil compaction, salinization, and soil pollution, among other topics, and a book entitled The Soils of Slovenia [17], which includes a special chapter on soil degradation, covering soil sealing, contamination, and erosion, among other topics. Worth mentioning is also an e-publication of the Slovenian Environment Agency entitled Environmental Indicators in Slovenia [18], which covers various aspects of agriculture, including land take and soil contamination. Among the publications dealing exclusively with soil erosion, a book entitled Erosion Processes in Slovene Istria [19] and a publication entitled Erosion in Slovenia [20] should be mentioned, and among the publications dealing exclusively with soil pollution, a publication entitled Research on Soil Pollution in Slovenia [21].

2 Land-Use Change, Organic Matter, and Organic Carbon

The predominant *land use* in Slovenia is forests, which cover approximately 60% of the country. Meadows and pastures predominate among the types of agricultural land (they cover nearly a fifth of the country), followed by arable land (nearly a tenth of the country), orchards (just under 2%), and vineyards (just under 1%). Built-up land accounts for over 5.5% (Table 2; Fig. 1) [22]. Over the past two centuries, Slovenia's land use has changed significantly (Figs. 2 and 3). The forest area has increased by over 20%, whereas the share of meadows and pastures has decreased by just over 14% and the share of arable land has decreased by just over 6% (Fig. 4). The share of vineyards has also more than halved. The built-up land had the highest relative increase: in the early nineteenth century, it accounted for only 1.4% [4].

Slovenia is losing a significant share of agricultural land especially because of afforestation or overgrowth (Figs. 2 and 5). From 1825 to 2020, the share of forest increased from just under 40% to over 60% of all land (Fig. 3) [22, 24]. Afforestation is typical of areas with less favorable farming conditions, where farming is being abandoned and thus the cultural landscape is falling into decay [25]. This process is

Fig. 1 Land use in Slovenia, 2017 [23]

Fig. 2 Types of land-use changes between 1825 and 2016 [23]

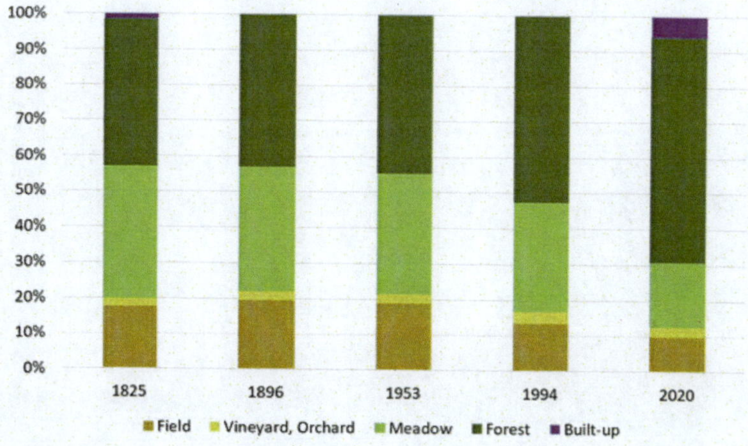

Fig. 3 Land use in Slovenia between 1825 and 2020 [22, 24]. The sources used do not provide information on built-up land for 1896, 1953, and 1994

distinctly spontaneous and one-directional [26]. On the other hand, afforestation is also strongly present in areas with a low inclination at an elevation below 500 m. In 2014, 32,916 ha or approximately 1.6% of all land was affected by afforestation; 23,374 ha of this area (71%) was located below 500 m [25]. In 2020, 24,390.9 ha, or approximately 1.2% of all land, was afforested [22].

Land use influences topsoil organic matter and the level of organic carbon. Slovenian soils are rich in *organic matter* [11], primarily because the predominant type of agricultural use is grassland (about two-thirds of all agricultural land) [27] and the large quantity of manure (solid and liquid manure) that is used in fields and areas of permanent crops [11].

In Slovenia, there is less topsoil organic matter in areas of intensive agriculture than in areas that are not plowed or trenched. In general, the soil is rich in organic matter, with as much as 86.2% of agricultural land containing over 2% and 30.9% containing over 4% organic matter [11]. Research conducted in 2005 and 2006 showed that the largest quantities of organic matter are in grassland soils (5.7–9.1% on average) and smaller quantities in tilled fields (3.4–3.7%) and areas of permanent crops (2.3–3.2%) [28]. Similar average values were recorded in 2008 and 2009 [17]: 6% in grasslands, 3.9% in tilled fields, gardens, and hop yards, 3.3% in orchards, 2.6% in vineyards, and 2.4% in olive groves, or an average of 4.3% for all types of agricultural land.

Despite the high topsoil organic matter levels, these may be decreasing in certain areas of northeastern Slovenia [13]. This is attributed to the local climate (less precipitation than in the rest of the country, temperature conditions, and problems with drought) and the predominance of Cambisols [17].

In Slovenia, topsoil *organic carbon* (0–30 cm) is estimated to be 0.2 Gt [11, 29]. The total carbon reserves in agricultural topsoil amount to 90 t/ha. Above-average reserves are found on land covered in trees and shrubs (118.1 t/ha)

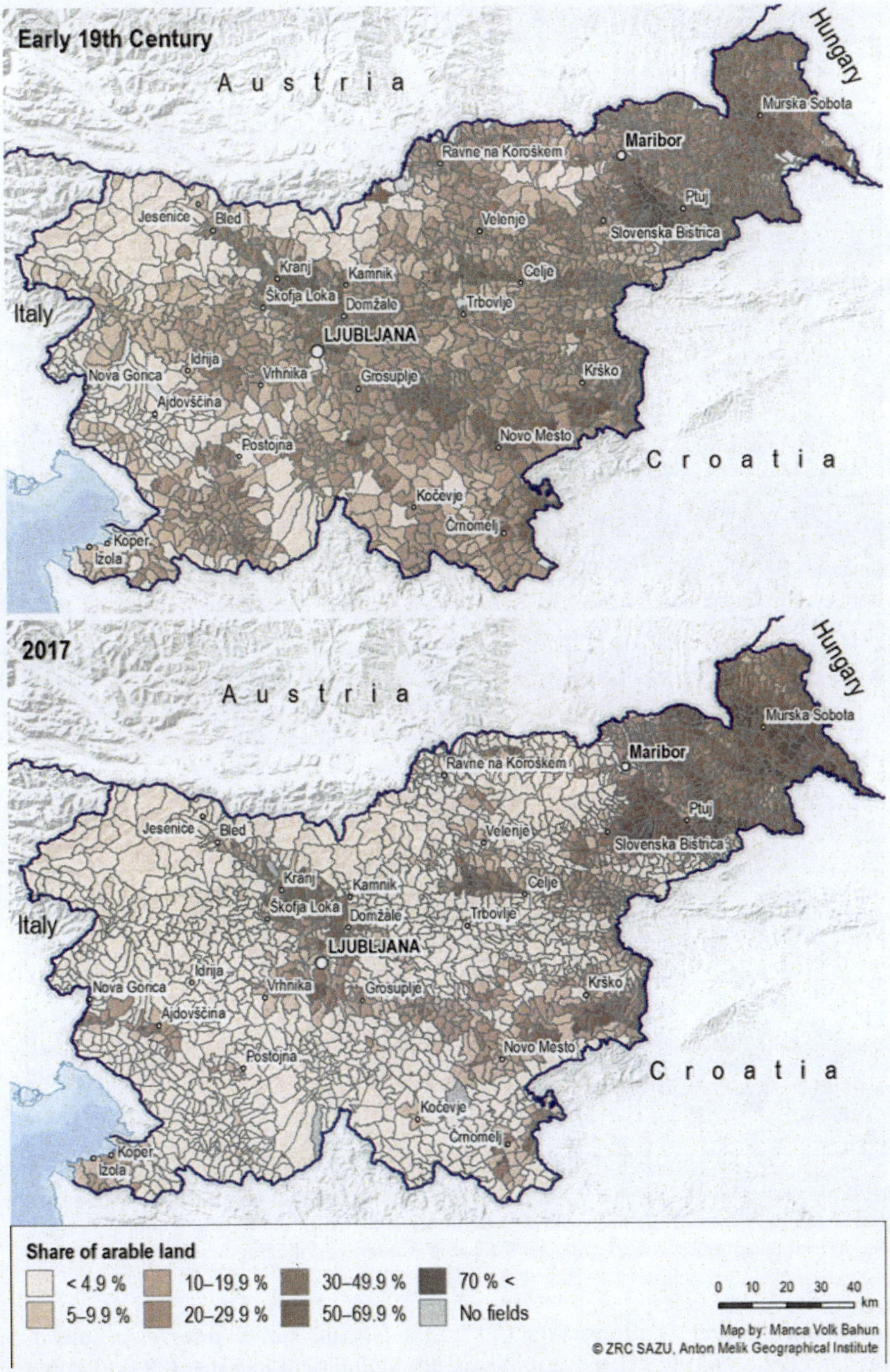

Fig. 4 Share of arable land in cadastral municipalities in the early nineteenth century and in 2017 [23]

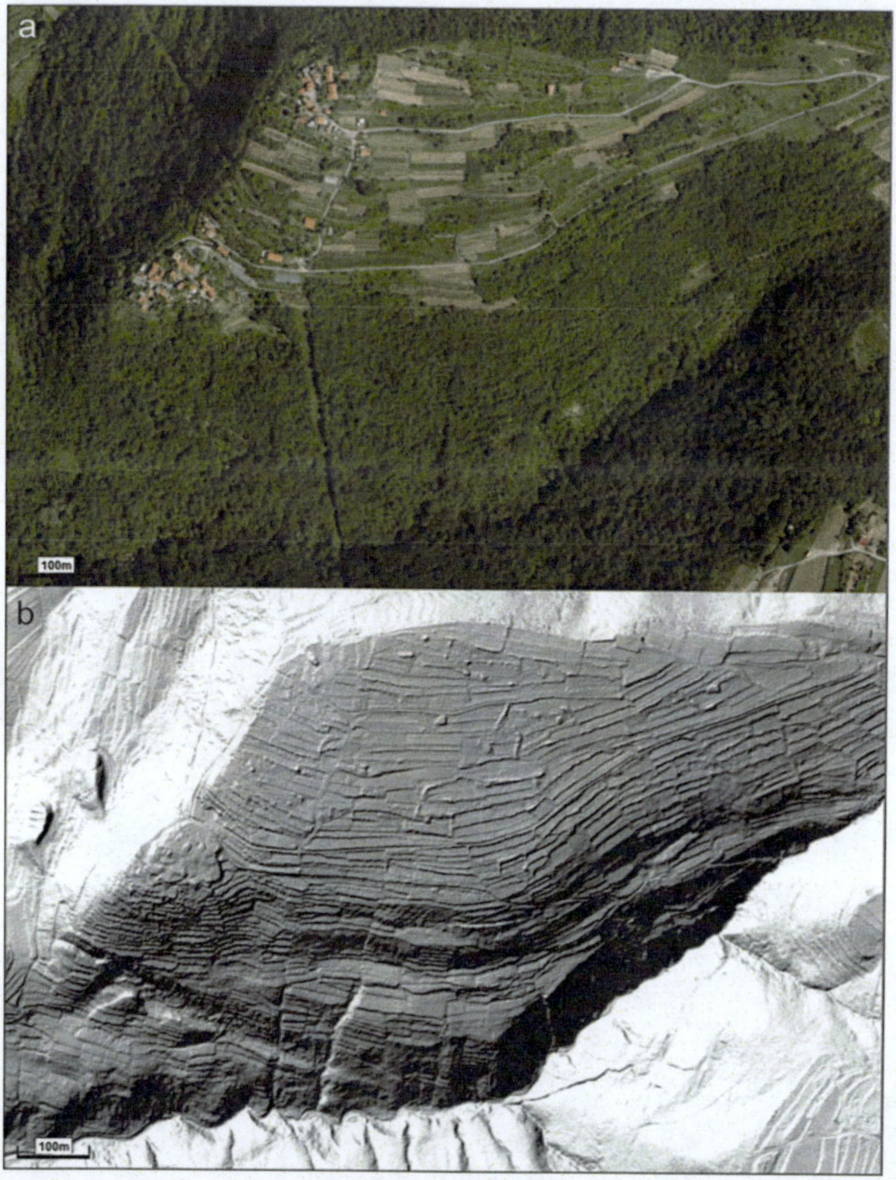

Fig. 5 Afforestation of agricultural land in Slovenia. An example of afforestation on cultivated terraces near the village of Glem in southwestern Slovenia: (**a**) a digital orthophoto from 2019, (**b**) a lidar-derived image (source: Surveying and Mapping Authority of the Republic of Slovenia)

and on land affected by afforestation (99.1 t/ha). Organic carbon reserves in soils in grassland (92.8 t/ha) in extensive orchards (90.5 t/ha), and in fields (89.8 t/ha) are close to the Slovenia average. Organic carbon in intensive orchards (71.5 t/ha) and vineyards (63.7 t/ha) are below the Slovenian average. The organic carbon levels decrease with depth, but this trend is not observed everywhere. The levels decrease the least in fields, which has to do with cultivation, in which the upper 20 to 30 cm of soil is mixed evenly every year, and they decrease the most on land affected by afforestation [30].

3 Legislation on Agricultural Land Degradation

Slovenia's awareness of agricultural land degradation problems is evident from various documents adopted at the national level [6]. The Development Strategy of Slovenian Agriculture [31] adopted in 1993 was the first document after the independence in 1991, setting the direction of agriculture development. Its long-term goals included the requirement to preserve agricultural land and protect it against pollution and unsustainable use. The 1999–2002 Agricultural Policy Reform Program [32] included the stopping or reducing of agricultural land shrinking among its goals. Protecting the best agricultural land from being permanently converted to other uses was also among the goals of the Resolution on the Strategic Guidelines for the Development of Slovenian Agriculture and Food Industry until 2020 [33]. Moreover, the protection and use of agricultural land is regulated by the Agricultural Land Act [34], which has been amended several times, also to improve the protection of agricultural land. The protection of agricultural land is also included in the Slovenian Constitution (Article 71) [35].

The current National Environment Protection Program (valid until 2030; [5]) has the following goals regarding agricultural soils: (1) increased capacity of the soil to perform ecosystem services achieved through managing the degradation processes related to a decline of organic matter in soil, preventing soil erosion and contamination, and restoring and revitalizing degraded areas, and through sustainable soil and land management, and reducing the net growth in built-up land by 25% by 2030, with the additional goal of achieving zero growth in built-up areas from 2050 onward; (2) acquisition of data on soil conditions; and (3) awareness raising. The program also specifies the measures to achieve these goals (Table 1), such as strengthening efforts to reduce soil sealing, preserving and increasing organic matter in agricultural soils, reducing soil erosion, safeguarding, preserving, and improving soil biodiversity, preventing contamination, restoring soils in polluted areas, including various aspects of soil protection into decision-making procedures at all levels and in various sectors. Nonetheless, soil as a natural resource is not covered by some other major national documents, such as the Spatial Planning Strategy of Slovenia [36], in which there is no mention of soil protection or soil loss prevention due to sealing. Something similar applies to the Resolution on the Transport Policy of the Republic of Slovenia [37], which does not mention soil protection in connection with transport infrastructure development or traffic pollution. Furthermore, soil protection is not covered in the National Mineral Resource Management Program [13, 38].

Table 1 Soil protection measures as defined in the National Environment Protection Program [5]

Type of measure	Measure	Measure indicator	Responsible body/participants	Implementation deadline
Land management	Soil restoration priority on degraded building land and reuse of this land	Area of degraded building land on which soil has been restored and land reused	Ministry of the Environment and Spatial Planning	Permanent task
	Recategorizing non-built-up land already zoned for construction back to agricultural land or forests where appropriate	Share of recategorized land	Ministry of the Environment and Spatial Planning, municipalities	Permanent task
	Improved land management policy for activating non-built-up building land to reduce net annual growth in area of built-up land	Net annual growth in area of built-up land	Ministry of the Environment and Spatial Planning, municipalities	Permanent task
	Setting up an intersectoral IT service providing data and information on present land use	IT system setup	Ministry of the Environment and Spatial Planning	From 2020 onward
Drawing up legislation	Amending spatial planning and construction regulations to incorporate subsurface management	Regulations adopted and amended	Ministry of the Environment and Spatial Planning	2022
	Amending comprehensive environmental impact assessment regulations due to enhanced coverage of soil protection	Regulations adopted and amended	Ministry of the Environment and Spatial Planning	2020
Monitoring, soil data, and databases	Establishing and carrying out soil contamination monitoring: (1) Establishing a national monitoring network and carrying out monitoring (2) Establishing an intersectoral IT service for providing data and information on soil contamination and trends	Monitoring established	Slovenian Environmental Agency	2021
	Producing legal bases, establishing and carrying out monitoring of organic matter and nutrients in soils: (1) Specifying requirements, method, and frequency of monitoring and reporting on organic matter in agricultural and forest soils in Slovenia to establish records on greenhouse gas emissions	Monitoring established	Ministry of Agriculture, Forestry, and Food	From 2020 onward

	(2) Specifying requirements, method, and frequency of monitoring and reporting on organic matter in agricultural soils (3) Establishing an intersectoral IT service for providing data and information on organic matter and nutrients in soils and trends			
	Producing legal bases, establishing and carrying out monitoring of agricultural soil erosion: (1) Specifying requirements, method, and frequency of monitoring and reporting on agricultural soil erosion (2) Establishing intersectoral IT service for providing data and information on agricultural soil erosion and trends	Monitoring of agricultural soil erosion established	Ministry of Agriculture, Forestry, and Food	2021
	Establishing and maintaining databases on: (1) Areas allegedly affected by soil contamination (2) Soil biodiversity in relation to conservation of protected species and habitat types	Database established and maintained	Slovenian Environmental Agency	2020/2022 (establishment), then permanent task
Soil management	Efforts to ensure environmentally sustainable use of pesticides, fertilizers, and farming techniques (producing legal bases and improving those already in force, and producing or improving expert instructions)	Water and soil conditions improved	Ministry of Agriculture, Forestry, and Food	Permanent task
	Efforts to ensure soil-friendly use of forest machinery in forest management plans and awareness raising	Guidelines produced and awareness-raising activities implemented	Ministry of Agriculture, Forestry, and Food, Slovenian Forest Service	Permanent task
	Examining suitability of introducing new (legislative, financial, and surveillance) measures to improve soil relocation system and introducing such measures	Analysis conducted and measures introduced	Slovenian Environmental Agency, Ministry of Agriculture, Forestry, and Food	2022

(continued)

Table 1 (continued)

Type of measure	Measure	Measure indicator	Responsible body/participants	Implementation deadline
	Efforts to provide areas for temporary storage of non-contaminated soil removed	Areas defined in spatial planning acts	Ministry of the Environment and Spatial Planning	2023
	Efforts to provide areas for temporary storage of contaminated soil removed	Areas defined in spatial planning acts	Ministry of the Environment and Spatial Planning	2023
	Improved soil relocation inspection		Ministry of the Environment and Spatial Planning	Permanent task
Awareness raising	Holding educational workshops and youth research camps, producing brochures, and providing online information	Number of events	Ministry of the Environment and Spatial Planning, Ministry of Agriculture, Forestry, and Food	Permanent task
Stakeholder collaboration	Active operation of Slovenian Soil Partnership	Partnership activities	Ministry of the Environment and Spatial Planning, partnership members	Permanent task

4 Soil Sealing

Soil sealing is defined by the European Environment Agency (EEA) as "the destruction or covering of the ground by an impermeable material" [39]. The EEA considers soil sealing to be one of the main causes of soil degradation in the EU, which also applies to Slovenia. The current National Environment Protection Program [5] identifies soil sealing (e.g., with asphalt or concrete) and soil compaction as the greatest threats to Slovenian soils. There are hardly any data available on *soil compaction*, except for observations in permanent crop plantations, where tractor trafficking is common [40], and in forests due to the construction of forest roads and the use of heavy logging machinery [17]. In contrast, there is a very noticeable problem of *soil sealing* especially due to the construction of major infrastructure (e.g., highways) [41] and urban sprawl (Fig. 6), including the needs of industry (e.g., mining, Fig. 7) [17, 42–44]. Due to the country's polycentric settlement pattern, urbanization is very dispersed (Fig. 8) [45].

Attention has been drawn to the urban sprawl into agricultural land and the subsequent soil degradation since at least the 1970s [43]. Important in terms of legislation was the 1982 adoption of the Agricultural Land Protection Act [47] and the harmonization of municipal spatial planning acts with the Slovenian republic-level (Slovenia was one of Yugoslav republics) agricultural land protection regulations. This act introduced compensation for recategorizing agricultural land to other uses, depending on the land quality [43]. From 1958 to 1988, Slovenia lost 17,033 ha of agricultural land (just under 1% of all land) due to urbanization [25].

Before 1982, or the adoption of aforementioned act, the annual volume of agricultural land recategorized to other uses was between 900 and 1200 ha, and from 1982 to 1989 it decreased to 400 to 500 ha (i.e., it more than halved). In the 1990s, the pressure to recategorize agricultural land to other uses again grew stronger. In 1995, municipalities proposed the recategorization of 141 ha of agricultural land, whereas such proposals covered 2,384 ha in 2002, 3,777 ha in 2009, and 1,142 ha in 2010 [6], excluding major national infrastructure (e.g., highways). It is alarming that the tendencies for these land-use changes are often connected with real state profit [25, 48, 49].

In the 1990s, agricultural land management was regulated anew by the Agricultural Land Act [50], which was updated in 2003 [51]. Indirectly, agricultural and forest land management is also covered by other laws, such as the Act on Forests [52], the Nature Conservation Act [53], and the Water Act [54]. Agricultural land management is also governed by the Construction Act [55] and the Spatial Planning Act [56], which, among other things, directs the spatial development of settlements onto vacant land within settlements or less important land [43].

From 1993 to 1997, 4,078 ha was newly built up and the volume of agricultural land decreased by 81,092 ha (approximately 4% of all land in Slovenia). Agricultural land decreased from 38% of Slovenia's total area in 1993 [10] to 31% in 2001 [12] due to urbanization, infrastructure development, and afforestation, and built-up land grew from 2.9% to 3.8% [9].

Fig. 6 The spread of Ljubljana, Slovenia's capital, onto the former outskirts and arable land (sources: Surveying and Mapping Authority of the Republic of Slovenia; ZRC SAZU Anton Melik Geographical Institute Archive)

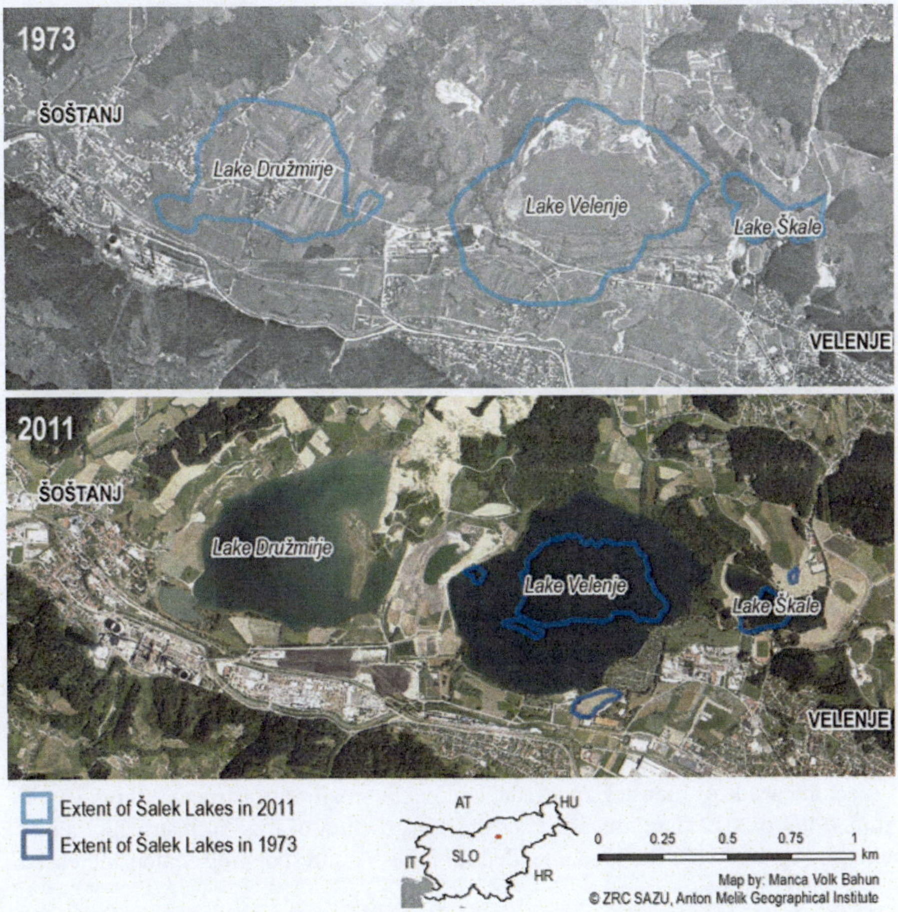

Fig. 7 Mining also causes agricultural land to disappear. In the Velenje area, the ground is subsiding due to underground mining activity, and the subsided areas have filled with water. The resulting lakes cover over 2.5 km^2, including former agricultural land (sources: Surveying and Mapping Authority of the Republic of Slovenia; [46])

From 2002 to 2007, 2.97% (14,121.7 ha) of agricultural land, or 7.99% of premium-quality agricultural land, was urbanized [42] (according to another source [6] 12,863 ha was urbanized during this period) and the share of urbanized land increased by 22.5%. Based on rough estimates, 11 ha of agricultural land was lost daily during this period [17, 57]. The area of fields and gardens decreased by 35,035 ha (approximately 1.7% of all land in Slovenia) during this period. These lands predominantly changed to meadows, and 6% of it was built up [6].

From 2008 to 2012, 13,024 ha of land was newly built up (just over 0.6% of all land in Slovenia), or 8.9 ha per day, whereby the volume of built-up land increased the most on agricultural land. During this period, the largest share of newly built-up

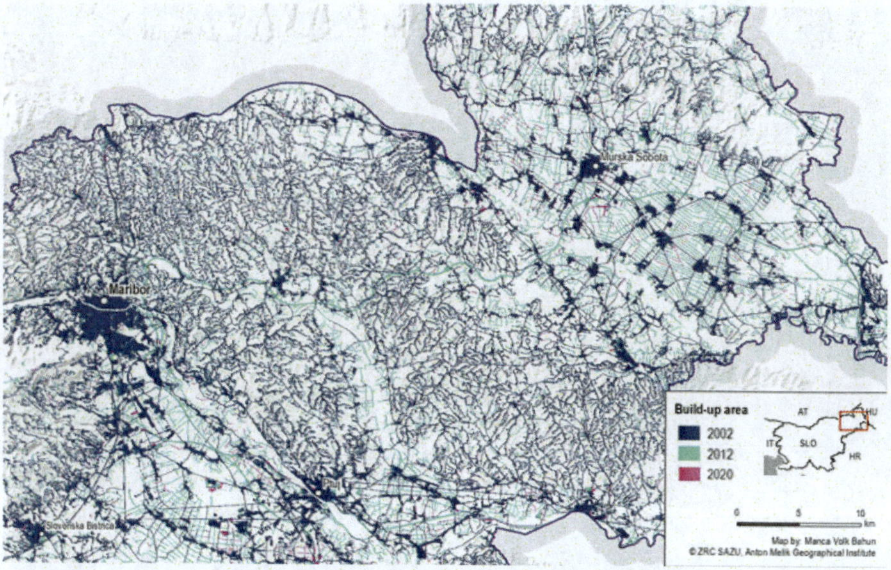

Fig. 8 Built-up areas in northeastern Slovenia in 2002, 2012, and 2020 [22]. Major infrastructure (highways) was built on agricultural land between 2002 and 2012, and industrial plants between 2012 and 2020 (source: Ministry of Agriculture, Forestry and Food)

land was recorded in grasslands (6,208.1 ha), forests (3,274.2 ha), and fields (1,550.5 ha) [13]. The largest changes occurred especially on the edges of towns to meet the needs of industry and trade, and next to major infrastructure (highways). From 1996 to 2012, land intended for the construction of road infrastructure alone increased by 1,090 ha, and land intended for industry and shopping centers increased by 228 ha [13].

The volume of built-up land continued to increase even after 2012; it accounted for 5.6% among the various land-use types in 2019. During this period, built-up land primarily spread to grassland (47%), forests (21%), and permanent crop areas (13%). From 2012 to 2019, the total volume of built-up land increased by 3,966 ha [58].

According to CORINE Land Cover, which uses a slightly different methodology, in 2018 Slovenia had 3.52% built-up land, ranking twenty-fourth among the thirty-nine European states [58].

Based on the latest national data [22], Slovenia had 5.72% built-up land in 2020, which is an increase of 0.24%, or 4,898.3 ha compared to 2002. In the same period, arable land decreased by 0.44%, or 7,854.5 ha (Table 2).

Most built-up land can be found on the best agricultural soil (Table 3). By far the largest share of fields in Slovenia can be found on Dystric Cambisol (17.9% of all fields) and Eutric Cambisol (19.1% of all fields), and most built-up land can also be found on these soils (17.6% of all built-up land on Dystric Cambisol and 19.7% on Eutric Cambisol).

Table 2 Land use in Slovenia between 2002 and 2020 according to data provided by the Ministry of Agriculture, Forestry, and Food [22]

Land use	2002		2012		2020		2012–2002		2020–2012		2020–2002	
	Hectares	%	Hectares	%	Hectares	%	Hectares	%	Hectares	%	Hectares	%
Field	216,976.2	10.96	191,200.7	9.62	195,801.5	9.88	−25,775.5	−1.34	4,600.8	0.26	−21,174.7	−1.08
Vineyard	25,411.4	1.28	21,531.0	1.08	17,890.5	0.90	−3,880.5	−0.20	−3,640.5	−0.18	−7,520.9	−0.38
Orchard	26,109.3	1.32	30,781.2	1.55	35,340.6	1.78	4,671.9	0.23	4,559.4	0.23	9,231.3	0.46
Meadow	351,770.3	17.77	378,225.6	19.04	363,380.1	18.34	26,455.3	1.26	−14,845.5	−0.70	11,609.8	0.56
Agricultural land	**620,267.2**	**31.34**	**621,738.5**	**31.29**	**612,412.7**	**30.90**	**1,471.3**	**−0.05**	**−9,325.8**	**−0.39**	**−7,854.5**	**−0.44**
Forest	1,250,277.6	63.18	1,256,284.2	63.23	1,255,850.4	63.37	6,006.7	0.05	−433.9	0.14	5,572.8	0.20
Built-up	**108,472.4**	**5.48**	**108,832.0**	**5.48**	**113,370.7**	**5.72**	**359.6**	**0.0**	**4,538.6**	**0.24**	**4,898.3**	**0.24**

Table 3 Land use in Slovenia by soil type [22, 59]

	Field		Vineyard		Orchard		Meadow		Forest		Built-up		Soil type (total)	
Soil type	Hectares	%	Hectares	%	Hectares	%	Hectares	%	Hectares	%	Hectares	%	Hectares	%
Aric Anthrosol	2,217.61	1.14	**2,579.42**	**14.66**	1,191.48	3.39	4,121.62	1.14	6,420.10	0.51	1,390.47	1.24	17,920.70	0.91
Aric Anthrosol with Eutric and Dystric Cambisol	64.35	0.03	127.32	0.72	81.69	0.23	188.57	0.05	441.50	0.04	122.64	0.11	1,026.07	0.05
Calcaric Cambisol	735.35	0.38	686.66	3.90	1,361.17	3.87	1,202.15	0.33	8,222.33	0.66	866.11	0.77	13,073.77	0.66
Calcaric Fluvisol	7,426.32	3.81	20.85	0.12	288.59	0.82	3,469.27	0.96	5,535.01	0.44	2,602.28	2.31	19,342.32	0.98
Calcaric Regosol	0.84	0.00	0.38	0.00	2.75	0.01	13.87	0.00	43.26	0.00	5.77	0.01	66.87	0.00
Chromic Cambisol	5,766.70	2.96	862.05	4.90	**2,300.98**	**6.54**	**40,976.47**	**11.33**	**188,369.82**	**15.09**	**7,628.27**	**6.78**	**245,904.29**	**12.48**
Chromic Cambisol, Chromic Cambisol and Eutric Cambisol	0.33	0.00	0.00	0.00	1.14	0.00	2.47	0.00	13.30	0.00	27.82	0.02	45.07	0.00
Chromic Cambisol, Eutric Cambisol and Rendzic Leptosol	0.01	0.00	0.00	0.00	0.00	0.00	4.54	0.00	15.64	0.00	0.21	0.00	20.41	0.00
Chromic Cambisol, Rendzic Leptosol and Dystric Cambisol	2.32	0.00	0.00	0.00	7.09	0.02	59.39	0.02	159.42	0.01	15.57	0.01	243.79	0.01
Dystric Cambisol	**34,940.82**	**17.94**	**1,385.64**	**7.87**	**7,734.48**	**22.00**	**73,641.01**	**20.35**	**269,105.50**	**21.56**	**19,819.97**	**17.62**	**406,627.42**	**20.64**
Dystric Cambisol and Dystric Planosol	19.96	0.01	0.03	0.00	11.06	0.03	212.04	0.06	521.65	0.04	32.99	0.03	797.74	0.04
Dystric Cambisol and Eutric Cambisol	452.72	0.23	49.80	0.28	121.15	0.34	1,020.13	0.28	2,922.95	0.23	213.00	0.19	4,779.75	0.24
Dystric Cambisol, Dystric Leptosol and Eutric Cambisol	63.56	0.03	0.07	0.00	80.86	0.23	701.60	0.19	2,274.09	0.18	92.42	0.08	3,212.60	0.16
Dystric Cambisol, Rendzic Leptosol and Eutric Cambisol	0.00	0.00	0.00	0.00	0.00	0.00	1.23	0.00	124.98	0.01	0.22	0.00	126.43	0.01
Dystric Fluvisol	**15,106.30**	**7.76**	29.21	0.17	396.25	1.13	1,922.57	0.53	6,898.28	0.55	2,218.69	1.97	26,571.32	1.35
Dystric Fluvisol and Dystric Gleysol	1500.31	0.77	14.58	0.08	45.72	0.13	337.48	0.09	590.79	0.05	170.74	0.15	2,659.63	0.14
Dystric Gleysol	**14,103.97**	**7.24**	30.07	0.17	278.87	0.79	3,384.04	0.94	9,650.89	0.77	2,308.27	2.05	29,756.12	1.51
Dystric Leptosol	1,138.89	0.58	88.61	0.50	516.42	1.47	4,548.32	1.26	32,653.30	2.62	1,129.99	1.00	40,075.53	2.03
Dystric Leptosol and Dystric Cambisol	6.88	0.00	0.33	0.00	7.04	0.02	94.05	0.03	329.27	0.03	16.45	0.01	454.02	0.02
Dystric Planosol	**13,313.57**	**6.84**	387.35	2.20	1,101.84	3.13	6,570.42	1.82	16,818.72	1.35	3,128.04	2.78	41,319.93	2.10
Eutric Cambisol	**37,196.24**	**19.10**	**7,556.10**	**42.94**	**10,533.83**	**29.96**	**77,032.14**	**21.29**	**144,979.77**	**11.61**	**22,142.12**	**19.68**	**299,440.20**	**15.20**
Eutric Cambisol and Dystric Cambisol	3,668.62	1.88	393.81	2.24	519.58	1.48	3,162.47	0.87	6,191.37	0.50	1,014.97	0.90	14,950.81	0.76

Soil type	Field		Vineyard		Orchard		Meadow		Forest		Built-up		Soil type (total)	
	Hectares	%	Hectares	%	Hectares	%	Hectares	%	Hectares	%	Hectares	%	Hectares	%
Eutric Cambisol and Eutric Planosol	0.12	0.00	0.00	0.00	0.00	0.00	6.88	0.00	94.21	0.01	0.02	0.00	101.23	0.01
Eutric Cambisol and Urbic Anthrosol	26.83	0.01	0.02	0.00	21.20	0.06	142.29	0.04	267.35	0.02	135.31	0.12	593.00	0.03
Eutric Cambisol, Dystric Cambisol and Dystric Leptosol	784.17	0.40	77.24	0.44	139.04	0.40	480.41	0.13	1,647.95	0.13	165.23	0.15	3,294.04	0.17
Eutric Fluvisol	**10,860.12**	**5.58**	**1,092.24**	**6.21**	1,463.58	4.16	12,074.83	3.34	14,701.23	1.18	5,291.83	4.70	45,483.83	2.31
Eutric Fluvisol and Dystric Fluvisol	192.71	0.10	0.00	0.00	31.46	0.09	578.72	0.16	642.71	0.05	193.13	0.17	1638.73	0.08
Eutric Gleysol	**16,028.02**	**8.23**	302.86	1.72	895.00	2.55	15,223.42	4.21	11,466.44	0.92	5,163.99	4.59	49,079.74	2.49
Eutric Gleysol and Dystric Gleysol	61.71	0.03	0.00	0.00	0.00	0.00	33.44	0.01	1.53	0.00	2.44	0.00	99.12	0.01
Eutric Leptosol	81.36	0.04	5.42	0.03	30.22	0.09	149.87	0.04	1,440.36	0.12	42.19	0.04	1,749.41	0.09
Eutric Planosol	7,519.13	3.86	238.18	1.35	656.51	1.87	5,745.09	1.59	9,466.84	0.76	2,696.23	2.40	26,321.98	1.34
Eutric Planosol and Dystric Planosol	2,352.07	1.21	110.45	0.63	104.34	0.30	1109.45	0.31	2,749.32	0.22	374.17	0.33	6,799.79	0.35
Eutric Regosol	16.49	0.01	0.09	0.00	2.37	0.01	31.88	0.01	98.09	0.01	4.47	0.00	153.38	0.01
Feric Podzol	0.00	0.00	0.00	0.00	0.00	0.00	5.43	0.00	137.31	0.01	1.37	0.00	144.11	0.01
Fibric Histosol	1.46	0.00	0.00	0.00	0.03	0.00	8.47	0.00	660.04	0.05	2.75	0.00	672.75	0.03
Haplic Luvisol	4,824.84	2.48	253.30	1.44	818.57	2.33	9,555.06	2.64	29,442.94	2.36	2,708.63	2.41	47,603.36	2.42
Lithic Leptosol	11.52	0.01	0.00	0.00	4.09	0.01	2,240.55	0.62	6,614.03	0.53	26.21	0.02	8,896.41	0.45
Mollic Gleysol	2,224.94	1.14	0.00	0.00	47.65	0.14	3,967.52	1.10	910.40	0.07	492.57	0.44	7,643.08	0.39
Mollic Leptosol	1,419.34	0.73	627.54	3.57	1,027.65	2.92	12,170.52	3.36	47,764.84	3.83	2,650.42	2.36	65,660.31	3.33
Mollic Leptosol, Chromic Cambisol and Eutric Cambisol	0.24	0.00	0.00	0.00	0.49	0.00	5.75	0.00	10.14	0.00	2.11	0.00	18.72	0.00
Mollic Leptosol, Eutric Cambisol and Chromic Cambisol	6.70	0.00	0.00	0.00	12.76	0.04	140.17	0.04	594.88	0.05	17.06	0.02	771.56	0.04
Mollic Leptosol, Eutric Cambisol and Dystric Cambisol	0.99	0.00	0.00	0.00	4.72	0.01	91.74	0.03	171.84	0.01	8.76	0.01	278.05	0.01
Rendzic Leptosol	3,036.26	1.56	367.04	2.09	**1,800.80**	**5.12**	**55,193.60**	**15.26**	**356,690.01**	**28.57**	8,258.62	7.34	**425,346.32**	**21.59**
Rendzic Leptosol and Chromic Cambisol	1,986.59	1.02	159.04	0.90	682.70	1.94	13,393.03	3.70	55,190.49	4.42	2,258.50	2.01	73,670.34	3.74
Terric Histosol	1,095.19	0.56	0.00	0.00	47.90	0.14	2,309.67	0.64	519.72	0.04	301.07	0.27	4,273.56	0.22
Urban and degraded areas	3,850.34	1.98	137.21	0.78	676.98	1.93	3,741.41	1.03	3,554.01	0.28	**16,027.34**	**14.25**	27,987.29	1.42
Urbic Anthrosol	33.27	0.02	0.00	0.00	2.15	0.01	68.51	0.02	32.14	0.00	57.96	0.05	194.03	0.01
Water bodies	635.72	0.33	12.78	0.07	106.67	0.30	668.02	0.18	1116.12	0.09	664.65	0.59	3,203.97	0.16

Soil types covering more than 5% of land use are in bold.

Soil types covering more than 5% of land use are in bold

Table 4 Main soil pollutants and causes of soil pollution [21]

Pollutants (for locations see Fig. 10)		Cause of pollution
Inorganic	Cadmium	Mining, smelting, industry
	Chromium	Metallurgic activity
	Cooper	Metallurgic activity
	Mercury	Mining, smelting
	Nickel	Metallurgical industry
	Lead	Mining, smelting, industry
	Zinc	Mining, smelting, industry
Organic	Pesticides	Intensive agriculture

5 Soil Contamination

Most soils in Slovenia are not contaminated, however some areas are polluted with certain inorganic pollutants (e.g., cadmium, lead, arsenic, and copper) and organic pollutants (e.g., pesticides) (Table 4). Inorganic pollutants primarily predominate in areas with past or ongoing mining and smelting or metallurgic activity (Figs. 9 and 10) [11]. These areas tend to exceed the alert thresholds, and in some places even the critical values of soil contamination [13].

Key in terms of assessing the contamination of agricultural land and soil in Slovenia is the Environmental Protection Act [60], which, among other things, regulates the monitoring and control of soil quality and forms the basis for soil protection measures against contamination and other degradation processes [27] defined in the EU's Thematic Strategy for Soil Protection [61]. Also relevant is the Decree on Limit Values, Alert Thresholds, and Critical Levels of Dangerous Substances in the Soil [62], which prescribes the limit, alert, and critical values for certain hazardous substances. The systematic study of soil contamination was defined in the National Environment Protection Action Program [63] and in the later Resolution on National Environmental Protection Action Plan 2005–2012 [64], but the envisaged extent of sampling in both cases proved to be unrealistic [27]. Monitoring is planned to be launched again in 2021 based on the current National Environment Protection Program [5].

The heaviest soil contamination with *inorganic pollutants* is observed in the Meža Valley, the Celje Basin, Jesenice, and Idrija (Fig. 10) [13]. In the Meža Valley, the soil is contaminated with lead, zinc, and cadmium as a result of five hundred years of mining and smelting. The soil in the Idrija area is contaminated with mercury, also as a result of a five-hundred-year mining and ore smelting history. In both sites, contamination can also be traced downstream on agricultural land on the river terraces. It is estimated that, in the case of the Idrija mine, about 2,000 t of mercury has been deposited downstream in the Idrijca River alluvial sediments. The sediment contamination area along the Idrijca and Soča Rivers is estimated to be more than 100 km up to the Gulf of Trieste. In the Idrijca Valley, there is up to 300 mg of Hg/kg or more deposited in the sediments [65]. In the Upper Meža Valley,

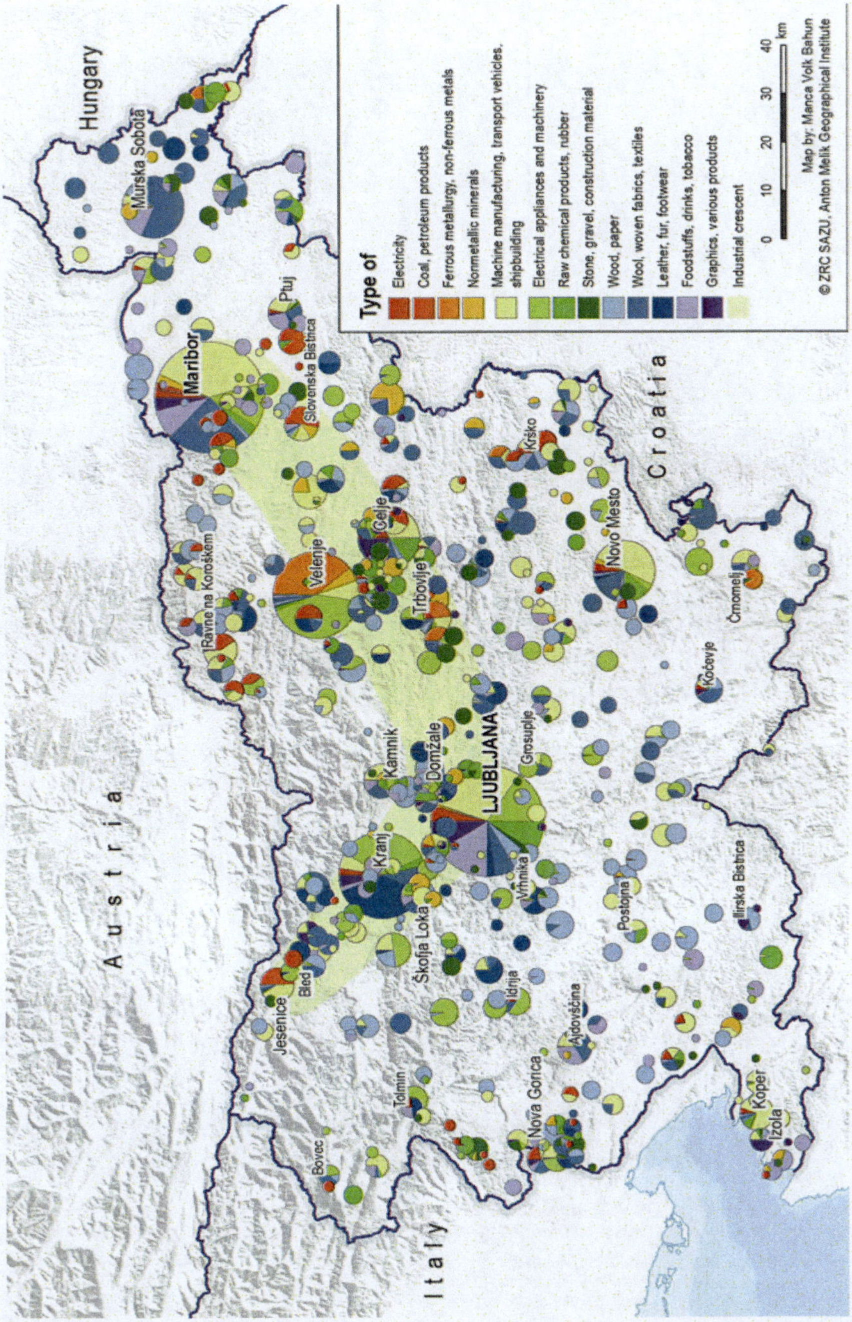

Fig. 9 Agricultural land degradation due to industrial activity is primarily present in the country's industrial crescent [67]

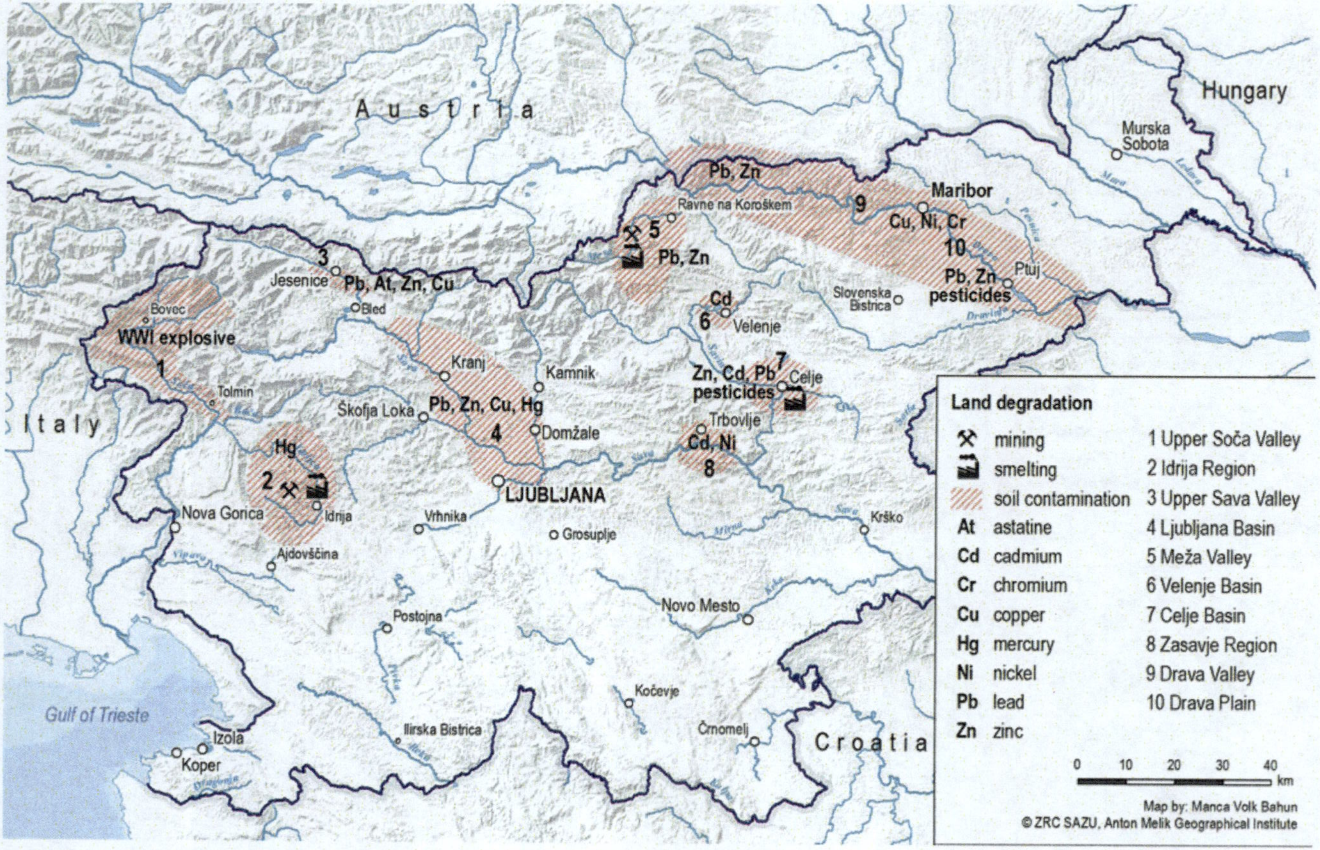

Fig. 10 Areas in Slovenia with the most heavily contaminated agricultural land and soil [10, 27, 65]

the topsoil is heavily contaminated within a radius of approximately 23 km^2, where the lead and zinc levels exceed 400 mg/kg. Levels even significantly higher than that have been measured in local gardens (lead: over 2,300 mg/kg, zinc: over 1,500 mg/kg). Alluvial sediments continue to be contaminated for several dozen kilometers downstream, and the total area of contaminated soil is estimated at about 60 km^2 [65]. Due to industrial activity, soils in the Celje area are contaminated with zinc (which was smelted for one hundred years) in an area of about 17 km^2, and also with cadmium and lead [10, 65]. Zinc and lead contamination is also present on the Drava River alluvial plain because of the historical lead and zinc mining upstream in the Meža Valley, and in Austria and Italy. Soils are contaminated on flood plains covering an area of 88 km^2. This corresponds to the largest contaminated area in Slovenia [65]. In the 1990s, elevated lead content was also detected in the soils along the main roads across the country [66]. In addition, soils are contaminated in the Velenje Basin (cadmium), the Zasavje region (cadmium and nickel), the Ljubljana Basin (lead, zinc, copper, and mercury), the Upper Sava Valley (lead, astatine, zinc, and cadmium), and the Maribor area (copper, nickel, and chrome). In some individual points located in western Slovenia, the values are also elevated due to the ammunition used on the Soča/Isonzo Front during the First World War (Fig. 10) [10].

Alert values of arsenic, nickel, and chrome appear at individual points in the soil, primarily due to releases from illegal dumps and/or increased levels in the parent rock, such as nickel in southwestern Slovenia because of the predominantly flysch rocks there [10, 27].

In Slovenia, contamination of soil with *organic pollutants* is less problematic because their levels do not exceed the alert thresholds [13]. Most often detected hazardous organic compounds were polycyclic aromatic hydrocarbons [21]. In certain areas of intensive agriculture, the limit values for pesticides or their degradation products were slightly exceeded [13, 21].

Mineral fertilizers began to be used more intensively in the 1950s, increasing to about 160 kg per hectare by the end of the century [68]. They are mainly used to add nitrogen, and phosphorus and potassium fertilizers are combined with manure to prepare the soil for planting [11].

The consumption of mineral fertilizers in Slovenia decreased 31% from 1992 to 2017 (Fig. 11). A decline in consumption can be observed after 1999, when it amounted to over 180,000 tons. Consumption recorded after 2008 has been below 140,000 tons, with 130,000 tons recorded in 2016 and 2017. During the same period (1992–2017), the consumption of mineral fertilizers per hectare of utilized agricultural area also declined by 21%, from 342 kg/ha to 271 kg/ha. The reduced consumption of mineral fertilizers can be attributed to the requirements imposed by the EU Nitrates Directive [69] and the agricultural fertilization best practices [70].

The period from 1992 to 2017 also saw the consumption of *nutrients* (nitrogen, phosphorus, potassium) decrease by 27%, from 135 kg/ha to 98 kg/ha (Fig. 12). The average consumption per hectare of agricultural land was 62 kg of nitrogen, 27 kg of phosphorus, and 34 kg of potassium. The predominant nutrient in mineral fertilizers is nitrogen (51%), followed by potassium (27%) and phosphorus (22%). From 2007

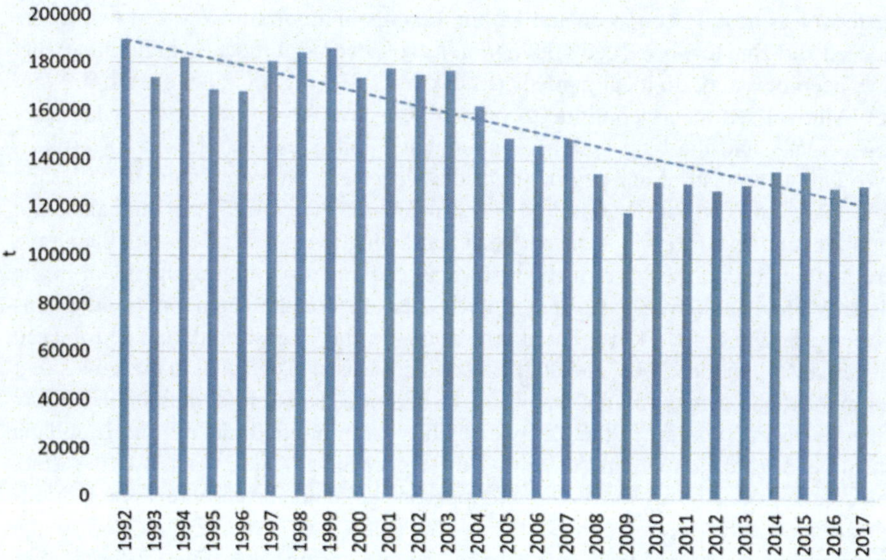

Fig. 11 Total consumption of mineral fertilizers in Slovenia from 1992 to 2017 [70]. The dotted line shows the trend

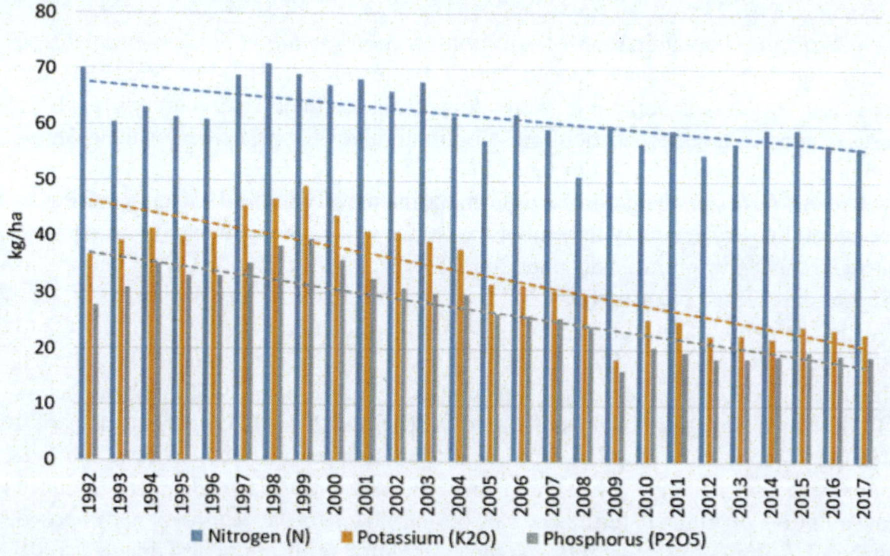

Fig. 12 Consumption of nutrients (N, P_2O_5, K_2O) per hectare of utilized agricultural area in Slovenia from 1992 to 2017 [70]. The dotted lines show the trends

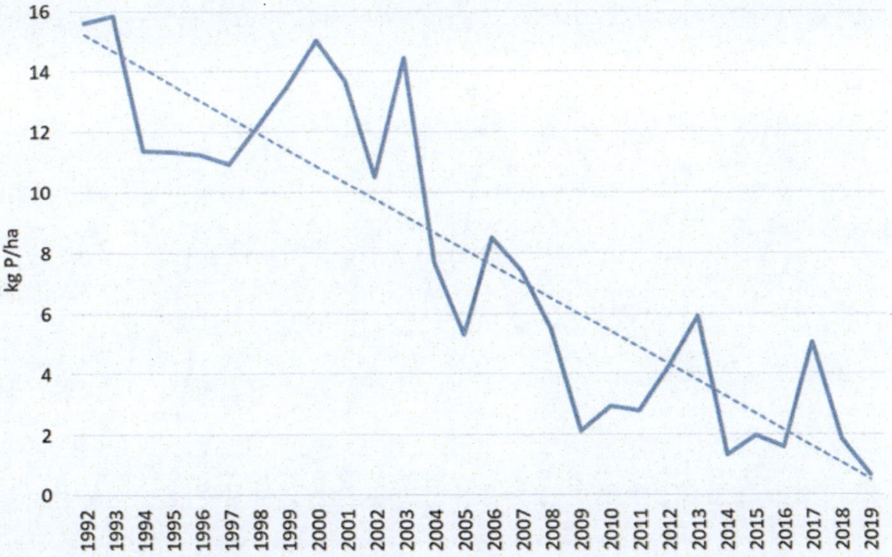

Fig. 13 Phosphorus budget in Slovenian agriculture from 1992 to 2019 [71]. The dotted line shows the trend

to 2019, the consumption of nitrogen per hectare of utilized agricultural area in Slovenia (57 kg/ha) was slightly below the EU average (60 kg/ha). In contrast, the consumption of phosphorus per hectare of utilized agricultural area (20 kg/ha) was higher than the EU average (14 kg/ha) [70]. The consumption of mineral fertilizers and nutrients on agricultural soil continued to decrease after 2017 [11].

From 1992 to 2019, the phosphorus use decreased by 97% (Fig. 13). A total of 10 to 15 kg/ha was typical up to 2005 and after decreased to below 5 kg/ha. From 2004 to 2015, the phosphorus budget in Slovenia (4.5 kg/ha) was above the EU average (2.2 kg/ha) [71].

Before the introduction of mineral fertilizers, soils were usually poor in nutrients, especially phosphorus and potassium. After more than half a century of using mineral fertilizers, the phosphorus and potassium levels are higher but vary greatly. In general, potassium levels are higher than phosphorus. Meadows and pastures, which account for two-thirds of Slovenian agricultural land, are less fertilized and contain less phosphorus and potassium than fields [27, 68]. Especially in areas where crops have a high productivity (hops, vineyards, intensive orchards, market gardening, and potatoes) fields were usually over-fertilized, which is now reflected in excessive phosphorus and potassium levels in the soil [68].

From 1992 to 2019, the nitrogen budget also decreased significantly: the gross nitrogen budget decreased by 50%, or 1.6 kg/ha per year, and the net nitrogen budget decreased by 81%, or 1.5 kg/ha per year (Fig. 14). It varied between 42 and 112 kg/ha in this period. Between 2004 and 2015, the gross nitrogen budget was slightly higher (54 kg/ha) than the EU average (50 kg/ha) [72]. Based on two decades' older

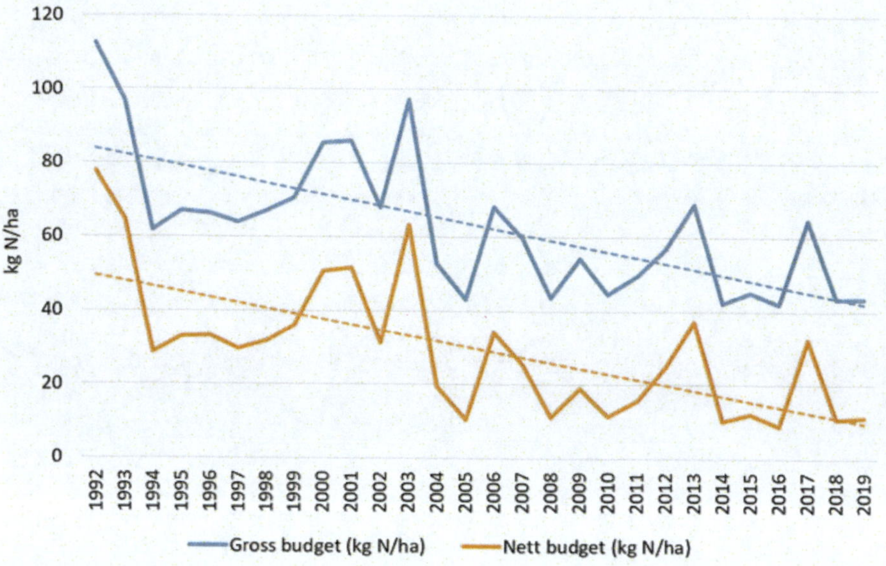

Fig. 14 Nitrogen budget in Slovenian agriculture from 1992 to 2019 [72]. The dotted lines show the trends

data, the nitrogen budget was >60 kg/ha [73]. According to the OECD data [73], the average for Slovenia was 36 to 42 kg of nitrogen per hectare. The main nitrogen source is manure (50% on average) and mineral fertilizers (36% on average). The reduced consumption of mineral fertilizers contributed the most to decreasing the use of nitrogen [11].

Overall, agriculture does not excessively burden the soil and underground water with nitrogen through the use of fertilizers or manure, but its consumption is not distributed evenly across the country [27]. The highest nitrogen budget can be observed in northeastern Slovenia [11], where agriculture is more intensive.

The limit values for *pesticides* residues were exceeded in some areas in Celje Basin in eastern Slovenia and on the Drava Plain in northeastern Slovenia (Fig. 10) [74]. Based on sales data, the consumption of these substances halved: from 1992 to 2015, there was a decrease from 2,031 tons 1,102 tons, respectively. Between 2000 and 2015, the average consumption was 6.2 kg/ha, or between 7.4 kg/ha in 2002 and 4.7 kg/ha in 2013 (Fig. 15). The use of pesticides per hectare of arable land is higher than in most EU member states. Slovenia is characterized by a high share of areas with permanent crops (orchards, vineyards, and hop yards) where the consumption of pesticides (especially fungicides) per hectare is higher than with grains and most root vegetables. Fungicides account for over two-thirds of all the pesticides used; among them, inorganic sulfur-based fungicides, which are less detrimental to the environment, account for the largest share. The smaller consumption of pesticides

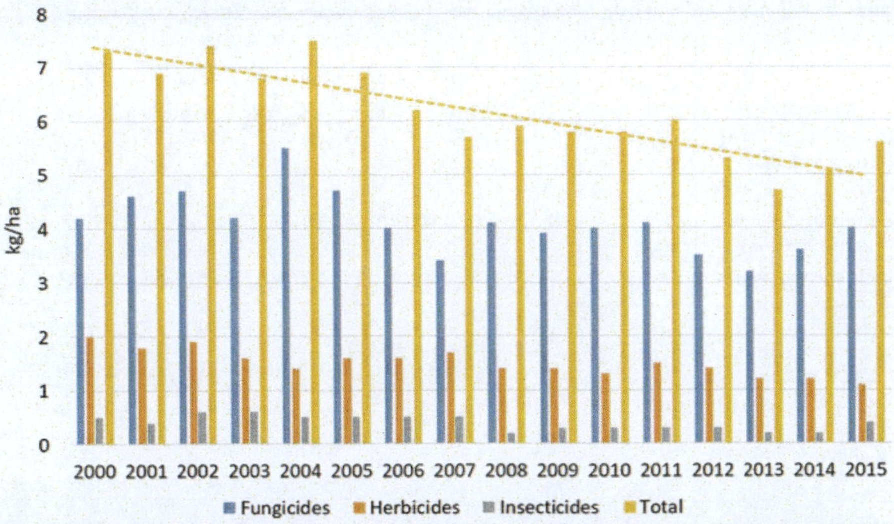

Fig. 15 Use of pesticides per hectare of arable land [75]. The dotted line shows the trend

was primarily influenced by the modernization of agricultural machinery and a more effective method of usage [11, 75].

6 Soil Erosion

Water soil erosion predominates on Slovenian agricultural land, but wind erosion is also common [17, 19, 76, 77]. Soil erosion is most intense in permanent crops fields [11, 79]. It is assumed that in the past soil erosion played a more important role because share of arable land was higher and there was also significantly less forest (Figs. 3 and 4) [12, 80].

Using the RUSLE model to estimate soil loss by *water erosion* across the EU, Panagos et al. [81] observed that the highest mean annual soil loss rate (at the country level) was in Italy (8.46 t/ha), followed by Slovenia (7.43 t/ha; 1.49% of the total EU soil loss) and Austria (7.19 t/ha). This results from a combination of high rainfall erosivity and steep topography [81]. Slovenian estimates show lower mean soil loss rates, between 3.70 and 4.52 t/ha per year [19, 79] or 3.68 t/ha per year [20, 82]. An even lower mean annual soil loss (1.2 t/ha; 0.4% of the total EU soil loss) was identified by Cerdan et al. [83], whose estimates for all of Europe were based on measurements and land-use data (Table 5).

Slovenia belongs to the EU countries with the highest rainfall erosivity, with maximum annual values exceeding 10,000 MJ mm ha^{-1} h^{-1} $year^{-1}$ [20, 82, 84]. According to Panagos et al. [85], the mean soil erodibility in Slovenia,

Table 5 Water soil erosion data for Slovenia based on selected sources (t/ha)

Method	Slovenia	Arable land	Grassland	Forest	Source
Literature review (measurements and different models) (Fig. 16)	3.70–4.52	0.86–1.93	0.79–0.81	0.32–0.28	Komac and Zorn [79]
Plot measurements (interrill erosion)	–	90–118[a]	0.90–1.68[b]	3.91–4.21[c]	Zorn [19, 77]; Zorn and Petan [87]
Field measurements and land use data	1.2	–	–	–	Cerdan et al. [83]
RUSLE	7.43	4.63	–	–	Panagos et al. [81]
RUSLE	3.13–3.68	7.64	6.82	0.29–0.43	Vrščaj et al. [20, 82]

[a] Inclination 9.4°, bare soil in an olive grove
[b] Inclination 5.5°
[c] Inclination 7.8–21.4°

estimated based on the K factor in the RUSLE model, is between 0.0282 and 0.0313 t ha h ha^{-1} MJ^{-1} mm^{-1}. According to Vrščaj et al. [82], it is 0.026 t ha h ha^{-1} MJ^{-1} mm^{-1}, whereby it can range between 0.001 and 0.048 t ha h ha^{-1} MJ^{-1} mm^{-1}. Therefore, soil erodibility in Slovenia is below the EU mean of 0.032 t ha h ha^{-1} MJ^{-1} mm^{-1} [85].

According to Panagos et al. [81], the mean annual soil loss rate on arable land in Slovenia is 4.63 t/ha per year (ranking third in the EU) and on arable land without Good Agricultural Environmental Condition (GAEC) it is 5.33 t/ha per year (fifth in the EU). Based on this model, 269,900 ha, or 37.1%, of agricultural land in Slovenia is exposed to strong or moderate soil erosion [11, 13].

According to Slovenian RUSLE-based estimates [20, 82], the mean soil loss rates on arable land are higher than identified by Panagos et al. [81]: 7.58 t/ha per year in fields and gardens. The highest soil loss rates were identified for permanent crops: 17.79 t/ha per year (olive groves: 41.62 t/ha per year, vineyards: 34.97 t/ha per year, and intensive orchards: 11.91 t/ha per year). Based on this model, the soil loss rate in grasslands amounts to 6.82 t/ha per year, and in forests to 0.43 t/ha per year (Table 5) [20, 82]. However, these model-based data still need to be validated through field measurements [86]. These are generally rare in Slovenia [19, 77], but they show higher values than the ones estimated based on the RUSLE model for bare agricultural land (90–118 t/ha per year on bare soil in an olive grove) and forests (3.91–4.21 t/ha per year), and lower values for grasslands (0.90–1.68 t/ha per year). Because the measured values apply to interrill soil erosion, which based on the simultaneous measurements of rill and interrill erosion (Fig. 17) accounts for only 10 to 25% of the total soil loss, the actual soil loss is significantly higher [19, 77].

Wind soil erosion is even less visible in Slovenia than water soil erosion. According to Borelli et al. [88], Slovenia belongs to the EU countries with the lowest susceptibility to wind erosion, recording only 0.1% of land with very low

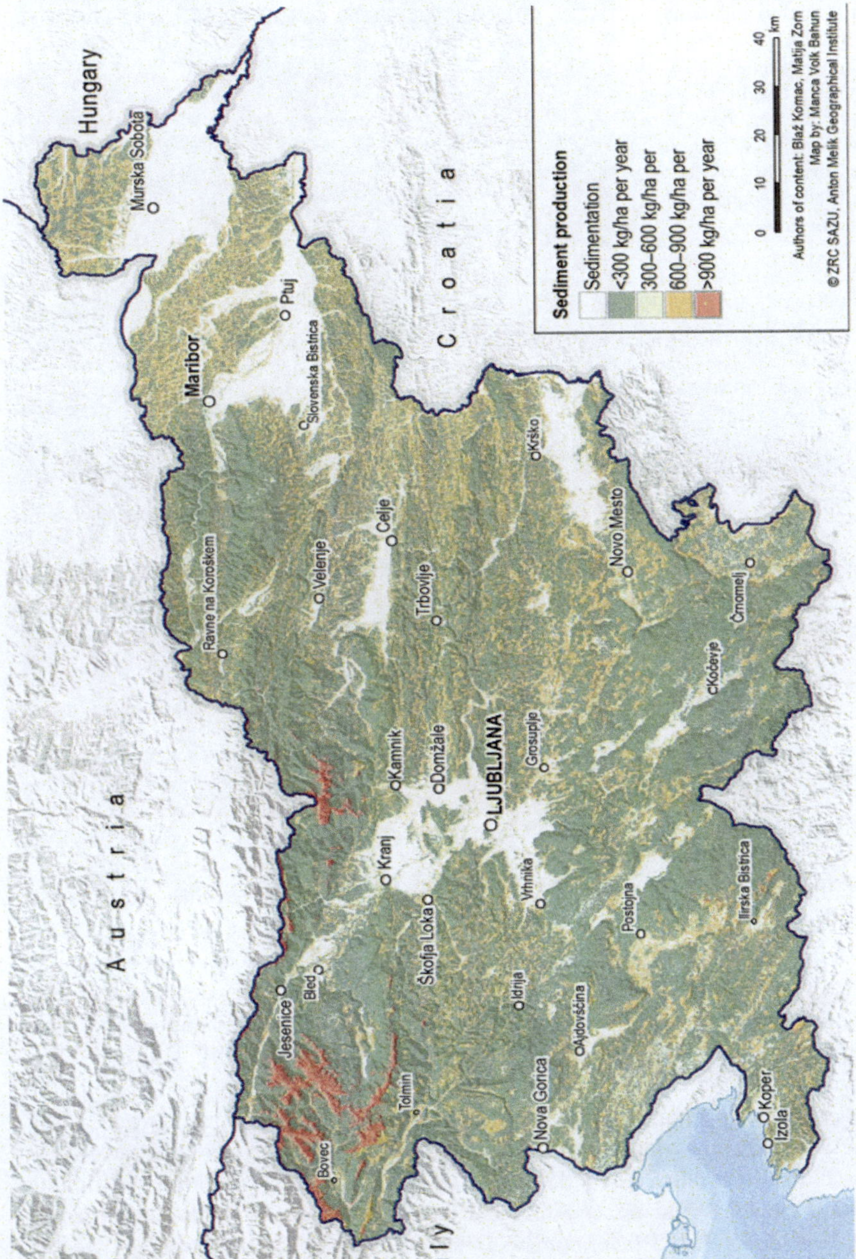

Fig. 16 Sediment production in Slovenia according to literature review in Komac and Zorn [79]

Fig. 17 (**a**) Rill soil erosion on bare agricultural land; (**b**) in heavy rainfall with daily erosivity of 1,110.5 MJ mm ha^{-1} h^{-1} $year^{-1}$, 2.67 kg/m^2 of soil was lost on a 1 m^2 bare erosion plot [19, 77] (photo: Matija Zorn, ZRC SAZU Anton Melik Geographical Institute Archive)

susceptibility to wind erosion. According to the same source, Slovenia's wind-erodible fraction of soil is among the lowest in Europe. Nonetheless, significant wind erosion can also occur locally in Slovenia. This is especially common in southwestern Slovenia, where the bora wind is typical. This is a cold, gusty, and mostly dry northeastern wind, which is most common between November and April and can reach well over 100 km/h (and even over 200 km/h). However, few quantitative data are available for southwestern Slovenia. In February 1954, in the Koper countryside a bora with a maximum speed of 23.7 m/s removed up to 10 cm of soil. Due to strong wind erosion, fields in this part of Slovenia were cultivated in the past mainly on the leeward slopes, whereas pastures predominated on the windward slopes. Wind erosion in the Koper countryside was also observed in November 2005. With a maximum wind speed of 24 m/s and an average weekly maximum speed of 13.5 m/s, approximately 64.28 g of soil per m^2 was eroded in a week [19, 77].

Fig. 18 Wind erosion in the Vipava Valley in March 2010 (photo: Karel Natek, ZRC SAZU Anton Melik Geographical Institute Archive)

Because of the bora wind, today wind erosion is primarily a problem on bare plowed fields in the Vipava Valley (Fig. 18). Wind erosion was especially strong after extensive drainage or amelioration in the first half of the 1980s, when many hedges (Fig. 19) were removed and plowed fields were left bare over the winter [66, 89]. According to the Wind Erosion Equation (WEQ), the annual soil loss by wind in the Vipava Valley is approximately 0.83 t/ha [90], but it can be much higher in extreme conditions. It was especially intense (with the bora reaching up to 130 km/h) [91] in early 2012 (Fig. 20), when up to 10 cm of topsoil, or a total of approximately 600,000 tons of soil (530 t/ha), was removed from 1,200 ha of agricultural land [78]. However, in southwestern Slovenia soil erosion by the bora wind also occurred in previous centuries [92].

However, wind erosion is also common in northeastern Slovenia, where agriculture is more intensive [20].

In Slovenia soil erosion has often been overlooked [93], the main reason being the lack of data on its intensity and extent. Soil erosion is not covered in the Slovenian umbrella law on natural disasters – Protection Against Natural and Other Disasters Act [94] or the umbrella law on agriculture – Agriculture Act [95]. The establishment of systemic agricultural soil erosion monitoring is only envisaged in the National Environment Protection Program [5] adopted in 2020 (Table 1). However,

Fig. 19 Agricultural land near Ajdovščina in the Vipava Valley before (**a**) and after (**b**) extensive amelioration. The main problem in preventing wind erosion is the lack of planted windbreaks (source: Surveying and Mapping Authority of the Republic of Slovenia)

people were already aware of the soil erosion threats in the past, which can be seen from the widespread construction of cultivated terraces as a traditional anti-erosion measure [82] across most of Slovenia (Fig. 21).

7 Conclusion

Slovenia is among the EU countries with the lowest share of agricultural and arable land, which continues to decrease further every year, especially due to continuous afforestation (the consequence of land abandonment) and urbanization. The disregard for soil or agricultural land as an important natural resource is reflected in siting major infrastructure and commercial structures, taking the highest-quality arable land. A further problem is the low value of agricultural land, purchases of which are often connected with its intended recategorization into building land, which is then sold at a significantly higher price. To improve the current situation, spatial planning should identify soil as a limited natural resource and the foundation for providing key ecosystem services.

Soil sealing as a result of the continuous expansion of built-up areas is the most important process of agricultural land degradation in Slovenia. The other forms of agricultural land degradation are soil contamination and soil erosion. Soil contamination with inorganic pollutants predominates mainly in areas with past or ongoing mining, smelting or metallurgic activity. Soil contamination with organic pollutants is concentrated in areas with intensive agriculture. On agricultural land soil erosion by water predominates, but wind erosion is also common. It is assumed that soil erosion played a more important role in the past because the share of arable land was significantly higher and there was also significantly less forest.

Fig. 20 The consequences of wind erosion in the Vipava Valley in February 2012: (**a**) up to 10 cm of soil was removed from a corn field; (**b**) over 10 cm of soil was deposited in a vineyard on the opposite side of the valley (photo: Matija Zorn, ZRC SAZU Anton Melik Geographical Institute Archive)

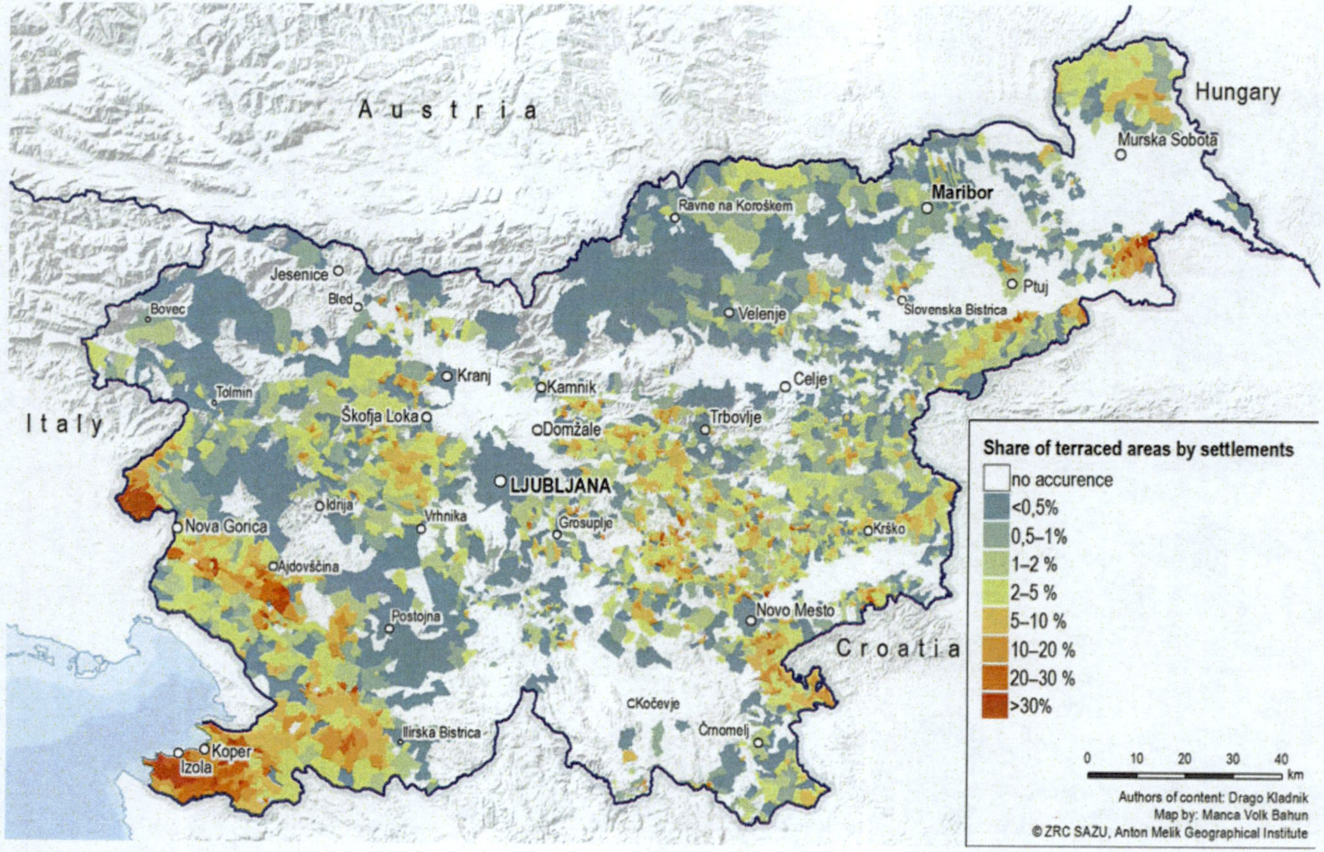

Fig. 21 Widespread construction of cultivated terraces in Slovenia as a traditional anti-erosion measure [96, 97]

Acknowledgments This study was conducted as part of the research program The Geography of Slovenia (P6-0101), financed by the Slovenian Research Agency and within the framework of the Commission on Land Degradation and Desertification (COMLAND) of the International Geographical Union.

References

1. Perko D, Ciglič R, Zorn M (2020) Slovenia: A European landscape hotspot. In: Perko D, Ciglič R, Zorn M (eds) The geography of Slovenia: small but diverse. Springer Nature, Cham, pp 1–20. https://doi.org/10.1007/978-3-030-14066-3_1
2. Hrustel Majcen M (2004) Trajnostni razvoj in kmetijstvo [Sustainable development and agriculture]. In: Lah A (ed) Sonaravno uravnoteženi razvoj Slovenije. Svet za varstvo okolja Republike Slovenije, Ljubljana, pp 99–102
3. Ciglič R, Hrvatin M, Komac B, Perko D (2012) Karst as a criterion for defining areas less suitable for agriculture. Acta Geogr Slov 52(1):61–98. https://doi.org/10.3986/AGS5210310
4. Gabrovec M, Kumer P (2019) Land-use changes in Slovenia from the Franciscean Cadaster until today. Acta Geogr Slov 59(1):63–81. https://doi.org/10.3986/AGS.4892
5. Resolucija o Nacionalnem programu varstva okolja za obdobje 2020–2030 [National environment protection programme with programmes of measures until 2030] (2020) Uradni list Republike Slovenije 31/2020. Ljubljana
6. Revizijsko poročilo: Uspešnost varovanja kmetijskih zemljišč kot pogoj za samooskrbo [Audit report: Effectiveness of agricultural land protection as a condition for self-sufficiency] (2013) Računsko sodišče Republike Slovenije, Ljubljana
7. Bedrač M, Cunder T (2018) Agricultural area - arable land per capita. In: Kazalci okolja. Agencija Republike Slovenije za okolje, Ljubljana. http://kazalci.arso.gov.si/en/content/agricultural-area-arable-land-capita. Accessed 25 Mar 2021
8. Plut D (2012) Prehranska varnost sveta in Slovenije [Food security in the world and in Slovenia]. Dela 38:5–23. https://doi.org/10.4312/dela.38.1.5-23
9. Suhadolc M (2005) Implementation of the Convention to combat desertification/land degradation in Slovenia. The Regional Environmental Center for Central and Eastern Europe, Ljubljana
10. Grčman H, Hudnik V, Lobnik F, Mihelič R, Prus T, Vrščaj B, Zupan M (2004) Tla [Soil]. In: Zych B, Mihelač Š (eds) Narava Slovenije. Mladinska knjiga, Ljubljana, pp 147–165
11. Program razvoja podeželja RS za obdobje 2014–2020 [Rural development program of the Republic of Slovenia for the period 2014–2020] (2020) Ministrstvo za kmetijstvo, gozdarstvo in prehrano, Ljubljana
12. Ogrin D, Plut D (2012) Aplikativna fizična geografija Slovenije [Applied physical geography of Slovenia]. Znanstvena založba Filozofske fakultete, Ljubljana
13. Poročilo o okolju v Republiki Sloveniji 2017 [Report on the environment in the Republic of Slovenia 2017] (2017) Vlada Republike Slovenije, Ljubljana
14. Tla [Soil] (2016) Okolje v Sloveniji 1996. Agencija Republike Slovenije za okolje, Ljubljana, pp 111–122. https://www.arso.gov.si/varstvo%20okolja/poro%c4%8dila/poro%c4%8dila%20o%20stanju%20okolja%20v%20Sloveniji/008f.pdf. Accessed 25 Mar 2021
15. Revizijsko poročilo: Zagotavljanje prehranske varnosti s pomočjo prehranske samooskrbe [Audit report: Ensuring food security through nutritional self-sufficiency] (2021) Računsko sodišče Republike Slovenije, Ljubljana
16. Knapič M (ed) (2007) Strategija varovanja tal v Sloveniji [Soil protection strategy in Slovenia]. Pedološko društvo Slovenije, Ljubljana
17. Vrščaj B, Repe B, Simončič P (2017) The soils of Slovenia. Springer, Dordrecht. https://doi.org/10.1007/978-94-017-8585-3

18. Kazalci okolja [Environmental indicators in Slovenia] (2021) Agencija Republike Slovenije za okolje, Ljubljana. http://kazalci.arso.gov.si/en. Accessed 10 Nov 2021
19. Zorn M (2008) Erozijski procesi v slovenski Istri [Erosion processes in Slovene Istria]. Geografija Slovenije 18. Založba ZRC, Ljubljana. https://doi.org/10.3986/9789612545482
20. Vrščaj B, Bergant J, Kastelic P, Šinkovec M (2020) Erozija v Sloveniji: Kratka predstavitev in ocena pomembne degradacije tal [Erosion in Slovenia: a brief presentation and assessment of significant soil degradation]. Kmetijski inštitut Slovenije, Ljubljana
21. Zupan M, Grčman H, Lobnik F (2008) Raziskave onesnaženosti tal Slovenije [Research on soil pollution in Slovenia]. Agencija Republike Slovenije za okolje, Ljubljana
22. Evidenca dejanske rabe kmetijskih in gozdnih zemljišč [Records of actual use of agricultural and forest land] (2021) Ministrstvo za kmetijstvo, gozdarstvo in prehrano, Ljubljana. https://rkg.gov.si/vstop/. Accessed 25 Mar 2021
23. Gabrovec M, Kumer P, Ribeiro D, Šmid Hribar M (2020) Land use in Slovenia. In: Perko D, Ciglič R, Zorn M (eds) The geography of Slovenia: small but diverse. Springer Nature, Cham, pp 279–290. https://doi.org/10.1007/978-3-030-14066-3_18
24. Bičík I, Gabrovec M, Kupková L (2019) Long-term land-use changes: a comparison between Czechia and Slovenia. Acta Geogr Slov 59(2):91–105. https://doi.org/10.3986/AGS.7005
25. Samec R (2014) Globalni pogledi in trajnostno upravljanje kmetijskih zemljišč v Sloveniji [Global view and sustainable management of agricultural land in Slovenia]. M.Sc. Thesis. Fakulteta za kmetijstvo in biosistemske vede, Maribor
26. Cunder T (1999) Zaraščanje kmetijskih zemljišč v slovenskem alpskem svetu [Abandoning of agricultural land in the Slovenian Alps]. Dela 13:165–175. https://doi.org/10.4312/dela.13.165-175
27. Grčman H, Vidic NJ, Zupan M, Lobnik F, Jones A, Montanarella L (eds) (2015) Soils of Slovenia with soil map 1:250,000. European Commission Joint Research Centre, Luxembourg. https://doi.org/10.2788/88750
28. Sušin J, Vrščaj B (2007) Količina organske snovi v zgornjem sloju kmetijskih tal Slovenije ter zasnova monitoringa organske snovi [The quality of organic matter in agricultural topsoil and the organic matter monitoring scheme]. In: Knapič M (ed) Strategija varovanja tal v Sloveniji. Pedološko društvo Slovenije, Ljubljana, pp 247–258
29. Yigini Y, Panagos P (2016) Assessment of soil organic carbon stocks under future climate and land cover changes in Europe. Sci Total Environ 557–558:838–850. https://doi.org/10.1016/j.scitotenv.2016.03.085
30. Šinkovec M, Bergant J, Mali B, Grčman H, Vrščaj B (2021) Zaloge oranskega ogljika v tleh kmetijskih zemljišč Slovenije – preliminarno poročilo večletnega projekta [Soil organic carbon stocks in agricultural land-uses of Slovenia – the multiannual project preliminary report]. Novi izzivi v agronomiji 2021:19–26
31. Strategija razvoja slovenskega kmetijstva [Development strategy of Slovenian agriculture] (2013) Državni zbor Republike Slovenije, Ljubljana
32. Program reforme kmetijske politike 1999–2002 [Agricultural policy reform program 1999–2002] (1998) Ministrstvo za kmetijstvo, gozdarstvo in prehrano, Ljubljana
33. Resolucija o strateških usmeritvah razvoja slovenskega kmetijstva in živilstva do leta 2020 – “Zagotovimo.si hrano za jutri” [Resolution on the strategic guidelines for the development of Slovenian agriculture and food industry until 2020 – “Ensuring Food for Tomorrow”] (2011) Uradni list Republike Slovenije 25/2011. Ljubljana
34. Zakon o kmetijskih zemljiščih [Agricultural land act] (2011) Uradni list Republike Slovenije 71/2011. Ljubljana
35. Ustava Republike Slovenije [Constitution of the Republic of Slovenia] (1991) Uradni list Republike Slovenije 33/1991. Ljubljana
36. Odlok o strategiji prostorskega razvoja Slovenije [Ordinance on spatial planning strategy of Slovenia] (2004) Uradni list Republike Slovenije 76/2004. Ljubljana
37. Resolucija o prometni politiki Republike Slovenije [Resolution on the transport policy of the Republic of Slovenia] (2006) Uradni list Republike Slovenije 58/2006. Ljubljana

38. Državni program gospodarjenja z mineralnimi surovinami [National mineral resource management program] (2009) Vlada Republike Slovenije, Ljubljana
39. What is soil sealing and why is it important to monitor it? (2021) European Environment Agency, Copenhagen. https://www.eea.europa.eu/help/faq/what-is-soil-sealing-and. Accessed 10 Nov 2021
40. Prus T (2007) Zbijanje in zaslanjevanje tal v Sloveniji [Soil compaction and salinization in Slovenia]. In: Knapič M (ed) Strategija varovanja tal v Sloveniji. Pedološko društvo Slovenije, Ljubljana, pp 103–110
41. Magyar A (2005) Soil sealing – highways in Slovenia. Academia Danubiana 1:8–9
42. Vrščaj B (2007) Urbanizacija tal v Sloveniji [Soil urbanisation in Slovenia]. In: Knapič M (ed) Strategija varovanja tal v Sloveniji. Pedološko društvo Slovenije, Ljubljana, pp 263–280
43. Lisec A, Lobnik F (2007) Spreminjanje rabe kmetijskih zemljišč kot posledica urbanizacije v Sloveniji [The change of the agricultural land use as the consequence of the urbanisation process in Slovenia]. In: Knapič M (ed) Strategija varovanja tal v Sloveniji. Pedološko društvo Slovenije, Ljubljana, pp 307–318
44. Repe B (2009) Izguba rodovitnih prsti Prekmurja zaradi trajnih sprememb rabe tal [Soil loss in Prekmurje due to the permanent land changes in the last 15 years]. In: Kikec T (ed) Pomurje: trajnostni regionalni razvoj ob reki Muri. Društvo geografov Pomurja, Murska Sobota, pp 127–142
45. Bole D, Goluža M, Tiran J, Kumer P, Topole M, Nared J (2020) The settlement system in Slovenia. In: Perko D, Ciglič R, Zorn M (eds) The geography of Slovenia: small but diverse. Springer Nature, Cham, pp 171–180. https://doi.org/10.1007/978-3-030-14066-3_11
46. Zorn M, Tiran J, Breg Valjavec M (2020) Pokrajinska preobrazba Velenjske kotline zaradi pridobivanja lignita [Landscape transformation of the Velenje Basin due to lignite mining]. In: Bole D (ed) Velenje, industrijsko mesto v preobrazbi. CAPACities 4. Založba ZRC, Ljubljana, pp 199–212. https://doi.org/10.3986/9789610502623_16
47. Zakon o varstvu kmetijskih zemljišč [Agricultural land protection act] (1982) Uradni list Socialistične republike Slovenije 44/1982. Ljubljana
48. Pihlar T (2017) V Sloveniji se je razpaslo lovljenje dobičkov pri preprodaji in pozidavi kmetijskih zemljišč [Profit hunting has disintegrated in Slovenia in the resale and sealing of agricultural land]. Dnevnik (June 3 2017). https://www.dnevnik.si/1042773890. Accessed 25 Mar 2021
49. Daugul L (2018) Še tisto malo rodovitne zemlje, ki jo imamo, slabo varujemo [Even that little bit of fertile land we have is poorly protected]. MMC RTV SLO (Nov 16 2018). https://www.rtvslo.si/okolje/novice/se-tisto-malo-rodovitne-zemlje-ki-jo-imamo-slabo-varujemo/472052. Accessed 25 Mar 2021
50. Zakon o kmetijskih zemljiščih [Agricultural land act] (1996) Uradni list Republike Slovenije 59/1996. Ljubljana
51. Zakon o kmetijskih zemljiščih [Agricultural land act] (2003) Uradni list Republike Slovenije 55/2003. Ljubljana
52. Zakon o gozdovih [Act on forests] (1993) Uradni list Republike Slovenije 30/1993. Ljubljana
53. Zakon o ohranjanju narave [Nature conservation act] (2004) Uradni list Republike Slovenije 96/2004. Ljubljana
54. Zakon o vodah [Water act] (2002) Uradni list Republike Slovenije 67/2002. Ljubljana
55. Zakon o graditvi objektov [Construction act] (2004) Uradni list Republike Slovenije 110/2004. Ljubljana
56. Zakon o prostorskem načrtovanju [Spatial planning act] (2007) Uradni list Republike Slovenije 33/2007. Ljubljana
57. Vrščaj B (2008) Strukturne spremembe kmetijskih zemljišč, njihova urbanizacija in kakovost v obdobju 2002–2007 [The structural changes of agricultural land, their quality and urbanization between 2002–2007]. Hmeljarski bilten 15:73–84
58. Lampič B, Rebernik L (2020) Land take. In: Kazalci okolja. Agencija Republike Slovenije za okolje, Ljubljana. http://kazalci.arso.gov.si/en/content/land-take-0. Accessed 25 Mar 2021

59. Grafični in pisni podatki Pedološke karte in pedoloških profilov [Graphic and written data on soil map and soil profiles] (2007) Ministrstvo za kmetijstvo, gozdarstvo in prehrano, Ljubljana. http://rkg.gov.si/GERK/. Accessed 25 Mar 2021
60. Zakon o varstvu okolja [Environmental protection act] (2004) Uradni list Republike Slovenije 41/2004. Ljubljana
61. Thematic Strategy for Soil Protection [SEC(2006)620] [SEC(2006)1165] (2006) COM(2006) 231 final. https://eur-lex.europa.eu/legal-content/EN/TXT/?uri=CELEX:52006DC0231. Accessed 25 Mar 2021
62. Uredba o mejnih opozorilnih in kritičnih imisijskih vrednostih nevarnih snovi v tleh [Decree on limit values, alert thresholds and critical levels of dangerous substances into the soil] (1996) Uradni list Republike Slovenije 68/1996. Ljubljana
63. Nacionalni program varstva okolja [National environment protection action programme] (1999) Uradni list Republike Slovenije 83/1999. Ljubljana
64. Resolucija o Nacionalnem programu varstva okolja 2005–2012 [Resolution on National environmental protection action plan 2005–2012] (2006) Uradni list Republike Slovenije 2/2006. Ljubljana
65. Žibret G, Gosar M, Miler M, Alijagić J (2018) Impacts of mining and smelting activities on environment and landscape degradation - Slovenian case-studies. Land Degrad Dev 29(12): 4457–4470. https://doi.org/10.1002/ldr.3198
66. Okolje v Sloveniji 1996 [Environment in Slovenia 1996] (2018) Agencija Republike Slovenije za okolje, Ljubljana. https://www.arso.gov.si/varstvo%20okolja/poro%C4%8Dila/poro%C4%8Dila%20o%20stanju%20okolja%20v%20Sloveniji/PSO1996.html. Accessed 25 Mar 2021
67. Smrekar A, Breg Valjavec M, Polajnar Horvat K (2020) Human-induced degradation in Slovenia. In: Perko D, Ciglič R, Zorn M (eds) The geography of Slovenia: small but diverse. Springer Nature, Cham, pp 303–312. https://doi.org/10.1007/978-3-030-14066-3_20
68. Leskošek M, Mihelič R, Grčman H, Pavlič E (1998) Oskrbljenost kmetijskih tal s fosforjem in kalijem v Sloveniji [Agricultural soil supply with phosphorus and potassium in Slovenia]. Novi izzivi v poljedeljstvu 1998:37–41
69. Council Directive of 12 December 1991 concerning the protection of waters against pollution caused by nitrates from agricultural sources (91/676/EEC) (1991). https://eur-lex.europa.eu/legal-content/EN/TXT/PDF/?uri=CELEX:01991L0676-20081211&from=EN. Accessed 25 Mar 2021
70. Sušin J (2018) Consumption of mineral fertilisers. In: Kazalci okolja. Agencija Republike Slovenije za okolje, Ljubljana. http://kazalci.arso.gov.si/en/content/consumption-mineral-fertilisers-0. Accessed 25 Mar 2021
71. Verbič J, Sušin J (2021) The phosphorus budget in agriculture. In: Kazalci okolja. Agencija Republike Slovenije za okolje, Ljubljana. http://kazalci.arso.gov.si/en/content/phosphorus-budget-agriculture?tid=1. Accessed 25 Mar 2021
72. Sušin J, Verbič J (2021) The nitrogen budget in agriculture. In: Kazalci okolja. Agencija Republike Slovenije za okolje, Ljubljana. http://kazalci.arso.gov.si/en/content/nitrogen-budget-agriculture-0?tid=1. Accessed 25 Mar 2021
73. Lobnik F, Vrščaj B, Prus T (2003) Land degradation in Slovenia. In: Jones RJA, Montanarella L (eds) Land degradation. Joint Research Centre, Ispra, pp 290–302
74. Poročilo o okolju v Sloveniji 2009 [Report on the environment in Slovenia 2009] (2010) Agencija Republike Slovenije za okolje, Ljubljana
75. Simončič A, Leskovšek R (2016) Consumption of pesticides. In: Kazalci okolja. Agencija Republike Slovenije za okolje, Ljubljana. http://kazalci.arso.gov.si/en/content/consumption-pesticides-2. Accessed 25 Mar 2021
76. Hrvatin M, Komac B, Perko D, Zorn M (2006) Slovenia. In: Boardman J, Poesen J (eds) Soil erosion in Europe. Wiley, Chichester, pp 297–310. https://doi.org/10.1002/0470859202.ch25
77. Zorn M (2009) Erosion processes in Slovene Istria – part 1: soil erosion. Acta Geogr Slov 49(1): 39–87. https://doi.org/10.3986/AGS49102

78. Zorn M, Breg Valjavec M, Komac B, Volk Bahun M, Hrvatin M (2020) Soils of Slovenia. In: Perko D, Ciglič R, Zorn M (eds) The geography of Slovenia: small but diverse. Springer Nature, Cham, pp 91–107. https://doi.org/10.1007/978-3-030-14066-3_6
79. Komac B, Zorn M (2005) Soil erosion on agricultural land in Slovenia – measurements of rill erosion in the Besnica valley. Acta Geogr Slov 45(1):53–86. https://doi.org/10.3986/AGS45103
80. Gabrovec M, Komac B, Zorn M (2012) Vpliv sprememb rabe tal na geomorfne procese v zadnjih stoletjih na primeru Zgornjega Posočja [The influence of land-use changes on geomorphic processes during recent centuries on the example of Zgornje Posočje (the Upper Soča Valley)]. In: Andrič M (ed) Dolgoročne spremembe okolja 1. Opera Instituti Archaeologici Sloveniae 25. Založba ZRC, Ljubljana, pp 101–109. https://doi.org/10.3986/9789612545925
81. Panagos P, Borrelli P, Poesen J, Ballabio C, Lugato E, Meusburger K, Montanarella L, Alewell C (2015) The new assessment of soil loss by water erosion in Europe. Environ Sci Policy 54: 438–447. https://doi.org/10.1016/j.envsci.2015.08.012
82. Vrščaj B, Kastelic P, Bergant J, Šinkovec M (2021) Ocena erozije na kmetijskih zemljiščih v Sloveniji [Assessment of agricultural soil erosion in Slovenia]. Novi izzivi v agronomiji 2021: 35–42
83. Cerdan O, Govers G, Le Bissonnais Y, Van Oost K, Poesen J, Saby N, Gobin A, Vacca A, Quinton J, Auerswald K, Klik A, Kwaad FJPM, Raclot D, Ionita I, Rejman J, Rousseva S, Muxart T, Roxo MJ, Dostal T (2010) Rates and spatial variations of soil erosion in Europe: a study based on erosion plot data. Geomorphology 122:167–177. https://doi.org/10.1016/j.geomorph.2010.06.011
84. Mikoš M, Bezak N (2020) Precipitation and soil erosion – Slovenia case. In: Oskoruš D, Rubinić J (eds) Okrugli stol s međunarobnim sudjelovanjem Nanos u vodnim sustavima - stanje i trendovi. Hrvatsko hidrološko društvo, Zagreb, pp 143–154
85. Panagos P, Meusburger K, Ballabio C, Borrelli P, Alewell C (2014) Soil erodibility in Europe: a high-resolution dataset based on LUCAS. Sci Total Environ 479–480:189–200. https://doi.org/10.1016/j.scitotenv.2014.02.010
86. Evans R, Boardman J (2015) The new assessment of soil loss by water erosion in Europe. Panagos P. et al., 2015 Environmental Science & Policy 54, 438–447—A response. Environ Sci Policy 58:11–15. https://doi.org/10.1016/j.envsci.2015.12.013
87. Zorn M, Petan S (2008) Interrill soil erosion on flysch soil under different types of land use in Slovenian Istria. IOP Conference Series: Earth and Environmental Science 4. https://doi.org/10.1088/1755-1307/4/1/012045
88. Borrelli P, Panagos P, Ballabio C, Lugato E, Weynants M, Montanarella M (2016) Towards a pan-European assessment of land susceptibility to wind erosion. Land Degrad Dev 27(4): 1093–1105. https://doi.org/10.1002/ldr.2318
89. Čarman M, Mikoš M, Pintar M (2007) Različni vidiki erozije tal v Sloveniji [Different aspects of soil erosion in Slovenia]. In: Knapič M (ed) Strategija varovanja tal v Sloveniji. Pedološko društvo Slovenije, Ljubljana, pp 39–50
90. Pliberšek R (2015) Ocena učinkov vetrne erozije prsti s pomočjo geoinformacijskih orodij na primeru Vipavske doline [Assessment of the effects of wind erosion of soil by using geoinformation tools in the case of the Vipava Valley]. B.Sc. Thesis. Filozofska fakulteta, Ljubljana
91. Bertalanič R (2013) Viharni vetrovi v Sloveniji leta 2012 [Storm winds in Slovenia in 2012]. Ujma 27:78–93
92. Panjek A (2006) Človek, zemlja, kamen in burja: zgodovina kulturne krajine Krasa (oris od 16. do 20. stoletja) [Man, soil, stone and bora: the history of the cultural landscape of the Karst (outline from the 16th to the 20th century)]. Založba Annales, Koper
93. Zorn M (2015) Erozija prsti – prezrt okoljski problem [Soil erosion – overlooked environmental problem]. Geografski obzornik 63(2–3):29–39

94. Zakon o varstvu pred naravnimi in drugimi nesrečami [Protection against natural and other disasters act] (2006) Uradni list Republike Slovenije 51/2006. Ljubljana
95. Zakon o kmetijstvu [Agriculture act] (2008) Uradni list Republike Slovenije 45/2008. Ljubljana
96. Kladnik D, Ciglič R, Geršič M, Komac B, Perko D, Zorn M (2016) Diversity of terraced landscapes in Slovenia. Annales: Series historia et sociologia 26(3):469–486. https://doi.org/10.19233/ASHS.2016.38
97. Perko D, Ciglič R, Geršič M, Kladnik D (eds) (2017) Terraced landscapes. Založba ZRC, Ljubljana. https://doi.org/10.3986/9789610500193

Agricultural Land Degradation in Spain

Natalia Rodríguez-Berbel, Rocío Soria, Raúl Ortega, Manuel Esteban Lucas-Borja, and Isabel Miralles

Contents

1 Introduction 264
2 Geographic Characterization 265
2.1 Main Climatic Characteristics 266
2.2 Main Soil Types 268
2.3 Outlines on Soil Degradation 269
3 Soil Erosion 270
4 Overgrazing 272
5 Soil Contamination 273
6 Salinity 276
7 Agrochemical Use 280
8 Microplastics 283
9 Conclusions 285
References 286

Abstract Soil degradation is a global problem and in Spain is especially important because of its climatic, geographical, and socioeconomic particularities. A large surface of the Spanish territory is occupied by agricultural activities, constituting one of the main land's uses due to its importance for the country's economy. Land use linked to agriculture, livestock, and forestry has been present historically, which has generated an impact on the soil that continues today because of the intensification of agricultural activities and inadequate soil management. This chapter brings combined the information on the current state of knowledge of soil degradation in Spain

N. Rodríguez-Berbel, R. Soria, R. Ortega, and I. Miralles (✉)
Department of Agronomy, Centre for Intensive Mediterranean Agrosystems and Agri-food Biotechnology (CIAIMBITAL), University of Almeria, Almería, Spain
e-mail: nrodfer@ual.es; rocio.soria@ual.es; rortega@ual.es; imiralles@ual.es

M. E. Lucas-Borja
Higher Technical School of Agricultural and Forestry Engineering, University of Castilla-La Mancha, Albacete, Spain
e-mail: manuelesteban.lucas@uclm.es

Paulo Pereira, Miriam Muñoz-Rojas, Igor Bogunovic, and Wenwu Zhao (eds.),
Impact of Agriculture on Soil Degradation II: A European Perspective,
Hdb Env Chem (2023) 121: 263–298, DOI 10.1007/698_2022_924,

Published online: 16 December 2022

caused by agricultural activities. First, the state of the climate and the main existing soil types are contextualized. Subsequently, in different sections, a review of the main factors causing soil degradation, such as soil erosion, overgrazing, soil contamination, salinization, agrochemical use, and the presence of microplastics is given. The deficit of available information on the degradation rates of Spanish agricultural soils has been highlighted and the processes of desertification have become evident, predicting a more pronounced acceleration of degradation with fatal consequences for these soils. Likewise, the several changes in soil use have caused large areas to be affected by salinity, owing to the use of poor-quality irrigation water and the overexploitation of aquifers and the excessive application of agrochemicals, and soil contamination, have resulted in a reduction in fertility, and have modified the physical, chemical, and biological soil properties. In view of the above, adequate land management and land-use planning is necessary to improve the state of Spanish soils, as well as the remediation of existing degradation problems and the transformation of agricultural productive activities to sustainability. It is essential to comply with existing policies strictly and rapidly and implement adequate management aimed at mitigating their degradation, as their current and future state could compromise the provision of recourses and services by Spanish agricultural ecosystems, exceeding their resilience capacity and implying a negative impact on food production and human health.

Keywords Land-use changes, Microplastics, Overgrazing, Soil contamination, Soil degradation

1 Introduction

Soil is a very slow-forming, complex, dynamic, and living resource, non-renewable on a human scale [1]. It is subject to increasing pressures due to anthropocentric activities [2]. Its degradation limits their provision of benefits and services [3] and implies the decrease in quality, limiting agricultural production [1]. Furthermore, land degradation is considered one of the triggers of desertification, especially in vulnerable areas [4–6]. In these areas, changes in land use, climatic restrictions, loss of organic matter, soil erosion, forest fires, or intensification of agriculture could be factors that accelerate degradation and desertification processes [7].

Land-use changes together with climate change are considered the main drivers of global change, and both are major contributors to land degradation processes [8]. Recently, the Intergovernmental Panel on Climate Change (IPCC) provided new estimates about the probabilities of exceeding global warming of at least 1.5°C over the next 20 years [9]. The relevant evidence of the impacts of climate change over the last 40 years places Spain in a compromising position. These are confirmed by visible effects such as the expansion of semi-arid climates, the elongation of summers, or the increase in the surface temperature of the Mediterranean Sea of

0.34°C per decade [5]. Therefore, due to its geographical location and socioeconomic characteristics, Spain confronts important risks derived from climate change, because some key economic sectors of its economy are closely linked to the climate, such as agriculture and forestry [5].

Agriculture occupies almost half of Spain's territory. According to the National Statistics Institute (INE), the usable agricultural area in Spain is more than 23 million ha (76% dedicated to rainfed and 24% to irrigated crops), while forestry covers 18 million ha [10, 11]. Throughout history, agriculture has contributed to the creation and conservation of a wide variety of landscapes and habitats. However, inappropriate agronomic practices could also have negative environmental effects [3, 12, 13]. An example of this is the intensification of irrigated crops in areas of Spain with lower rainfall, resulting in land degradation due to overexploitation of aquifers, causing problems of water scarcity and salinization of agricultural soils [14]. Soil is continuously exposed to degradation processes, some of which are directly linked to agriculture, such as erosion, compaction, salinization, and contamination (by heavy metals, agrochemicals, or microplastics) [7, 15, 16]. This fact, together with the forecast of desertification due to climate change [9], demonstrates the threats to the Spanish territory, which could lead to the decrease of soil, water, and air quality, as well as to the fragmentation of habitats and the depletion of wildlife.

Hence, through the contextualization of the country's geography, climatology, and soil types, together with a review of the main drivers of soil degradation in Spain, the aim of this chapter is to unify the current information in order to provide a general vision of the vulnerable situation of the territory.

2 Geographic Characterization

Spain is located in the southeast of the European continent at a latitude between 43° 47′ 25″ (Estaca de Bares, La Coruña) and 35° 59′ 50″ (Isla de Tarifa, Cádiz). It covers an area of 506,030 km^2, of which 493,514 km^2 are on the Iberian Peninsula. The archipelagos of the Balearic Islands (Mediterranean Sea) and the Canary Islands (Atlantic Ocean) cover 4,992 km^2 and 7,492 km^2, respectively. Additionally, 32 km^2 belong to the cities of Ceuta and Melilla in North Africa.

The orography in Spain is diverse, with an average altitude of 650 m above sea level. Large mountains are situated in a west-east direction: the Cantabrian Mountains and the Pyrenees in the north, the Iberian System in the northwest, the Central System and the Toledo Mountains in the middle, and the Sierra Morena and Sierra Nevada in the south. Additionally, it is also made up of two large plateaus: the Castilian-Leonese Plateau in the north and the Castilian-Manchegan Plateau in the south. Finally, it has two depressed areas between the edges of the plateaus known as the North Valley of the Ebro River (NE Spain) and the South Valley of the Guadalquivir River (SW Spain) [17].

The Spanish littoral has 5,755 km^2 surrounded by two large bodies of water, the north and southwest by the Atlantic Ocean and the east and southeast coast by the Mediterranean Sea [18]. The different thermal characteristics provided by the interaction of air and water masses give the Iberian Peninsula varied thermal and dynamic peculiarities [19].

2.1 Main Climatic Characteristics

According to the Spanish National Geographic Institute [20], the average annual temperatures fluctuate between 0°C and 22°C, with the lowest temperatures recorded in the Pyrenees, Cantabrian, and Iberian mountain ranges and the Sierra Nevada, and the highest are observed in the extreme south of the country. Winter can be described as cold in the interior of the northern half of the peninsula, where temperatures do not exceed 6°C, while in the south they are milder (≈12°C). However, north of the peninsula has summers with pleasant temperatures (<18°C), and as the latitude decreases, the summer temperature increases progressively until it reaches an average temperature of over 26°C in Andalusia (Fig. 1).

Annual solar radiation in Spain varies from 1,800 sunshine hours per year in the north of the peninsula through values of 2,000–2,600 sunshine hours per year in the

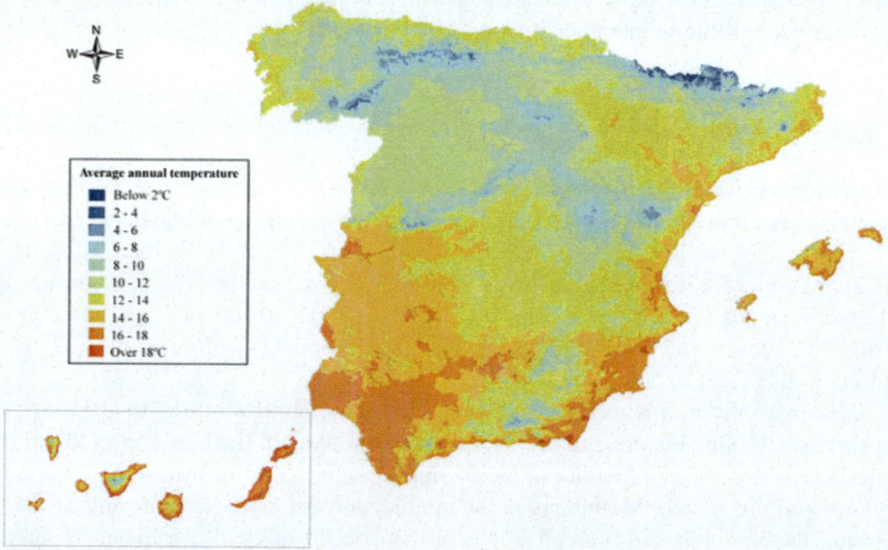

Fig. 1 Average annual temperature obtained from information generated by the Ministry of Agriculture, Fisheries and Food using geostatistical interpolation methods (kriging) from 1803 stations belonging to the Spanish State Meteorological Agency (AEMET). Source: Infrastructure for Spatial Data in Spain (IDEE), Infrastructure for Spatial Information in Europe (INSPIRE), Ministry of Agriculture, Fisheries and Food (MAPAMA)

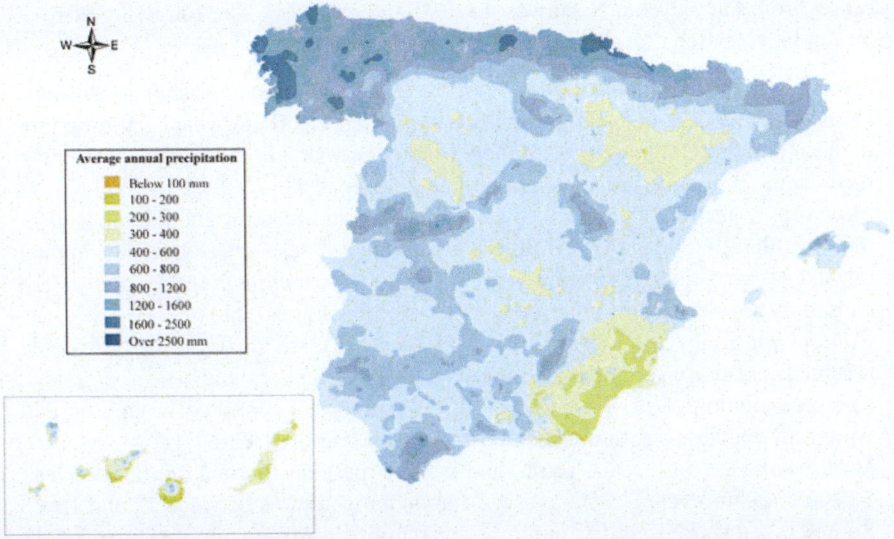

Fig. 2 Average annual rainfall obtained from information generated by the Ministry of Agriculture, Fisheries and Food using geostatistical interpolation methods (kriging) from 4189 stations belonging to the Spanish State Meteorological Agency (AEMET). Source: Infrastructure for Spatial Data in Spain (IDEE), Infrastructure for Spatial Information in Europe (INSPIRE), Ministry of Agriculture, Fisheries and Food (MAPAMA)

centre of the peninsula (southern Galicia, Castilla and León, La Rioja, Navarra, Aragón, Catalonia, and northern areas of the Canary Islands) to more than 2,600 insolation hours per year in the southern half of the peninsula (the Duero and Ebro basins, the Balearic Islands, and the rest of the Canary Islands). The maximum sunshine (>3,000 h per year) is recorded in Cadiz, Huelva, Lanzarote, and Fuerteventura [20].

In general, rainfall is scarce and irregularly distributed (Fig. 2). The annual average is 650 mm per year [21], so there are areas, such as Galicia and north of the peninsula, with values of more than 1,800 and 2,000 mm of rain per year, placing them among the most humid areas in Europe, while 32% of the country receives an average rainfall of 300–500 mm per year. Additionally, there are other areas where values of less than 300 mm year^{-1} are recorded, such as southwest of the Iberian Peninsula (Almería) or some areas of the Canary Islands, with values of less than 200 or 150 mm per year [20].

The heterogeneity of the relief, together with thermodynamic factors, is the driving factor of climate on the peninsula and on the islands, resulting in a great diversity of climates in Spain. Three major climatic zones can be delimited. Predominantly, there is a Mediterranean climate characterized by hot, dry summers and moderate, rainy winters. In the north, we find an oceanic climate, and in the southeast of Spain, we find a semi-arid climate, especially in the Region of Murcia [22]. The aspect of the mountain ranges combined with the Foëhn effect results in semi-arid or arid climates in the southern and eastern slopes of the mountain systems. The aridity increases from the extreme Northwest to the Southeast [21]. The following distinct areas can be considered accordingly [17]:

1. *Humid Atlantic and Cantabrian regions,* from the west (Galicia) to the east (Western Pyrenees). It has high rainfall (more than 1,200 mm year^{-1}) throughout the year and mild temperatures. If there is a dry period, it is short. Most of the year the oceanic influence prevents temperatures below 0°C.
2. *Mountain zones*: the Pyrenees, Sistema Central, and Sistema Ibérico, as well as the Cordillera Bética and Cordillera Penibética. These areas have an annual rainfall of more than 1,000 mm year^{-1}. In summer, there is a mild water deficit caused by regular storms coupled with low temperatures (<11°C).
3. *Continental zones* (plains northern and southern) are distinguished by wide seasonal and daily temperature ranges, with cold winters and warm summers. The precipitation (400–600 mm year^{-1}) is mainly concentrated in autumn and winter, in which temperatures oscillate between 9 and 12°C.
4. *Mediterranean zones* (east, south, and west) experience annual rainfall between 500 and 800 mm year^{-1} with average annual temperatures between 12 and 15°C. Summers are warm and dry, autumns are rainy, and winters are short and mild.
5. *Semi-arid and arid Mediterranean zones* (Central Ebro Valley and south-eastern Spain) – generally torrential rainfall events occur mainly in autumn (150 and 350 mm year^{-1}), with long, dry summers and mild winters, which provide long periods of water stress.
6. *Island areas* (Canary and Balearic Islands) – the Canary Islands are characterized by moderate temperatures (the mean annual temperature does not exceed 20°C) with rainfall mainly concentrated at the end of autumn and winter. Rainfall is scarce and irregular (<300 mm year^{-1}) in the lower areas and more abundant in the midlands (800–1,000 mm year^{-1}) [23]. The mean annual temperature recorded in the Balearic Islands is 14–15°C. On the island of Mallorca, the annual rainfall is around 1,000–1,500 mm, and Menorca is located in an intermediate position between Ibiza and Formentera (500–700 mm year^{-1}) [24].

2.2 *Main Soil Types*

Soil formation and distribution in Spain have been conditioned by the different climatic and geological factors to which the Spanish territory has been exposed, and it is mainly calcareous [22]. In general, soils are shallow in semi-arid areas and humid mountain hillslopes [17]. However, because of the long history of civilizations that have populated the Iberian Peninsula, it is difficult to find soils that have not been affected by human activity (see point 2.3).

According to the World Reference Base for Soil Resources [25], 10 different soil types are mainly found in Spain: Cambisols, Umbrisols, Leptosols, Paleosols, Luvisols, Gypsisols, Calcisols, Regosols, Solonchaks, and Solonetz.

Specifically, the north of Spain presents a humid and mountainous strip from west to east, where the most evolved soils are found, including Podzols, among which

Cambisols and Umbrisols dominate, while Leptosols are found on steep slopes. We find soils with mixed continental and Mediterranean characteristics, such as Paleosols, in the northern region of the Sierra del Sistema Central (Duero River basin) as well as Luvisols, Leptosols, and Cambisols. To the south of the central system, there are typical Mediterranean soils, such as Luvisols, Cambisols, Leptosols, and Gypsisols, which may be red on stable surfaces or white when carbonates are abundant. Vertisols predominate in the Andalusian valley bottoms, as well as saline soils in the areas that have not been drained in the marshes of the Guadalquivir River. Arid soils are found in the provinces of Almería and Murcia (SE Spain) and in the valley of the Ebro River. Additionally, Calcisols, Cambisols, and Regosols, and sometimes Fluvisols and Gypsisol, are found in the most arid areas and saline soils [17].

2.3 Outlines on Soil Degradation

Land degradation can be defined as the long-term decline in ecosystem functionality and productivity [26]. In recent years, this has become an issue that needs to be addressed by all nations [27]. Deterioration of soils leads to a loss of ecosystem productivity, compromising the provision of resources and services and causing a serious impact on biodiversity as well as the functioning of natural systems [28]. According to the Food and Agriculture Organization (FAO) [29], 95% of food is produced directly or indirectly in soils and, as a consequence, one-third of soils worldwide are moderately or highly degraded because of erosion, loss of organic matter, salinization, acidification, compaction, fire, or contamination (heavy metals, agrochemicals, medicines, etc.) [30].

Socioeconomic conditions have proved to be important potential drivers of degradation [28] which, together with the fragility of Spanish ecosystems, have led to excessive transformation and degradation of the original environment [21]. Numerous changes in land use, such as deforestation for the cultivation of cereals and livestock (transhumance) between the sixteenth and nineteenth century, the largest deforestation in the mid-nineteenth century, and government aid for the afforestation in the mid-twentieth century, caused severe deterioration in the soils of the nation [31]. At the beginning of the twentieth century, with the entry into the European Union and the implementation of the Common Agricultural Policy (CAP), the rate of degraded soils increased because of the change in the orientation to a greater market profitability, which promoted irrigated crops [32]. Therefore, problems, such as the intensification of agriculture [31] and livestock farming [33], the increase of urban pressure [32], and industrial pollution [34], combined with frequent forest fires favoured by periods of drought [21], have positioned Spain among the countries with the highest degradation rates (12.5% of the territory), with the 0.2% (63,266 km^2) of the total degraded soils in the world [26]. As a result, Spain is in a compromised situation, given that soil degradation, as a consequence of changes in land use and loss of vegetation cover, combined with desertification stimulated by climate change, is accelerating the loss of land productivity [3].

3 Soil Erosion

Soil erosion refers mainly to the removal of soil particles at the surface or at shallow depths by water (water erosion) or wind (wind erosion). Soil erosion on a geological scale (a natural phenomenon that is involved in the modelling of the landscape) is different compared to the human scale, in which erosion is generally compensated by natural rates of soil formation. However, anthropogenic soil erosion or accelerated erosion mainly lies in the inappropriate use of natural resources by humans, with marked environmental, economic, and social consequences.

Soil erosion is a key factor of land degradation in Spain and is directly and indirectly related to lithology, topography, land use, plant cover, soil management, and climatology. According to one study, human activity over time explains eroded landscapes and sedimentary structures in Spain [31]. According to the most recent data from the National Inventory of Soil Erosion (INES), which is a homogeneous geographic information system on erosion processes, almost 30% of the total surface area of Spain suffers from medium to high erosion processes (soil loss of more than 10 tons per hectare per year). The average annual soil loss in Spain is around 14.2 tons per hectare per year, which varies according to the different autonomous communities. The most affected regions are Catalonia, Andalusia, and Cantabria, with losses of more than 21 tons per hectare per year. With less than 5 tons per hectare per year, Castilla and León is the autonomous community with the least soil loss rate. Aragon, Madrid, Extremadura, and the Canary Islands also show moderate average losses (less than 10 t ha^{-1} $year^{-1}$). Regarding the classification of soil according to its level of erosion, there is a predominance of areas subject to moderate erosion processes in all the autonomous communities. In Andalusia and Catalonia, up to one-fifth of the soil is subject to high erosion processes (greater than 25 t ha^{-1} $year^{-1}$). Table 1 shows the percentage of surface area affected by erosion in the different autonomous communities.

In relation to factors affecting soil erosion in Spain, there are numerous studies about this topic [31, 35–37]. According to them, soil erosion problems in Spain vary depending on soil types, climatic conditions, soil cover, soil management, or topography. In general, erosion is more frequent in steeper slopes, and where rain erosivity is higher and the climate has large seasonal differences. Moreover, the degradation of vegetation and soil cover, associated with various perturbations (e.g. agriculture, forest fires, and flooding), is one of the main causes of soil loss in Spain [38]. The most affected Spanish regions tend to coincide with areas with a Mediterranean climate, mainly in the mid-mountains, and with strong human pressure [39].

Forecasts of climate change in Spain indicate scenarios that are more favourable to desertification processes (increasing aridity and temperatures). In fact, the report "Impacts of Climate Change in the Processes of Desertification in Spain" estimates that the risk of desertification will increase by about 22% in Spain as a consequence of changes in aridity [40]. This fact would change the percentages of arid, semi-arid, and dry sub-humid areas (Fig. 3). We can conclude that soil erosion is a serious problem that threatens the soil and, by extension, the human activities supported by it [41].

Table 1 Annual soil loss by sheet and runoff erosion calculated by the INES with the Revised Universal Soil Loss Equation (RUSLE) model, expressed in Tn ha^{-1}, in reference to the total geographical area of autonomous community

Region	% of land with erosive processes		
	Moderate intensity	Medium intensity	High intensity
Andalucía	57.61	19.76	22.63
Aragón	81.51	12.83	5.66
Asturias	61.92	21.67	16.42
I. Baleares	76.62	13.69	9.7
C. Valenciana	70.13	16.04	13.83
Canarias	69.25	21.86	8.89
Cantabria	59.91	22.39	17.7
Castilla y León	89.13	7.77	3.1
Cataluña	54.41	24.86	20.74
Extremadura	83.75	9.81	6.44
Galicia	74.34	13.06	12.61
La Rioja	65.84	20.43	13.72
Madrid	81.28	10.89	7.83
Murcia	66.41	18.13	15.46
Navarra	65.64	18.79	15.57

Source: Ministry of Agriculture, Fisheries and Food. Data for Castilla La Mancha and País Vasco not shown

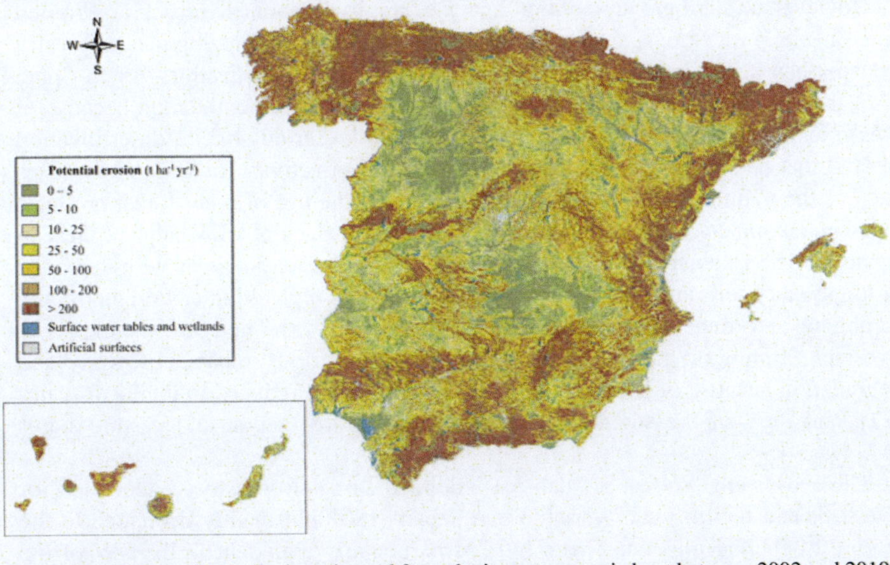

Fig. 3 Potential erosion in Spain estimated from the inventory carried out between 2002 and 2019 using the RUSLE (Revised Universal Soil Loss Equation) model (rainfall erosion index, soil erodibility, and topography). Source: Infrastructure for Spatial Data in Spain (IDEE), Infrastructure for Spatial Information in Europe (INSPIRE), Ministry of Agriculture, Fisheries and Food (MAPAMA)

4 Overgrazing

Overgrazing is still a problem for soils in different European Mediterranean regions [42]. *In future climate change scenarios, overgrazing with increased extreme weather events may reduce primary production and ecosystem organic content, accelerate soil erosion and alter soil chemical properties* [43]. Continuous trampling also causes serious soil compaction problems, thus increasing bulk density and decreasing infiltration capacity [15]. Pastoralism has been a historical activity in Spain [44] causing problems such as reduction of plant density and soil erosion because of heavy grazing [21]. Livestock currently accounts for approximately 40% of the total production of the country's agrosilvopastoral sector [45], while the remaining 60% is mainly in agriculture and, to a lesser extent, forestry. Combination of intensive livestock farming with the use of permanent pastures, as well as other extensive land-based activities linked to the use of natural pastures, accounts for the 19% of the total area of the territory, according to the National Geographic Institute [45]. Livestock stocking rates have been stabilized because of reforms due to the CAP in recent years, with the intention of minimizing the impact on threatened Spanish ecosystems, particularly in the arid and semi-arid regions of south-eastern Spain, which are at risk of desertification and are highly sensitive to climate change [21]. Extensive and semi-extensive livestock farming is widespread in grassland and pasture areas throughout Spain, but it is mainly located in the western part of the country, Galicia, and the Cantabrian coast, and is also present with varying intensities in mountain areas and fallow lands [45].

However, in national agricultural analyses, registers about livestock are limited and especially for extensive management [43]. Although overgrazing in agroforestry systems has been cited as a problem associated with desertification risk in Spain [21], there is not an inventory of soils affected or at risk of degradation because of overgrazing. Theoretical studies have highlighted the serious risk of desertification of land that has been intensively exploited by human activity for various purposes such as the maintenance of livestock, hunting, and the use of other forest products (firewood, cork, etc.), also known as "dehesas" and the lack of knowledge of these in Spain [4, 46]. Extensive livestock farming is mainly located in areas with degraded or low-quality soils and areas of complex orography or high altitude, with moorland, scrubland, and "dehesas" areas being the most affected [45]. Therefore, the impact of livestock farming on these fragile soils could have a serious impact. However, land degradation because of overgrazing on the soil has been reduced to the local scales [21], and most of the studies carried out focus on vegetation and pasture status [44, 47–49].

The ecosystems known as "dehesas", defined as a agroforestry system used for livestock and hunting with variable tree density [50] and mainly dedicated to the rearing of Iberian pigs and sheep [15], is particularly abundant in the SW of the country and in the Andalusian Sierras Béticas. Physical and biological degradation has been observed in these soils because of overgrazing, and the type of farm management increases the risk of erosion [15]. Trampling and frequent rainfall events favour ruts and gullies formation with considerable soil losses [51]. One study reported the compaction of dehesa soils in a depth of 5–10 cm in soils of 10 farms distributed in the Extremadura region (SW, Spain) under intensive grazing [15]. The authors found a positive and significant relationship between animal stocking (mainly pigs and sheep and, to a lesser extent, goats and cows) and the area of bare soil exposed to increased degradation, as it affected pasture composition and productivity due to increased bulk density, and decreased infiltration capacity and water retention. The highest soil organic carbon content in dehesas is found in the first 5 cm of the topsoil [52]. In addition, a thinning of horizons due to the impact of livestock pressure and a lower organic matter content with depth have both been shown in the bare soil between trees [53]. Other authors pointed out that topography could be a limiting factor in the impacts of extensive grazing, which could improve the content of nutrients such as soil nitrogen and potassium content related to organic matter inputs through livestock excrement and induce changes in the composition of plant communities in low-lying areas [54].

On the other hand, in mountain ecosystems livestock activity is being abandoned due to economic losses and European Union subsidies for agriculture [55]. As a result, a decrease in soil compaction in the first 10 cm of the soil and changes in plant communities in the mountains of northern Spain (Cantabria region) have been reported in just 5 years after the number of animals has been reduced, as well as the pressure on grazing areas used for extensive livestock farming [56]. In another area, on the Pyrenees, problems due to grazing and transhumance abandonment have also been reported. Historically, high mountain areas were deforested to generate larger areas of pasture for animals during the summer, changing the hydrological dynamics and favouring erosive processes which are currently declining due to a decrease in livestock farming activity [57]. This abandonment has favoured the development of scrublands [58, 59] but also the increase of wildfire risk [57]. In recent years, controlled grazing to prevent fires has gained significance as a management tool in Mediterranean forests [60–62], and in some cases, grazing activities are combined with low-intensity controlled burns [63], but there is still a lack of information on the impact on soils.

In any case, minimizing the impact of soil degradation due to overgrazing requires addressing the proper management of extensive livestock farming, especially in arid and semi-arid areas [21].

5 Soil Contamination

Both local and diffused soil pollution is a serious problem resulting from anthropogenic activities on a global scale [29], compromising food production and security, soil and water quality, and in general the environment [64].

Soil contamination is regulated by Spanish Government Legislation (Law 22/2011. Waste and Contaminated Soils [65]). It defines contaminated soil as those whose characteristics have been negatively altered by the presence of hazardous chemical components derived from anthropogenic activities and in concentrations that pose a risk to the environment or human health, and have been declared as contaminated by regional authorities [64]. An inventory of economic activities considered potentially soil contaminators, as well as the standards and criteria for declaration of contaminated soil, was set out by the Royal Decree 9/2005 [66]. In addition to other activities, in which the obligation of decontamination by the causers or owners of the affected soils is indicated, these edicts establish a voluntary regime of decontamination that must be communicated to the autonomous communities by means of a formal declaration. Regarding the latest information about the status of European soil contamination [67], in Spain, only 50% of the 19 autonomous communities provided updated information, and the rest was estimated by the Ministry of Agriculture and Fisheries, Food and Environment. Regional authorities are responsible for keeping a record of voluntary remediation and for the management of contaminated sites throughout the country. Even though the law requires the causers to take responsibility for decontaminating the affected soils, according to the last European report, 100% of the investment for the remediation of contaminated areas came from public funds [67]. At the end of 2018, Spain reported 133,344 sites that were estimated as engaging in potential soil-contaminating activities, with 4,924 estimated to require investigation and of which 270 were under investigation [68].

Despite the risks of soil degradation because of pollution in the country, it is difficult to find precise data on the area or concentration of soil pollutants. An estimate of the possible risks of soil contamination that could be associated with potentially polluting activities is available from the State Register of Polluting Emissions and Effluents of the Ministry for Ecological Transition and the Demographic Challenge (MITECO) (https://prtr-es.es). The main sources of soil contamination evaluated in Spain come from waste or accidents related to industrial activity according to the Spanish Geographical Institute [69]. Among the latest notable accidents is the Aznalcollar mine spill (Seville, SW Spain), which affected agricultural and natural soils with acid waste and sludge [70–72]. The mining is the main activity which generates contaminated soil across the peninsula [17], and Andalucía is the regional area with the highest production value [73]. There are other localized sources, such as thermal, nuclear power plants, and military sites of special importance. Practically, the whole territory engages in potential soil-contaminating activities, although the mountainous areas, such as the Pyrenees, the Iberian and Central System, Castilla and León, Aragón, and the Balearic Islands, have a smaller number of contaminating sources [45]. On the contrary, the coastal areas of the east of the country, the industrial areas of the Basque Region and Asturias, the Galician estuaries, and Madrid are the most affected [45]. Besides, soil contamination derived from the use of fertilizers in agricultural activities, irrigation with wastewater, application of sewage sludge, and pesticides and fertilizers play an important role in soil degradation in Spain. The additional importance lies in the large area devoted to agricultural production due to the fact that 40% of Spanish territory is suitable for crop production and because it is a historical activity linked to land use [74]. Like that, Spain is one of the largest agricultural countries in the European Union and one of the main consumers of fertilizers and pesticides [75]. The abusive use of fertilizers mainly causes nitrate contamination of groundwater as they are leached after application to the soil. In the Survey of Commercialization of Phytosanitary Products, conducted in 2019 by the Ministry of Agriculture, Fisheries and Food of the Government of Spain, a total quantity of 75,397 Tn of pesticides were commercialized [76], positioning Spain as one of the main consumers of pesticides in the European Union [75]. Fungicides and bactericides were the most used, accounting for 45.2% of total sales, followed by herbicides (22.6%), molluscicides and growth regulators (22.8%), and, to a lesser extent, insecticides (10.4%).

The concentration of heavy metals in natural soil depends on the geochemical composition of the geological parent material; however, agricultural management can modify this concentration [77]. High heavy metals content in soils could reduce crop productivity and the seed germination rate by up to 30% [78]. Furthermore, these pollutants are capable of entering in the trophic chain, being a serious health risk associated with deaths from digestive tumours in Spain [79, 80]. High contents of heavy metals in rice crops irrigated with wastewater have been detected in Valencia (E, Spain) [81]. In the NE of the country there are important agricultural areas, such as the Ebro River basin, with 4.2 million ha of intensively cultivated land and important industrial activity. The use of pesticides, fungicides, and fertilizers increases the concentrations of Zn, Cu, and Pb from agricultural activity, as well as the deposition of Hg generated by industrial emissions [82]. In the SE, the Mediterranean region of Almería, dominated by greenhouse agriculture, heavy metal concentrations were found to be 88% higher than the natural baseline for geochemical concentrations [83]. In 2018, the same greenhouse soils were evaluated 20 years later and the concentration of Cu, Ni, Pb, Cd, Cu, and Zn had increased, indicating that there is contamination associated with agricultural practices [84]. Results with a similar trend were found for soil concentrations of Hg and Cr [79]. A high Hg soil content has also been associated with coal-fired power generation, and its average content in Spanish soils is 67.22 $\mu g\ kg^{-1}$ [74].

The use of sludge, animal manure, and contaminated water for irrigation in agriculture has also been referenced in the Spanish scientific literature as a source of soil contamination by pharmaceutical products. Different geographical locations and types of crops, such as rice fields and citrus and vegetable fields in Valencia, barley and wheat crops in Segovia (Castilla and León), and citrus, vegetables, and cereals in the Region of Murcia, are locations where traces of paracetamol or carbamazepine, among others, have been found [85]. The presence of antibiotics in soils has also been found in Spanish agricultural soils in different geographical areas, such as Galicia (NW) [86] and Catalonia (NE) [87], with the use of irrigation water or organic amendments from animal waste as the main drivers. The application of sewage sludge has also been responsible for the contamination of agricultural soils with polybrominated diphenyl ethers (PBDEs) [88]. Additionally, pesticides and organic contaminant compounds associated with the management practices in intensively cultivated soils with horticultural and olive crops, such as dichloro diphenyl trichloroethane (DDT) and endosulfan, polycyclic aromatic hydrocarbons, and other organochlorine compounds, have been found in Spanish soils [89–

92]. The use of other types of amendments, such as the phosphogypsum (PG) applied in agriculture, is another cause for concern in western Andalucía [93, 94]. PG is obtained as a by-product in the manufacture of phosphate fertilizers located in Huelva province [95]. This residue contains mainly gypsum, but also heavy metals and radionuclides that are highly contaminating. Its use was limited in 2001 due to public health concerns; however, its use as a soil amendment was allowed by the Spanish Government Legislation (Royal Decree 824/2005 of July 2005 [96]) obviating its content of radioactive isotopes [95]. Currently, PG wastes management must comply with Spanish Government Law 22/2011 on waste and contaminated soil [65], Decree 18/2015, of January 27 guidelines, which approves the regulations governing the regime applicable to contaminated soils, of Andalusian Government [97], and Spanish Government Law 25/1964, of April 29 which establishes the criteria for radioactive waste management [98].

Finally, another major concern corresponds to an important and general problem associated with agriculture management – contamination by nitrates, especially on groundwaters. Vulnerable zones cover an area of 110,482.22 km^2 in Spain, which corresponds to 21.8% of the national territory and continues to increase [99].

In any case, soil contamination problems in Spain are diverse, varied, and widespread, posing a serious threat to human health and the goods and services of agricultural ecosystems. The availability of a national inventory of contaminated soils and the ease with which it can be consulted could facilitate the detection of problems and prevent – as far as possible – their impact.

6 Salinity

Soil salinization is one of the problems that contribute most to the loss of soil quality and degradation. The accumulation of salts generated by natural processes and as a consequence of inadequate irrigation management in agriculture is a worldwide issue that particularly affects areas susceptible to desertification, such as arid and semi-arid regions [100]. Mostly, salinization is caused by the deposit of soluble sodium, calcium, and magnesium in the soil [101], which can be the result of primary processes (from natural sources) and secondary processes (caused by human activity) that alter the physicochemical soil properties and lead to their degradation [102]. Most often, soil salinization is associated with irrigation, particularly in areas with low rainfall and high evapotranspiration as is the case of arid and semi-arid ecosystems [103]. Additionally, soil salinization is influenced by soil texture, which can limit the washing of salts. The aforementioned factors, together with the use of poor quality water or water rich in salts, increase their accumulation in the surface layers of soils, thus reducing their fertility [101]. Soil salinization is favoured in coastal areas because of the wind and rain deposit oceanic salt in the environment [104], a fact that, together with the excessive use of groundwater, causes saltwater intrusion into aquifers. Spain, together with Portugal, Greece, and

Italy, is one of the most affected countries by salinization [105], causing the remediation of saline soils large economic costs [106].

Over 20% of the Spanish territory is desert or severely degraded [105], with about 840,000 ha affected by salinization processes [13]. Soil deterioration by salt accumulation is one of the greatest difficulties faced in agriculture because it limits its potential and is closely related to poor management and the overexploitation of water resources, especially in arid climates [16, 107]. It is estimated that about 30% of agricultural soils applying irrigation systems are rapidly degraded by salinization [108]. This phenomenon occurs frequently when water evaporates on the surface depositing salts from parent material and subsoil layers. The Spanish Ministry of Agriculture, Fishing and Food [109] carried out a classification of these areas, among which the following were highlighted in the northwest of the peninsula: the Ebro Valley (Bárdenas reales, Monegros, Llanos de Urgell, and Ebro delta, among others); the Mediterranean coast [Segura River lowland (Alicante and Murcia) or the Guadalentín Valley and Mazarrón, among others]; and Andalusia in the south of the peninsula (middle and lower course of the Guadalquivir River, Campo de Níjar, Campo de Dalias, Barbate-La Janda, Guadalhorce, and the lower course of the Odiel and Tinto Rivers, among others). Saline intrusion has also affected other areas, such as the Campo de Cartagena in Murcia, Castellón, the Alt Empordá in Gerona, or the deltas of the Besós, Llobregat, and Marenases rivers in Catalonia. Likewise, this problem also seriously affects island areas, such as the southern area of Mallorca (Balearic Islands) or the Canary Islands, where Fuerteventura stands out with 54% of its territory affected, as well as Lanzarote (30%), Gran Canaria (12%), La Gomera (10%), and Tenerife (9%; www.agrosal.ivia.es). For this reason, water planning criteria in the Canary Islands include restrictions on the intensive use of groundwater to mitigate aquifer depletion [110].

Soil deterioration by salinization is one of the greatest difficulties faced in agriculture because it limits its potential and is closely related to poor management and the overexploitation of water resources, especially in arid climates [16]. It is estimated that about 30% of agricultural soils applying irrigation systems are rapidly degraded by salinization [108]. Saline soils, from an agricultural point of view, are those with an electrical conductivity of saturation soil extract of more than 4 dS m^{-1} at 25°C [111]; that is, they contain sufficient soluble salts to adversely affect the growth of most crop plants [112], thus causing yield decline [113]. An example of this was the drastic reduction in yield and the crop growth of adult mandarin trees after periods of more than 2 years of water and salt stress [114]. However, there are sensitive crops that are even affected at lower values [115]. The reduction of agricultural profitability leads to excessive fertilization, which in turn can lead to increased soil salinity, heavy metal accumulation, water eutrophication, and nitrate accumulation, thus becoming a threat to soil conservation [34]. Salt accumulation and translocation is mainly determined by the irrigation method and water quality and quantity. Poor agricultural management has led to an overexploitation of aquifers. An example of that is observed in the Guadalentín Valley (Murcia), where groundwater has been contaminated by the abuse of irrigated crops inducing soil salinization [102]. Similarly, another study conducted on Haplic Calcisol soils in

Campo de Cartagena (Murcia, Spain) revealed that fluctuations in soil salinity over 4 years depended on plot management and, particularly, on the irrigation water used [116]. This problem occurs in irrigated crops and in rainfed crops, such as olive groves. Additionally, it was observed that soil salinity was conditioned by the soil washing capacity, increased linearly with irrigation water conductivity, and affected the deeper layers [117]. The increase of electrical conductivity in agricultural soils derived from the addition of salt-rich water generates a negative effect on biological fertility and soil biochemistry [118].

At the beginning of the twentieth century, the Spanish government promoted the transformation of rainfed agriculture into irrigated agriculture, mainly in the arid extensions of the central Ebro Valley [119], without knowing the serious consequences, such as the economic losses caused after the irrigation of Bardenas semi-arid area in Zaragoza (NE Spain) [120]. This agricultural management transition, in addition to generating economic losses, also affected the agricultural and environmental functionality of the area. In the central Ebro Basin, geological materials rich in salts were favoured in their weathering by irrigation, which promoted soil salinization [121]. Similarly, the relief and low-quality water caused salinization problems in the upper basin of the Guadiana River in central Spain [122]. Currently, in Spain, rice cultivation is an important economic activity, forming part of the agroecosystems of natural parks such as the Ebro Delta (NE Spain), the Albufera de Valencia (E Spain), or the Doñana National Park, in the Guadalquivir Valley (S Spain) [123]. It has been observed that these Mediterranean wetlands are beneficial to the environment because crops can fix carbon on the surface [115] and act as a barrier to prevent the rise of phreatic water with high salt content [123].

In addition to the aforementioned studies, there are others about soil salinity in the east coast of the peninsula. For example, in the Vega Baja region of the Segura River (Alicante), spatial and temporal changes in salinity of Fluvisol soils were examined between 2002 and 2006, concluding that the area near the "El Hondo" lagoon had saline and sodic soils [108]. In a similar study in this same area, it was observed that irrigation with high salt content water drastically worsened the problem in Fluvisol soils, with coastal areas being more affected by salinization. This was caused by the overexploitation of groundwater by the demands of increasing urbanization, industry, and agriculture, leading to a reduction in the water table and a triggering of the intrusion of saltwater from the sea [124]. The soils of Murcia are heavily affected by salinization, because of the large area devoted to intensive agriculture. Soils in a highly productive agricultural area of this region were evaluated, confirming that the main origin of high electrical conductivity was mainly because of poor irrigation water quality as well as primary salinization [125]. Additionally, the semi-arid conditions of the area caused the highest concentrations in the month of July, concluding that the salt content depended on evaporation and capillary rise. However, the extensive use of saline or sodic water in Spain for crop irrigation [126] has caused the country to become an example of the application of desalinated water in irrigation – with 40% of the total existing desalination plants worldwide – placing Spain in first place for the use of desalinated water for agricultural purposes [110].

Soil degradation by salinization could be ameliorated by carrying out specific soil studies to improve soil management and the implementation of efficient irrigation systems. However, Spain has no legislation at national or autonomous levels that demand this type of studies to the potential contaminant activities [113]. One possibility for the recovery of saline degraded soils could be the implementation of salinity-resistant crops, although some authors suggest that in order to obtain successful results it is necessary to apply additional technologies including drainage systems that allow the elimination of salts, as observed in the Monegros-Flumen in Aragon (NE Spain) [127]. In semi-arid areas, where part of the irrigation is lost because of high evapotranspiration rates, irrigation optimization can be a solution to avoid salt accumulation in sandy loam soils [128]. An example of that is observed in the soils of the Flumen irrigation district (27,500 ha; NE Spain), where improved irrigation systems and agricultural management over 24 years lowered soil salinity [107]. Another example of how good agricultural practices can lead to success in non-salinization of irrigated land after 70 years is found in the Comarca de Violada, located between the provinces of Huesca and Zaragoza (NE Spain) [119]. Another technique to avoid or reverse this process could be to increase the irrigation dose with quality water to wash away surface salts that may affect vegetation [129]. Other authors propose that – in specific cases – a possible solution to this problem could be the optimization of irrigation management in combination with the use of amendments. In a trial conducted on a grapevine crop in which compost mulch was applied, a decrease in required irrigation and soil salinity was observed [130]. The reuse of residue for soil remediation could have great benefits, both for the circular economy and for significantly increasing crop yields [16].

In general, excessive salts induce the death or decrease the productivity of plants and soil organisms [131], and can lead to the displacement of autochthonous vegetation, thus favouring the introduction of the allochthonous species because it generates a toxic environment that slows the absorption of nutrients or decrease the availability of water for native plants [112]. In Murcia, increased salinity in soils has benefited the introduction of plant species such as *Nicotiana glauca*, which modifies the functionality of the soil microbial community [132]. Moreover, the effect of salinity on germination and seedling establishment has also been observed. It was concluded that salts significantly reduce vegetation germination and survival [133], but this phenomenon also affects halophytic plants, such as *Limonium tabernense*, endemic of the Tabernas Desert (Almería, SE Spain), and their germination is affected by salinity, temperature, and interaction between them [134]. Therefore, some authors propose using vegetation as a bioindicator of soil conditions to improve the environmental management of saline ecosystems [135].

It should be noted that the state of the soils for a large part of the Spanish territory is unknown, because there are few studies that have been carried out in localized areas. Thus, new research focused on studying the state of the soils is needed. Similarly, studies should explore the influence of salinization on the different ecosystems of Spain and the organisms that inhabit them. Ignorance in this area could lead to the degradation of unaffected soils or to the worsening of damaged soils, rendering them irrecoverable.

7 Agrochemical Use

Since ancient times, agriculture has been a key part of sustaining human life. In recent decades, new practices have been implemented that include the application of agrochemicals, such as fertilizers and pesticides to improve crop production. Nowadays, the addition of these substances has become a fundamental part of agricultural systems to meet the huge food demand [75]. However, the large quantities of agrochemicals used in modern agriculture to increase productivity and reduce crop losses have led to the acceleration of soil contamination in the last few decades [34] and, consequently, the deterioration of soil and water quality because of supplementation with nitrogen, phosphorus, and persistent pesticides [136].

Spain has a total of 50 million ha dedicated to agriculture [137]. In 2019, 75,400 tons of plant protection products were marketed (Fig. 4), 2.8% more than in 2018, including fungicides, bactericides, herbicides, insecticides, molluscicides, and growth regulators [76] – leading the pesticide sales in Europe. Despite the different state and European regulations for the sustainable use of pesticides (Directive 2009/128/EC, Regulation 2009/1107, or Regulation 2009/1185/EC, and others [138–140]) their excessive application has generated soils in which these products have prevailed [75, 90, 141]. Several authors have studied this fact and observed that the presence of rest of pesticides in soils affects agricultural production [142] and the physicochemical soil properties [143].

Pesticides are substances intended to prevent, destroy, repel, or mitigate any pest that causes damage or interferes with agricultural production [144]. Despite being considered environmental pollutants because of their potential toxicity and widespread use [145], their cost-effectiveness is positive in controlling pests to maintain the necessary production yields and economic viability [146]. These can be organic or inorganic synthetic molecules, which are classified based on their chemical structure, their mode of action, their form of entry into the organism, and their target

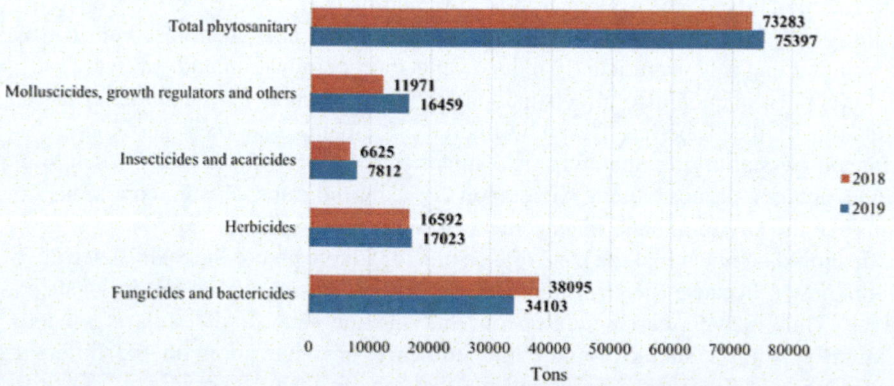

Fig. 4 Annual comparison between 2018 and 2019 of the marketing of plant protection products in Spain expressed in tons. Source: Ministry of Agriculture, Fisheries and Food [74]

organisms, and they can also be short-lived and not persist in the environment or be classified as persistent organic pollutants as is the case with organochlorine insecticides [145]. The toxicological effects of these substances on pests depend on their chemical composition, in turn affecting their interaction with soil components [147]. Because of the wide range of products existing, their persistence, behaviour, and mobility are enormously varied, so the mechanisms involved in their degradation and retention in soils are broad [146]. The overuse of fertilizers and manure for soil fertilization and the inefficient use of nutrients, such as nitrogen and phosphorus, are the primary constituents of the environmental problems caused by agriculture [148]. Furthermore, these fertilizers are also considered as sources of heavy metals (Hg, Cd, As, Pb, Cu, Ni, and Cu) [34], because excess of fertilization could affect other organisms in the ecosystem, in the case of agrochemicals residues that remain in the field affect rotational crops [149].

The behaviour of pesticides in soils is governed by complex physical, chemical, and biological dynamic processes, including sorption-desorption, volatilization, chemical, and biological degradation, plant uptake, runoff, and leaching, which are directly involved in the transport of pesticides within the soil and translocation from soil to water, air, or plants, driving these compounds into the food chain [150]. It has been estimated that less than 0.1% actually reaches the target pest [146], so proper management of these products is of vital importance to avoid the contamination of soils [151]. One of the major environmental problems with agrochemical use is the sorption and leaching of pesticides from soil to groundwater. These processes are determined mainly by the clay and organic matter content [152]. Examples include the widely applied herbicides for weed control, such as 6-chloro-N_2-ethyl-N4-isopropyl-1,3,5-triazine-2,4-diamine (atrazine) or 4-chloro-o-tolyloxyacetic acid (MCPA).

In southern Spain, 40,000 ha are dedicated to intensive agriculture under plastic greenhouses, where more than 2×10^9 kg of vegetables are produced [153]. The region of Almeria (SE, Spain) has 31,614 ha, the highest concentration of plastic greenhouses in the world. Given the poor quality of the soil in this area, it is common to add organic amendments rich in carbon, which, combined with the limestone nature of the substrate, causes the herbicides to be adsorbed and leached, contaminating the soil and, therefore, the water. In a study in Almería in which the synergy between the application of soil amendment and the excessive use of herbicides was studied, it was observed that there was a higher mobility of MCPA than atrazine, although at high concentrations of organic matter, atrazine can percolate to deeper layers with dissolved organic matter [154].

In Extremadura (W Spain), olive grove cultivation covers 265,000 ha, and the province of Badajoz accounts for 22% of this area [155]. It was observed there that the addition to the soil in two proportions (5% and 10% w/w) of solid olive-mill residue, named "*Alperujo*", derived on the first step of olive oil extraction, had a great influence on the fate of the herbicide simazine in Chromic Luvisol soils [156], greatly reducing its vertical movement and attenuating its leaching [157].

In Andalusia (Southern Spain), olive groves have become the most extensive crop in recent decades as a consequence of the high demand for olive oil, leading to an increase in the application of agrochemicals to make production more profitable [158]. These crops are normally located on sloping areas which, combined with the Mediterranean climate, favours soil degradation and the mobilization of the agrochemicals used. One study concluded that the accumulation of agrochemicals can lead to problems, such as the transportation of sediments containing simazine to herbaceous crops located at lower altitudes [159]. Similarly, the use of highly soluble fungicides, such as metalaxyl, or more hydrophobic fungicides, such as penconazole, on cucurbits, pome fruit trees, or grapevines can cause a large environmental impact because their behaviour in the soil and subsoil is influenced by the organic matter, texture, and calcium carbonate content of the soil [160]. La Rioja region is known worldwide for its vineyards, and the demand for wines from this area has led to an abuse of agrochemicals. In this region, soils were studied and up to 17 pesticides and high concentrations of the fungicides metalaxyl and triadimenol, the herbicides fluometuron and terbuthylazine, and the insecticide methoxyfenozide were found [145].

The excessive and insensitive use of these substances has become a danger to the environment and the ecosystems. Some of them remain for a long time in the soil of the fields, increasing in quantity each season [161]. In Spain, the use of artificial pesticides in intensive horticultural activities, such as organochlorines (suppress symbiotic nitrogen fixation) and organophosphates characterized by their high stability, persistence, and affinity for particulate materials [162], has generated contaminated soils as well as limitations in agricultural production [163].

The presence of pesticide residues in the environment makes monitoring programmes necessary to track the level of residue in soils, surface water, groundwater, and drinking water because agrochemicals are transported from the soil to nearby water sources as a result of rainfall or runoff. Thus, the agrochemical content of soils is reflected in river sediment. An example of this phenomenon is observed in the Ebro Delta (NE Spain), an important 320 km^2 international conservation wetland ecosystem. This area is dedicated to rice cultivation and receives large quantities of pesticides, mainly organochlorine herbicides, as well as industrial pollutants that are transported to the delta by the river. The excessive use of polychlorinated biphenyls (PCBs), PCB congeners, insecticides, and organophosphorus pesticides in crops, together with the dragging of these products by the river, has contaminated both soil and water, with strong consequences for the biodiversity of the delta [164, 165]. Similarly, the concentration of agrochemicals washed down from riverbeds can be observed in different areas in Spain, for example, the accumulation of organochlorine pesticides in the Albufera de Valencia (E Spain) [162], or also in coastal marine sediments such as traces of Aldrin along the coast of Alicante and PCBs and DDT in the Bay of Cadiz (SW Spain) [166, 167].

Thanks to Spanish legislation, the National Action Plan for the Sustainable Use of Plant Protection Products 2018–2020, and farmers' awareness of agrochemical use, the concentration of these products in soils is decreasing. Proof of this was observed in the monitoring conducted over 3 years in agricultural soils in Almeria, where the presence of pollutants such as DDT and endosulfan decreased during the study [90].

The controversy generated by the effects of commercial agrochemicals on human health and the ecosystem has prompted farmers to consider alternative options, such as the use of bio-based fertilizers and pesticides, as well as soil disinfection by biosolarization as an alternative to chemical spraying [168]. Therefore, further studies on the state of soils and the environmental effects of these pollutants are necessary for a more sustainable use of agrochemicals in agriculture.

8 Microplastics

From the middle of the twentieth century, microplastic (MP) residues are accumulating in soils, probably derived from petroleum and textile fibre products, among others [169]. MPs are defined as plastic polymer particles with diameters ranging from 5 mm to a few μm [170]. Although they have been poorly quantified in the terrestrial environment [171], this appears to be one of the most important long-term sinks for this waste because of their management on land [172]. Many soil properties are affected by MP contamination [173]. Although there are still few studies on the effects of MP on microbial activity, some studies have shown evidence about negative effects in the presence of polyacrylics, polystyrene, and polyester [174, 175]. However, it is known that soil pore space setup and its hydrological properties can affect the metabolic functions of microorganisms [176], and MPs reduce bulk density, increase water retention change, and can diminish the stability in water of soil aggregates, modifying soil structure [174]. In addition, MPs can accumulate in the food chain due to their small size if ingested by soil animals [177], such as earthworms, nematodes, isopods, and collembolan. Studies evidence lower life expectancy, changes in their intestinal microbiota, and deterioration of their immune and reproductive systems [178].

Agroecosystems are an important input of MPs to terrestrial compartments [179]. The frequent use and manipulation of plastics in modern agriculture facilitates the access to soils of plastics widely used in modern agriculture [172, 180]. In agricultural lands, input of plastic contaminants into orchards, horticulture, forest, and pasture soils is also caused by the application of irrigation with wastewater, water from contaminated rivers or water reservoirs [169], sewage sludge, and other organic amendments from organic waste composted [171, 181–184], among others.

Spain is ranked as the fourth country in Europe in demand for plastics according to data presented in Plastics Europe (2019), representing 7.6% (3.9 Mt) of total European consumption (51.2 Mt). Sixty-seven percent of the plastic waste removed in Spain is high-density polyethylene (HDPE), low-density polyethylene (LDPE), polypropylene (PP), and linear low-density polyethylene (LLDPE) polymers; 13.3% is polystyrene (PS); 10.3% is polyvinyl chloride (PVC); and 5.3% is polyethylene terephthalate (PET) [185]. Currently, there is no mapping of MP pollution in Spain. However, there are recent local studies focused on their presence in coastal sediment in areas bathed by the Mediterranean Sea and the Atlantic Ocean [186–188]; in river basins, such as the Llobregat in Catalonia [189]; the Ebro River [190]; or the Miño

River [191] as well as in wastewater effluents and sewage treatment plants [192, 193]. Because the water used to crop irrigation in Spain is usually taken from rivers, water reservoirs, wastewater, or desalinated water, the latter mainly in dry areas of the Mediterranean coast, it causes MP contamination in soils – an unavoidable risk for the time being.

Similarly, plastics play an important role in Spanish agricultural production, because they have economic and productive benefits [194] in horticultural and fruit crops in the territory, especially in water-scarce areas [195]. Among their main uses are the maintenance of specific microclimatic conditions, such as greenhouse crops, also known as "crops under plastic", which cover an area of 50,365 ha in Spain and are located mainly on the southern coast of the country [196]. Another common use is plastic mulch, which controls the growth of weeds while improving soil moisture conditions and improving physical and chemical properties, such as aggregate stability or nutrient retention [197], which are essential for crops in some areas of S and SE Spain where precipitation is low. For example, in Andalusia (S, Spain), different plastic cultivation systems are used for the cultivation of some types of vegetables and fruits such as greenhouses or plastic tunnels, the latter being widely used for strawberry crops [198]. Strawberry crops represent 97.3% of the total area, with Andalusia being the region with the largest area (96% of the total area), while only 2.7% are grown without plastic in the north of Spain where climatic conditions are more favourable [199]. As for the plastic use as mulch, it is widely employed in the Region of Murcia, where farmers have been practising this technique for 20 years for the cultivation of lettuce and escarole, for which Spain is one of the main producers worldwide [198]. Other plastic waste, such as those from packaging, containers used in agriculture, or silage, are also common sources of MP contamination [200]. Even though the use of plastic in Spain is widespread in agriculture, the impact of MPs on the soil has been scarcely studied, and the available literature is recent, so there are no long-term studies.

Another source of agricultural soils contamination is sewage sludge, given that, in Spain, 65% of the sludge generated is recycled through agricultural soils [170]. Among the studies carried out, the impact of MPs is highlighted because of the annual use of sewage sludge as fertilizer in rural areas of Valencia (E Spain) [170]. The authors found MPs in 97% of the samples analysed and reported that the successive application of sludge progressively increased the concentration of MPs threefold in 10 years, which was equivalent to a 256% higher MP content with respect to soils that were never treated [170]. The types of MPs found in treated soils came mainly from fragments, followed by synthetic fibres and films, and showed polypropylene (PP) and polyvinyl chloride (PVC) particles, with sizes of 150–250 μm as the most abundant. Information derived from studies in the last 20 years in the Region of Murcia (SE Spain) about MPs presence on agricultural soils cultivated using plastic mulches to prevent soil evapotranspiration has been mapped [201]. Additionally, it was considered that some of the plastic may have come from packaging or silage from the packaging of wind-dispersed produce or bags [201], which are often also consumed by sheep [201, 202]. All the analysed soil and livestock fecal samples contained MPs, with an average content of

2,116 ± 1,024 particles kg^{-1} in dry soil and 997 ± 971 particles kg^{-1} in dry feces. Although no bibliographic data quantifying the concentration of MPs in agricultural soils in the province of Almería (Andalusia, SE Spain) have been found, the incidence of plastics used in these crops on the seabed soils of the adjacent *Posidonia oceanica* coastal meadows bathed by the Mediterranean Sea has been evaluated [203]. In this study, the existence of strong correlations between seafloor MP pollution with the area of greenhouse crops and their production in a chrono sequence since the beginning of intensive agriculture in this area in 1975 was observed [203]. Thus, MP contamination in terrestrial soils could be significant. Moreover, other laboratory studies conducted with earthworm species widely used for vermicompost production (*Eisenia andrei* and *Eisenia foetida*) have shown that MPs can induce different metabolic stress problems on the biota, which could have an influence on soil productivity and its functions [204, 205]. However, research is still needed to understand the influence of MP contamination on soils and soil organisms, its effect on organic matter, soil microbiota and their metabolic activity, and vegetation, as well as the potential problems it may cause for agricultural production and food security in both the short and long term. Likewise, Spanish guidelines, standards, and policies should be adapted to prevent and reduce the impact of MPs on soils.

9 Conclusions

The geographical situation, climatic conditions, and the threat of climate change makes Spain vulnerable to soil degradation and desertification, mainly in its arid and semi-arid areas. This fact, added to erosion-derived problems and soil degradation caused by different human activities, such as changes in land use, industry, or the intensification of agriculture, has led to a large part of the territory being degraded or in high danger of degradation. Therefore, the Spanish state is in a compromised situation, given that one of its main socioeconomic activities is agriculture, which in turn is one of the main precursors of soil degradation in the country. It is consequently vitally important to make adjustments in the management of cultivated areas, predominantly in the proper use of water resources and the application of fertilizers and pesticides. Moreover, the presence of microplastics is another source of contamination in soils and how they can affect human health is largely unknown. Despite there being an important number of studies about problems derived from soil erosion, overgrazing, salinization, use of agrochemicals, and microplastics contamination, further research is still needed to make a statewide diagnosis and obtain more accurate information on the state of soil degradation. A thorough review of existing legislation and guidelines should be carried out to improve the prevention, mitigation, and remediation of the impacts that lead to its decline. Likewise, the uncertainty generated by global change forecasts requires an adaptive management of soils that could be a good management tool to guarantee their health and quality over time and, consequently, the ecosystem services provided by soils.

Acknowledgements This work was supported by the Spanish Ministry of Economy and Competitiveness (MINECO), AEI and FEDER funds, through the projects: CGL2017-88734-R (BIORESOC), PID2021-126946OB-I00, BIOQUALIRES (PID2021-1275910B-100) and CLIMARESTOR (TED2021-132687B-I00) as well as FEDER-Junta de Andalucía Research Projects RESTAGRO (UAL18-RNM-A021-B), Restoration of Abandoned Agricultural Soils in Semiarid Zones to Improve Productivity and Soil Quality and Enhance Carbon Sequestration (P18-RT-4112) and EVOCLIMED (UAL2020-RNM-A2063). Isabel Miralles is grateful for funding received from the Ramón y Cajal Research Grant (RYC-2016-21191) from the Spanish Ministry of Economy, Industry and Competitiveness (MINECO), Raúl Ortega thanks his postdoctoral contract HIPATIA of the University of Almería Research Plan, and Natalia Rodríguez-Berbel acknowledges a Ph.D. research grant from the Spanish government (PRE2018-084964).

References

1. Lal R (2015) Restoring soil quality to mitigate soil degradation. Sustain 7:5875–5895. https://doi.org/10.3390/SU7055875
2. Jónsson JÖG, Davídsdóttir B (2016) Classification and valuation of soil ecosystem services. Agr Syst 145:24–38. https://doi.org/10.1016/J.AGSY.2016.02.010
3. van Leeuwen CCE, Cammeraat ELH, de Vente J, Boix-Fayos C (2019) The evolution of soil conservation policies targeting land abandonment and soil erosion in Spain: a review. Land Use Policy 83:174–186. https://doi.org/10.1016/J.LANDUSEPOL.2019.01.018
4. Martínez Valderrama J (2005) Estudio de la desertificación por sobrepastoreo mediante un modelo de simulación dinámica. E.T.S.I. Agrónomos (UPM)
5. Gobierno de España. Ministerio para la Transición Ecológica y el Reto Demográfico (MITECO) (2020) Plan Nacional de Adaptación al Cambio Climático 2020–2030. Madrid
6. Carrión JS, Fernández S, Jiménez-Moreno G, Fauquette S, Gil-Romera G, González-Sampériz P, Finlayson C (2010) The historical origins of aridity and vegetation degradation in southeastern Spain. J Arid Environ 74:731–736. https://doi.org/10.1016/j.jaridenv.2008.11.014
7. Symeonakis E, Calvo-Cases A, Arnau-Rosalen E (2007) Land use change and land degradation in southeastern Mediterranean Spain. Environ Manag 40:80–94. https://doi.org/10.1007/s00267-004-0059-0
8. Castillo-Monroy AP, Maestre FT, Rey A, Soliveres S, García-Palacios P (2011) Biological soil crust microsites are the main contributor to soil respiration in a semiarid ecosystem. Ecosystems 14:835–847. https://doi.org/10.1007/s10021-011-9449-3
9. Intergovernmental Panel on Climate Change (IPCC) (2020) Climate change and land: an IPCC special report on climate change, desertification, land degradation, sustainable land management, food security, and greenhouse gas fluxes in terrestrial ecosystems. In: Shukla PR, Skea J, Buendia EC, Masson-Delmotte V, Pörtner H-O, Roberts DC, Zhai P, Slade R, Connors S, Diemen R van, Ferrat M, Haughey E, Luz S, Neogi S, Pathak M, Petzold J, Pereira JP, Vyas P, Huntley E, Kissick K, Belkacem M, Malley J (eds) Summary for policymakers. p 41
10. Instituto Nacional de Estadística (2019) Anuario Estadístico de España. Agricultura, silvicultura, ganadería y pesca
11. Gobierno de España. Ministerio para la Transición Ecológica y el Reto Demográfico (MITECO) (2020) Sector agrícola y ganadero. In: Minist. para la Transic. Ecológica y el Reto Demográfico. https://www.miteco.gob.es/es/cambio-climatico/temas/mitigacion-politicas-y-medidas/agricola.aspx. Accessed 20 Oct 2021
12. Comisión Europea (2009) Relaciones entre los procesos de degradación del suelo, las prácticas agronómicas no perjudiciales para el suelo y las medidas adoptadas en relación con ello. In: Agricultura sostenible y conservación de los suelos. pp 1–40

13. Szabolcs I (1989) Salt-affected soils. CRC Press Inc., Boca Raton
14. Alonso-Sarría F, Martínez-Hernández C, Romero-Díaz A, Cánovas-García F, Gomariz-Castillo F (2016) Main environmental features leading to recent land abandonment in Murcia Region (Southeast Spain). L Degrad Dev 27:654–670. https://doi.org/10.1002/LDR.2447
15. Pulido M, Schnabel S, Lavado Contador JF, Lozano-Parra J, González F (2018) The impact of heavy grazing on soil quality and pasture production in rangelands of SW Spain. L Degrad Dev 29:219–230. https://doi.org/10.1002/ldr.2501
16. Cuevas J, Daliakopoulos IN, Del Moral F, Hueso JJ, Tsanis IK (2019) A review of soil-improving cropping systems for soil salinization. Agronomy 9:295
17. Gallardo JF (2016) The soils of Spain. Springer
18. Gobierno de España La Moncloa. Geografía [España/País, Historia y Cultura/Geografía]. https://www.lamoncloa.gob.es/espana/paishistoriaycultura/geografia/Paginas/index.aspx. Accessed 31 May 2021
19. Capel Molina J (1978) Factores del Clima de la Península Ibérica. Paralelo 37(2):5–13
20. Instituto Geográfico Nacional (2021) Clima. http://atlasnacional.ign.es/wane/Clima. Accessed 31 May 2021
21. Gobierno de España. Ministerio de Medio Ambiente y Medio Rural y Marino (MAGRAMA) (2008) Programa de acción nacional contra la desertificación. Madrid
22. Rodríguez Martín JA, Álvaro-Fuentes J, Gonzalo J, Gil C, Ramos-Miras JJ, Grau Corbí JM, Boluda R (2016) Assessment of the soil organic carbon stock in Spain. Geoderma 264:117–125. https://doi.org/10.1016/j.geoderma.2015.10.010
23. López Gómez JYA (1979) El clima de Canarias según la clasificación de Köppen. In: Estudios Geográficos, 156th ed. Consejo Superior de Investigaciones Científicas. Instituto "Juan Sebastián Elcano", Madrid, p 321
24. Jansá A (1988) El clima de las baleares. Mediterraneidad y insularidad Treballs Geogr 39:39–43
25. IUSS Working Group WRB (2007) World reference base for soil resources 2006. World Soil Resources Report, Rome
26. Bai ZG, Dent DL, Olsson L, Schaepman ME (2008) Proxy global assessment of land degradation. Soil Use Manage 24:223–234
27. FAO-IUSS-ISRIC Working Group WRB 106 (2015) World Soil Resources Reports No. 106. World Reference Base for Soil Resources. Food and Agriculture Organization of the United Nations (FAO) Rome, p. 192
28. Martínez-Valderrama J, Ibáñez J, Del Barrio G, Sanjuán ME, Alcalá FJ, Martínez-Vicente S, Ruiz A, Puigdefábregas J (2016) Present and future of desertification in Spain: implementation of a surveillance system to prevent land degradation. Sci Total Environ 563–564:169–178. https://doi.org/10.1016/j.scitotenv.2016.04.065
29. FAO and ITPS (2015) Status of the world's soil resources (SWSR) – main report. Food and Agriculture Organization of the United Nations and Intergovernmental Technical Panel on Soils. Rome, Italy
30. Cantera X, Cánovas C, Garrido F, Mantel S (2015) El suelo, un paseo por la vida. Museo Nacional de Ciencias Naturales
31. García-Ruiz JM (2010) The effects of land uses on soil erosion in Spain: a review. Catena 81: 1–11
32. Guaita N, López I, Prieto F (2008) Cambios de ocupación del suelo en España: implicaciones para la sostenibilidad | Ciudad y Territorio Estudios Territoriales (CyTET). Ciudad y Territ 40: 235–260
33. Molinero Hernando F (2006) La evolución de La agricultura en españa: tradición, modernización y perspectivas. Norba Rev Geogr XI:213–3709
34. Rodríguez-Eugenio N, McLaughlin M, Pennock D (2019) La contaminación del suelo: Una realidad oculta. Rome
35. Cerdá A (2002) La erosión del suelo y sus tasas en España. Ecosistemas 10

36. Morales Gil A (2016) Paisaje, cultura territorial y vivencia de la Geografía. Libro homenaje al profesor Alfredo Morales Gil. Servicio de Publicaciones de la UA, San Vicente del Raspeig
37. Fernández Lorenzo C, Bueno Herrero D, García Ruiz O (2017) Efectos del uso del territorio en la erosión de suelos en España. MoleQla Rev Ciencias la Univ Pablo Olavide, vol 25, p 21, ISSN-e 2173-0903
38. López-Bermúdez F, García-Gómez J (2006) Desertification in the Arid and Semiarid Mediterranean Region. In: Kepner W, Rubio JL, Mouat DA, Pedrazzini F (eds) Desertification in the Mediterranean Region. A Security Issue, NATO Secur. Springer, Dordrecht, pp 401–428
39. López Bermúdez F (2009) Desertificación: Un desafío Ambiental, Económico y Social. In: Rubio JL, Ferri Avaria A (eds) Medio Ambiente: Un Medio de Oportunidades, vol 1. XVII Foro Universitario Juan Luis Vives. Ajuntament de València. XVII Foro Universitario Juan Luis Vives. Ajuntament de València. FIVEC, Valencia, pp 30–42
40. Gobierno de España. Ministerio de Agricultura Alimentación y Medio Ambiente (MAGRAMA). Oficina Española de Cambio Climático (2016) Impactos del cambio climático en los procesos de desertificación en España. Madrid, España
41. Gobierno de España. Ministerio para la Transición ecológica (MITECO) (2017) Plan ambiental de España. Madrid
42. Jones A, Panagos P, Barcelo S, Bouraoui F, Bosco C, Dewitte O, Gardi C, Hervás J, Hiederer R, Jeffery S, Montanarella L, Penizek V, Toth G, Van Den Eeckhaut M, Van Liedekerke M, Verheijen FGA, Yigini Y, Erhard M, Lukewille A, Petersen J, Marmo L, Olazabal C, Strassburger T, Viestova E (2012) The State of Soil in Europe. A contribution of the JRC to the European Environment Agency's Environment State and Outlook Report – SOER 2010. European Evironment Agency, Italy
43. Rubio A, Roig S (2017) Impactos, vulnerabilidad y adaptación al cambio climático en los sistemas extensivos de producción ganadera en España. Of Española Cambio Climático Minist Agric y Pesca, Aliment y Medio Ambient Madrid
44. San Miguel A, Roig S, Perea R (2017) The pastures of Spain Pastos 46:6–39
45. IGN (2019) Capítulo 10. Actividades agrarias y pesqueras. In: España M de FG de (ed) Atlas Nacional de España, Geoportal. Instituto Geográfico Nacional
46. Ibáñez J, Martínez J, Schnabel S (2007) Desertification due to overgrazing in a dynamic commercial livestock-grass-soil system. Ecol Model 205:277–288. https://doi.org/10.1016/j.ecolmodel.2007.02.024
47. Robles AB (2006) Pastos áridos y ganado del sudeste español
48. Bello F, Lepš J, Sebastià M (2007) Grazing effects on the species-area relationship: Variation along a climatic gradient in NE Spain. J Veg Sci 18:25–34. https://doi.org/10.1111/j.1654-1103.2007.tb02512.x
49. Robles AB, Ruiz-Mirazo J, Ramos ME, González-Rebollar JL (2008) Role of livestock grazing in sustainable use, naturalness promotion in naturalization of marginal ecosystems of Southeastern Spain (Andalusia). In: Agroforestry in Europe. Springer, pp 211–231
50. Pulido Fernández M (2014) Indicadores de calidad del suelo en áreas de pastoreo. Universidad de Extremadura
51. Schnabel S, Gómez Gutiérrez A, Lavado Contador JF (2009) Grazing and soil erosion in dehesas of SW Spain. In: Congreso Internacional sobre Desertificación, Murcia
52. Simón N, Montes F, Díaz-Pinés E, Benavides R, Roig S, Rubio A (2012) Spatial distribution of the soil organic carbon pool in a Holm oak dehesa in Spain. Plant and Soil 3661(366): 537–549. https://doi.org/10.1007/S11104-012-1443-9
53. Pulido-Fernández M, Schnabel S, Lavado-Contador JF, Miralles Mellado I, Ortega Pérez R (2013) Soil organic matter of Iberian open woodland rangelands as influenced by vegetation cover and land management. Catena 109:13–24. https://doi.org/10.1016/J.CATENA.2013.05.002
54. Peco B, Sánchez AM, Azcárate FM (2006) Abandonment in grazing systems: consequences for vegetation and soil. Agric Ecosyst Environ 113:284–294. https://doi.org/10.1016/J.AGEE.2005.09.017

55. Strijker D (2005) Marginal lands in Europe – causes of decline. Basic Appl Ecol 6:99–106. https://doi.org/10.1016/J.BAAE.2005.01.001
56. Aldezabal A, Moragues L, Odriozola I, Mijangos I (2015) Impact of grazing abandonment on plant and soil microbial communities in an Atlantic mountain grassland. Appl Soil Ecol 96: 251–260. https://doi.org/10.1016/J.APSOIL.2015.08.013
57. Nadal-Romero E, Lasanta T, Cerdà A (2016) Integrating extensive livestock and soil conservation policies in mediterranean mountain areas for recovery of abandoned lands in the Central Spanish Pyrenees. A long-term research assessment. L Degrad Dev 29:262–273. https://doi.org/10.1002/LDR.2542
58. García-Ruiz JM, Lana-Renault N (2011) Hydrological and erosive consequences of farmland abandonment in Europe, with special reference to the Mediterranean region – a review. Agric Ecosyst Environ 140:317–338. https://doi.org/10.1016/J.AGEE.2011.01.003
59. García-Martínez A, Olaizola A, Bernués A (2009) Trajectories of evolution and drivers of change in European mountain cattle farming systems. Animal 3:152–165. https://doi.org/10.1017/S1751731108003297
60. Ruiz-Mirazo J, Robles AB, González-Rebollar JL (2011) Two-year evaluation of fuelbreaks grazed by livestock in the wildfire prevention program in Andalusia (Spain). Agric Ecosyst Environ 141:13–22. https://doi.org/10.1016/j.agee.2011.02.002
61. Ruiz-Mirazo J (2008) La prevención de incendios forestales mediante pastoreo controlado: el estado del arte en Andalucía. Ganadería 6
62. Ruiz-Mirazo J, Robles AB, González-Rebollar JL (2009) Pastoralism in Natural Parks of Andalusia (Spain): a tool for fire prevention and the naturalization of ecosystems
63. Alcañiz M, Úbeda X, Cerdà A (2020) A 13-year approach to understand the effect of prescribed fires and livestock grazing on soil chemical properties in Tivissa, NE Iberian Peninsula. 11:1013. https://doi.org/10.3390/F11091013
64. Ramón F, Lull C (2019) Legal measures to prevent and manage soil contamination and to increase food safety for consumer health: The case of Spain. Environ Pollut 250:883–891. https://doi.org/10.1016/j.envpol.2019.04.074
65. Gobierno de España. Jefatura del Estado (2011) Ley 22/2011, de 28 de julio, de residuos y suelos contaminados. Boletín Oficial del Estado (BOE) n° 181, de 29/07/2011
66. Gobierno de España. Ministerio de la Presidencia (2005) Real Decreto 9/2005, de 14 de enero, por el que se establece la relación de actividades potencialmente contaminantes del suelo y los criterios y estándares para la declaración de suelos contaminados. Boletín Oficial del Estado (BOE) n° 15, de 18/01/2005
67. Pérez Payá A, Rodríguez Eugenio N (2018) Status of local soil contamination in Europe: Revision of the indicator "Progress in the management Contaminated Sites in Europe
68. Callaba De Roa A (2018) La Contaminación del Suelos en España
69. España a Través de los Mapas. https://www.ign.es/espmap/mapas_conta_bach/Contam_Mapa_05.htm. Accessed 3 May 2021
70. Simón M, Ortiz I, García I, Fernández E, Fernández J, Dorronsoro C, Aguilar J (1999) Pollution of soils by the toxic spill of a pyrite mine (Aznalcollar, Spain). Sci Total Environ 242:105–115. https://doi.org/10.1016/S0048-9697(99)00378-2
71. Vidal M, López-Sánchez JF, Sastre J, Jiménez G, Dagnac T, Rubio R, Rauret G (1999) Prediction of the impact of the Aznalcollar toxic spill on the trace element contamination of agricultural soils. Sci Total Environ 242:131–148. https://doi.org/10.1016/S0048-9697(99)00380-0
72. Romero-Baena AJ, González I, Galán E (2018) Soil pollution by mining activities in Andalusia (South Spain) – the role of Mineralogy and Geochemistry in three case studies. J Soil Sediment 18:2231–2247. https://doi.org/10.1007/s11368-017-1898-7
73. Gobierno de España. Ministerio para la Transición Ecológica y el Reto Demográfico (MITECO) (2019) Estadística minera de españa 2019 secretaría de estado de energía dirección general de política energética y minas. Madrid

74. Rodríguez Martín JA, Nanos N (2016) Soil as an archive of coal-fired power plant mercury deposition. J Hazard Mater 308:131–138. https://doi.org/10.1016/j.jhazmat.2016.01.026
75. Silva V, Mol HGJ, Zomer P, Tienstra M, Ritsema CJ, Geissen V (2019) Pesticide residues in European agricultural soils – a hidden reality unfolded. Sci Total Environ 653:1532–1545. https://doi.org/10.1016/J.SCITOTENV.2018.10.441
76. Gobierno de España. Ministerio de Agricultura Pesca y Alimentación (MAPA) (2019) Estadística anual de consumo de productos fitosanitarios en la Agricultura. Madrid
77. Rodríguez Martín JA, Arias ML, Grau Corbí JM (2006) Heavy metals contents in agricultural topsoils in the Ebro basin (Spain). Application of the multivariate geoestatistical methods to study spatial variations. Environ Pollut 144:1001–1012. https://doi.org/10.1016/j.envpol.2006.01.045
78. El-Alam I, Verdin A, Fontaine J, Laruelle F, Chahine R, Makhlouf H, Sahraoui ALH (2018) Ecotoxicity evaluation and human risk assessment of an agricultural polluted soil. Environ Monit Assess 190:1–17. https://doi.org/10.1007/s10661-018-7077-5
79. Ramos-Miras JJ, Gil C, Rodríguez Martín JA, Bech J, Boluda R (2020) Ecological risk assessment of mercury and chromium in greenhouse soils. Environ Geochem Health 42: 313–324. https://doi.org/10.1007/s10653-019-00354-y
80. Núñez O, Fernández-Navarro P, Martín-Méndez I, Bel-Lan A, Locutura Rupérez JF, López-Abente G (2017) Association between heavy metal and metalloid levels in topsoil and cancer mortality in Spain. Environ Sci Pollut Res 24:7413–7421. https://doi.org/10.1007/s11356-017-8418-6
81. Martínez-Cortijo J, Ruiz-Canales A (2018) Effect of heavy metals on rice irrigated fields with waste water in high pH Mediterranean soils: the particular case of the Valencia area in Spain. Agric Water Manag 210:108–123. https://doi.org/10.1016/J.AGWAT.2018.07.037
82. Rodríguez JA, Nanos N, Grau JM, Gil L, López-Arias M (2008) Multiscale analysis of heavy metal contents in Spanish agricultural topsoils. Chemosphere 70:1085–1096. https://doi.org/10.1016/j.chemosphere.2007.07.056
83. Gil C, Boluda R, Ramos J (2004) Determination and evaluation of cadmium, lead and nickel in greenhouse soils of Almería (Spain). Chemosphere 55:1027–1034. https://doi.org/10.1016/j.chemosphere.2004.01.013
84. Gil C, Boluda R, Rodríguez Martín JA, Guzmán M, del Moral F, Ramos-Miras J (2018) Assessing soil contamination and temporal trends of heavy metal contents in greenhouses on semiarid land. L Degrad Dev 29:3344–3354. https://doi.org/10.1002/ldr.3094
85. Aznar R, Sánchez-Brunete C, Albero B, Rodríguez JA, Tadeo JL (2014) Occurrence and analysis of selected pharmaceutical compounds in soil from Spanish agricultural fields. Environ Sci Pollut Res 21:4772–4782. https://doi.org/10.1007/s11356-013-2438-7
86. Conde-Cid M, Álvarez-Esmorís C, Paradelo-Núñez R, Nóvoa-Muñoz JC, Arias-Estévez M, Álvarez-Rodríguez E, Fernández-Sanjurjo MJ, Núñez-Delgado A (2018) Occurrence of tetracyclines and sulfonamides in manures, agricultural soils and crops from different areas in Galicia (NW Spain). J Clean Prod 197:491–500. https://doi.org/10.1016/j.jclepro.2018.06.217
87. Cerqueira F, Matamoros V, Bayona JM, Berendonk TU, Elsinga G, Hornstra LM, Piña B (2019) Antibiotic resistance gene distribution in agricultural fields and crops. A soil-to-food analysis. Environ Res 177:108608. https://doi.org/10.1016/j.envres.2019.108608
88. Eljarrat E, Marsh G, Labandeira A, Barceló D (2008) Effect of sewage sludges contaminated with polybrominated diphenylethers on agricultural soils. Chemosphere 71:1079–1086. https://doi.org/10.1016/j.chemosphere.2007.10.047
89. Plaza-Bolaños P, Padilla-Sánchez JA, Garrido-Frenich A, Romero-González R, Martínez-Vidal JL (2012) Evaluation of soil contamination in intensive agricultural areas by pesticides and organic pollutants: south-eastern Spain as a case study. J Environ Monit 14:1182–1189. https://doi.org/10.1039/c2em10993j
90. Padilla-Sánchez JA, Romero-González R, Plaza-Bolaños P, Garrido Frenich A, Martínez Vidal JL (2014) Residues and organic contaminants in agricultural soils in intensive

agricultural areas of spain: a three years survey. Clean Soil Air Water 43:746–753. https://doi.org/10.1002/clen.201300583
91. Hildebrandt A, Lacorte S, Barceló D (2009) Occurrence and fate of organochlorinated pesticides and PAH in agricultural soils from the Ebro River Basin. Arch Environ Contam Toxicol 57:247–255. https://doi.org/10.1007/s00244-008-9260-0
92. Cabrera A, Cox L, Koskinen WC, Sadowsky MJ (2008) Availability of triazine herbicides in aged soils amended with olive oil mill waste. J Agric Food Chem 56:4112–4119. https://doi.org/10.1021/jf800095t
93. Rodríguez-Jordá MP, Garrido F, García-González MT (2010) Potential use of gypsum and lime rich industrial by-products for induced reduction of Pb, Zn and Ni leachability in an acid soil. J Hazard Mater 175:762–769. https://doi.org/10.1016/j.jhazmat.2009.10.074
94. Tayibi H, Choura M, López FA, Alguacil FJ, López-Delgado A (2009) Environmental impact and management of phosphogypsum. J Environ Manage 90:2377–2386
95. Abril JM, García-Tenorio R, Enamorado SM, Hurtado MD, Andreu L, Delgado A (2008) The cumulative effect of three decades of phosphogypsum amendments in reclaimed marsh soils from SW Spain: 226Ra, 238U and Cd contents in soils and tomato fruit. Sci Total Environ 403:80–88. https://doi.org/10.1016/J.SCITOTENV.2008.05.013
96. Gobierno de España. Ministerio de la Presidencia (2005) Real Decreto 824/2005, de 8 de julio, sobre productos fertilizantes. Boletín Oficial del Estado (BOE) nº 171, de 19/07/2005
97. Junta de Andalucía (2015) Decreto 18/2015, de 27 de enero, por el que se aprueba el reglamento que regula el régimen aplicable a los suelos contaminados
98. Gobierno de España. Jefatura del Estado (1964) Ley 25/1964, de 29 de abril, sobre energía nuclear. Boletín Oficial del Estado (BOE) nº 107, de 04/05/1964
99. Gobierno de España. Ministerio de Transición Ecológica (MITECO) (2016) Dirección general del agua. Madrid
100. FAO (2021) Salinidad del suelo | Asociación Mundial por el Suelo | Organización de las Naciones Unidas para la Agricultura y la Alimentación. http://www.fao.org/global-soil-partnership/areas-of-work/soil-salinity/en/. Accessed 19 May 2021
101. Van-Camp L, Bujarrabal B, Gentile AR, Jones RJA, Montanarella L, Olazabal C, Selvaradjou S-K (2004) Reports of the technical working groups established under the thematic strategy for soil protection volume-II erosion
102. Schofield R, Thomas DSG, Kirkby MJ (2001) Causal processes of soil salinization in Tunisia, Spain and Hungary. L Degrad Dev 12:163–181. https://doi.org/10.1002/ldr.446
103. Luna L, Pastorelli R, Bastida F, Hernández T, García C, Miralles I, Solé-Benet A (2016) The combination of quarry restoration strategies in semiarid climate induces different responses in biochemical and microbiological soil properties. Appl Soil Ecol 107:33–47
104. Alonso Cabrera CA (2020) Técnicas de aprendizaje automático para el análisis de la salinidad de aguas en el valle del Guadalhorce. Universitat Oberta de Catalunya
105. Shahid SA, Zaman M, Heng L (2018) Soil salinity: historical perspectives and a world overview of the problem. In: Guideline for salinity assessment, mitigation and adaptation using nuclear and related techniques. Springer, pp 43–53
106. Qadir M, Quillérou E, Nangia V, Murtaza G, Singh M, Thomas RJ, Drechsel P, Noble AD (2014) Economics of salt-induced land degradation and restoration. Nat Resour Forum 38: 282–295. https://doi.org/10.1111/1477-8947.12054
107. Herrero J, Pérez-Coveta O (2005) Soil salinity changes over 24 years in a Mediterranean irrigated district. Geoderma 125:287–308. https://doi.org/10.1016/j.geoderma.2004.09.004
108. Ortiz Silla R, Maracute˜in Sanleandro P, Sánchez Navarro A, García Navarro A, Delgado Iniesta MJ (2009) Evaluation of salinity in the Fluvisols of the Vega Baja region, SE Spain. In: Land degradation and rehabilitation: dryland ecosystems. Papers presented at the Fourth International Conference on Land Degradation. Catena Verlag, Cartagena, Murcia, Spain, pp 63–70
109. Gobierno de España. Ministerio de Agricultura Pesca y Alimentación (MAPA) (2008) Programa de Vigilancia Ambiental del Plan Nacional de Regadíos. Madrid

110. FAO (2006) Water desalination for agricultural applications, Proceeding. Rome, Italy
111. Richards LA (1954) Diagnosis and improvement of saline alkali soils. In: Agriculture handbook no. 60. United States Salinity Laboratory Staff, Washington, p 154
112. Abrol IP, Yadav JSP, Massoud FI (1988) Salt-affected soils and their management. Food and Agriculture Organisation of United Nations (FAO), Rome
113. Gobierno de España. Ministerio de Agricultura y Pesca Alimentación y Medio Ambiente (MAPAMA) (2016) Análisis temático: Evaluación de aspectos ambientales. Madrid
114. Pagán E, Robles JM, Temnani A, Berríos P, Botía P, Pérez-Pastor A (2022) Effects of water deficit and salinity stress on late mandarin trees. Sci Total Environ 803:150109. https://doi.org/10.1016/J.SCITOTENV.2021.150109
115. Moreno-Ramón H, Zuzunaga-Rosas J, Ibáñez-Asensio S (2021) Acumulación de carbono en un humedal afectado por la salinidad en un contexto de cambio climático: La Albufera de Valencia. In: Almendro-Candel MB, Jordán-Vidal LM (eds) Libro de resúmenes del IX Simposio Nacional sobre el Control de la Degradación y Recuperación de Suelos. Elche, España, p 620
116. Sánchez-Navarro A, Girona-Ruiz A, Delgado Iniesta MJ (2020) Influence of intensive horticultural cultivation on soil salinity in campo de Cartagena (Murcia). Spanish J Soil Sci 10: 204–208. https://doi.org/10.3232/SJSS.2020.V10.N3.03
117. Mousa Yahia M (2007) Dinámica de la salinidad en un olivar bajo riego localizado: efecto de la precipitación invernal y respuesta del olivo. Universidad de Córdoba
118. Garcia C, Hernandez T (1996) Influence of salinity on the biological and biochemical activity of a calciorthird soil. Plant and Soil 178:255–263. https://doi.org/10.1007/bf00011591
119. Herrero J, Castañeda C (2018) The success story of irrigation against salinity in Violada, NE Spain. L Degrad Dev 29:3039–3049. https://doi.org/10.1002/ldr.3031
120. Zekri S, Albisu LM (1993) Economic impact of soil salinity in agriculture. A case study of Bardenas area. Spain Agric Syst 41:369–386
121. Castañeda C, Herrero J, Latorre B (2020) The vanishing legacy of soil salinity data from irrigated districts: a case study from Spain and a call for action. In: Advances in agronomy. Academic Press Inc., pp 325–355
122. Boukalfa A (1995) Influencia del relieve sobre la salinidad de los suelos en la llanura aluvial de la cuenca alta del Guadiana (centro de España). Universidad Politécnica de Madrid
123. Català M del M, Domingo C, Martínez-Eixarch M, Tomàs N, Pla E, Bertomeu A (2019) Impacto de la salinidad en las principales variedades de arroz cultivadas en España. Vida Rural 473:36–39
124. Garcia Navarro AF (2016) Caracterización y riesgos de salinización de los suelos de la red de riegos del Bajo Segura. Universidad de Murcia
125. Acosta JA, Faz A, Jansen B, Kalbitz K, Martínez-Martínez S (2011) Assessment of salinity status in intensively cultivated soils under semiarid climate, Murcia, SE Spain. J Arid Environ 75:1056–1066. https://doi.org/10.1016/j.jaridenv.2011.05.006
126. Mateo J, Burke J (2010) Agriculture and water quality interactions: a global overview SOLAW TR08
127. Isern JH, Ochoa RR (1989) Colmatación de drenes en suelos afectados por salinidad: finca experimental de San Juan de Flumen (Huesca). Institucio, Zaragoza
128. Caballero R, Bustos A, Román R (2001) Soil Salinity under Traditional and Improved Irrigation Schedules in Central Spain. Soil Sci Soc Am J 65:1210–1218. https://doi.org/10.2136/sssaj2001.6541210x
129. Zuzunaga-Rosas J, Moreno-Ramón H, Ibáñez-Asensio S (2021) 45 años de riego: Reevaluación de la aptitud de los suelos de la zona de la Pedrera (Alicante). In: Almendro-Candel MB, Jordán-Vidal LM (eds) Libro de resúmenes del IX Simposio Nacional sobre el Control de la Degradación y Recuperación de Suelos. Elche, España, p 620
130. Zribi W, Faci J, Aragüés R (2011) Efectos del acolchado sobre la humedad, temperatura, estructura y salinidad de suelos agrícolas. ITEA 107:148–162

131. Jordán MM, Navarro-Pedreño AJ, García-Sánchez AE, Mateu AJ, Juan AP (2004) Spatial dynamics of soil salinity under arid and semi-arid conditions: geological and environmental implications. Environ Geol 45:448–456. https://doi.org/10.1007/s00254-003-0894-y
132. Caravaca F, Díaz G, Torres P, Campoy M, Roldán A (2021) Efecto de la invasión de Nicotiana glauca sobre la funcionalidad de la comunidad microbiana del suelo en ecosistemas semiáridos mediterráneos. In: Almendro-Candel MB, Jordán-Vidal MM (eds) Libro de resúmenes del IX Simposio Nacional sobre el Control de la Degradación y Recuperación de Suelos. Elche, España, p 620
133. García-Fayos P, García-Ventoso B, Cerdà A (2000) Limitations to plant establishment on eroded slopes in southeastern Spain. J Veg Sci 11:77–86. https://doi.org/10.2307/3236778
134. Delgado Fernández IC, Giménez Luque E, Gómez Mercado F, Pedrosa W (2016) Influence of temperature and salinity on the germination of Limonium tabernense Erben from Tabernas Desert (Almería, SE Spain). Flora Morphol Distrib Funct Ecol Plants 218:68–74. https://doi.org/10.1016/j.flora.2015.12.001
135. González-Alcaraz MN, Jiménez-Cárceles FJ, Álvarez Y, Álvarez-Rogel J (2014) Gradients of soil salinity and moisture, and plant distribution, in a Mediterranean semiarid saline watershed: A model of soil-plant relationships for contributing to the management. Catena 115:150–158. https://doi.org/10.1016/j.catena.2013.11.011
136. Bhattacharyya C, Roy R, Tribedi P, Ghosh A, Ghosh A (2020) Biofertilizers as substitute to commercial agrochemicals. In: Agrochemicals detection, treatment and remediation. Elsevier, pp 263–290
137. Gobierno de España. Ministerio de Agricultura Pesca y Alimentación (MAPA) (2020) Encuesta sobre Superficies y Rendimientos Cultivos (ESYRCE). Encuesta de Marco de Áreas de España. https://www.mapa.gob.es/es/estadistica/temas/estadisticas-agrarias/agricultura/esyrce/. Accessed 11 May 2021
138. Official Journal of the European Union (2009) Directiva 2009/128/CE del Parlamento Europeo y del Consejo, de 21 de octubre de 2009, por la que se establece el marco de la actuación comunitaria para conseguir un uso sostenible de los plaguicidas
139. Official Journal of the European Union (2009) Regulation (EC) n° 1107/2009 of the European Parliament and of the Council of 21/10/2009 concerning the placing of plant protection products on the market and repealing Council Directives 79/117/EEC and 91/414/EEC
140. Official Journal of the European Union (2009) Regulation (EC) n° 1185/2009 of the European Parliament and of the Council of 25/11/2009 concerning statistics on pesticides
141. Belmonte Vega A, Garrido Frenich A, Martínez Vidal JL (2005) Monitoring of pesticides in agricultural water and soil samples from Andalusia by liquid chromatography coupled to mass spectrometry. Anal Chim Acta 538:117–127. https://doi.org/10.1016/j.aca.2005.02.003
142. AL-Ahmadi MS (2019) Pesticides, anthropogenic activities, and the health of our environment safety. Pestic – use misuse their impact environment https://doi.org/10.5772/INTECHOPEN.84161
143. Carpio MJ, Sánchez-Martín MJ, Rodríguez-Cruz MS, Marín-Benito JM (2021) Effect of organic residues on pesticide behavior in soils: a review of laboratory research environment 2021, vol 8, p 32. https://doi.org/10.3390/ENVIRONMENTS8040032
144. FAO (2006) International Code of conduct on the distribution and use of pesticides guidelines on efficacy evaluation for the registration of plant protection products food and agriculture organization of the United Nations
145. Pose-Juan E, Sánchez-Martín MJ, Andrades MS, Rodríguez-Cruz MS, Herrero-Hernández E (2015) Pesticide residues in vineyard soils from Spain: spatial and temporal distributions. Sci Total Environ 514:351–358. https://doi.org/10.1016/j.scitotenv.2015.01.076
146. Arias-Estévez M, López-Periago E, Martínez-Carballo E, Simal-Gándara J, Mejuto JC, García-Río L (2008) The mobility and degradation of pesticides in soils and the pollution of groundwater resources. Agric Ecosyst Environ 123:247–260
147. Singh DK (2012) Toxicology: agriculture and environment: pesticide chemistry and toxicology. Bentham Science Publishers

148. Kanter DR (2018) Nitrogen pollution: a key building block for addressing climate change. Clim Change 147:11–21. https://doi.org/10.1007/s10584-017-2126-6
149. Damalas CA, Eleftherohorinos IG (2011) Pesticide exposure, safety issues, and risk assessment indicators. Int J Environ Res Public Health 8:1402–1419
150. Malik Z, Ahmad M, Abassi GH, Dawood M, Hussain A, Jamil M (2017) Agrochemicals and soil microbes: interaction for soil health. Springer, Cham, pp 139–152
151. Stewart WM, Dibb DW, Johnston AE, Smyth TJ (2005) The contribution of commercial fertilizer nutrients to food production. Agron J 97:1–6
152. Hutson DH, Roberts TR (1990) Environmental fate of pesticides: progress in pesticide biochemistry and toxicology. Wiley, Chichester
153. Padilla-Sánchez JA, Romero-González R, Plaza-Bolaños P, Garrido Frenich A, Martínez Vidal JL (2015) Residues and organic contaminants in agricultural soils in intensive agricultural areas of spain: a three years survey. Clean Soil Air Water 43:746–753. https://doi.org/10.1002/clen.201300583
154. Socías-Viciana MM, Fernández-Pérez M, Villafranca-Sánchez M, González-Pradas E, Flores-Céspedes F (1999) Sorption and leaching of atrazine and MCPA in natural and peat-amended calcareous soils from Spain. J Agric Food Chem 47:1236–1241. https://doi.org/10.1021/jf980799m
155. Acopaex (2016) Los cultivos más rentables en Extremadura – Acopaex. https://acopaex.es/2016/08/18/cultivos-rentabilidad-extremadura/. Accessed 11 May 2021
156. Albarran A, Celis R, Hermosín M, López-Piñeiro A, Ortega-Calvo J, Cornejo J (2003) Effects of solid olive-mill waste addition to soil on sorption, degradation and leaching of the herbicide simazine. Soil Use Manage 19:150–156. https://doi.org/10.1079/SUM2002185
157. Albarrán A, Celis R, Hermosín MC, López-Piñeiro A, Cornejo J (2002) Effect of solid olive-mill waste amendment on pesticide sorption and leaching in soil. WIT Press
158. Wolpert F, Quintas-Soriano C, Plieninger T (2020) Exploring land-use histories of tree-crop landscapes: a cross-site comparison in the Mediterranean Basin. Sustain Sci 15:1267–1283. https://doi.org/10.1007/s11625-020-00806-w .
159. Jímenez-Hornero FJ, Giráldez JV, Laguna A (2001) Desplazamiento de suelo y sustancias agroquímicas por la erosión en olivares. In: V Jornadas sobre Investigación en la Zona no Saturada. Pamplona, España, pp 157–163
160. Marín-Benito JM, Sánchez-Martín MJ, Soledad Andrades M, Pérez-Clavijo M, Rodríguez-Cruz MS (2009) Effect of Spent Mushroom Substrate Amendment of Vineyard Soils on the Behavior of Fungicides: 1. Adsorption-Desorption of Penconazole and Metalaxyl by Soils and Subsoils. J Agric Food Chem 57:9634–9642. https://doi.org/10.1021/jf902108n
161. Rajput P, Thakur A, Devi P (2020) Emerging agrochemicals contaminants: current status, challenges, and technological solutions. In: Agrochemicals detection, treatment and remediation. Elsevier, pp 117–142
162. Peris E, Requena S, De La Guardia M, Pastor A, Carrasco JM (2005) Organochlorinated pesticides in sediments from the Lake Albufera of Valencia (Spain). Chemosphere 60:1542–1549. https://doi.org/10.1016/j.chemosphere.2005.02.043
163. Santos A, Flores M (1995) Effects of glyphosate on nitrogen fixation of free-living heterotrophic bacteria. Lett Appl Microbiol 20:349–522. https://doi.org/10.1111/j.1472-765X.1995.tb01318.x
164. Mañosa S, Mateo R, Guitart R (2001) A review of the effects of agricultural and industrial contamination on the Ebro delta biota and wildlife. Environ Monit Assess 71:187–205. https://doi.org/10.1023/A:1017545932219
165. Fernández MA, Alonso C, González MJ, Hernández LM (1999) Occurrence of organochlorine insecticides, PCBs and PCB congeners in waters and sediments of the Ebro River (Spain). Chemosphere 38:33–43. https://doi.org/10.1016/S0045-6535(98)00167-2
166. Lara-Martín PA, Gómez-Parra A, Petrovic M, Barceló D, González-Mazo E (2005) Distribución de contaminantes orgánicos en sedimentos costeros de la Bahía de Cádiz (SO de España). Cienc Mar 31:203–212. https://doi.org/10.7773/cm.v31i12.95

167. Linares-Mazariegos RM (2007) Evaluación ambiental de pesticidas organoclorados en sedimentos de la Laguna de Chantuto (Chiapas, México) y de la Bahía de Santander (Cantabria, España). Universidad de Cantabria
168. de los Santos B, Medina JJ, Miranda L, Gómez JA, Talavera M (2021) Soil disinfestation efficacy against soil fungal pathogens in strawberry crops in spain: an overview. Agronomy 11:526. https://doi.org/10.3390/agronomy11030526
169. Büks F, Kaupenjohann M (2020) Global concentrations of microplastics in soils – a review. Soil 6:649–662. https://doi.org/10.5194/soil-6-649-2020
170. van den Berg P, Huerta-Lwanga E, Corradini F, Geissen V (2020) Sewage sludge application as a vehicle for microplastics in eastern Spanish agricultural soils. Environ Pollut 261:114198. https://doi.org/10.1016/j.envpol.2020.114198
171. Bläsing M, Amelung W (2018) Plastics in soil: analytical methods and possible sources. Sci Total Environ 612:422–435
172. Yang L, Zhang Y, Kang S, Wang Z, Wu C (2021) Microplastics in soil: a review on methods, occurrence, sources, and potential risk. Sci Total Environ 780:146546
173. Kumar R, Sharma P (2021) Recent developments in extraction, identification, and quantification of microplastics from agricultural soil and groundwater. Springer, Singapore, pp 125–143
174. Machado AA d S, Lau CW, Till J, Kloas W, Lehmann A, Becker R, Rillig MC (2018) Impacts of microplastics on the soil biophysical environment. Environ Sci Technol 52:9656–9665. https://doi.org/10.1021/ACS.EST.8B02212
175. Awet TT, Kohl Y, Meier F, Straskraba S, Grün A-L, Ruf T, Jost C, Drexel R, Tunc E, Emmerling C (2018) Effects of polystyrene nanoparticles on the microbiota and functional diversity of enzymes in soil. Environ Sci Eur 301(30):1–10. https://doi.org/10.1186/S12302-018-0140-6
176. Six J, Frey SD, Thiet RK, Batten KM (2006) Bacterial and fungal contributions to carbon sequestration in agroecosystems. Soil Sci Soc Am J 70:555–569. https://doi.org/10.2136/SSSAJ2004.0347
177. Rilling M (2012) Microplastic in terrestrial ecosystems and the soil? Environ Sci Technol 46:6453–6454. https://doi.org/10.1021/ES302011R
178. Zhu F, Zhu C, Wang C, Gu C (2019) Occurrence and ecological impacts of microplastics in soil systems: a review. Bull Environ Contam Toxicol 1026(102):741–749. https://doi.org/10.1007/S00128-019-02623-Z
179. Nizzetto L, Futter M, Langaas S (2016) Are agricultural soils dumps for microplastics of urban origin? Environ Sci Technol 50:10777–10779
180. Liu EK, He WQ, Yan CR (2014) "White revolution" to "white pollution" – agricultural plastic film mulch in China. Environ Res Lett 9:091001
181. Weithmann N, Möller JN, Löder MGJ, Piehl S, Laforsch C, Freitag R (2018) Organic fertilizer as a vehicle for the entry of microplastic into the environment. Sci Adv 4:eaap8060. https://doi.org/10.1126/sciadv.aap8060
182. Zhang Z, Su Y, Zhu J, Shi J, Huang H, Xie B (2021) Distribution and removal characteristics of microplastics in different processes of the leachate treatment system. Waste Manag 120:240–247. https://doi.org/10.1016/j.wasman.2020.11.025
183. Bullard JE, Ockelford A, O'Brien P, McKenna Neuman C (2021) Preferential transport of microplastics by wind. Atmos Environ 245:118038. https://doi.org/10.1016/j.atmosenv.2020.118038
184. Nizzetto L, Bussi G, Futter MN, Butterfield D, Whitehead PG (2016) A theoretical assessment of microplastic transport in river catchments and their retention by soils and river sediments. Environ Sci Process Impacts 18:1050–1059. https://doi.org/10.1039/c6em00206d
185. Gala A, Guerrero M, Serra JM (2020) Characterization of post-consumer plastic film waste from mixed MSW in Spain: a key point for the successful implementation of sustainable plastic waste management strategies. Waste Manag 111:22–33. https://doi.org/10.1016/j.wasman.2020.05.019

186. Bayo J, Rojo D, Olmos S (2019) Abundance, morphology and chemical composition of microplastics in sand and sediments from a protected coastal area: the Mar Menor lagoon (SE Spain). Environ Pollut 252:1357–1366. https://doi.org/10.1016/j.envpol.2019.06.024
187. Alomar C, Estarellas F, Deudero S (2016) Microplastics in the Mediterranean Sea: deposition in coastal shallow sediments, spatial variation and preferential grain size. Mar Environ Res 115:1–10. https://doi.org/10.1016/j.marenvres.2016.01.005
188. Godoy V, Prata JC, Blázquez G, Almendros AI, Duarte AC, Rocha-Santos T, Calero M, Martín-Lara MÁ (2020) Effects of distance to the sea and geomorphological characteristics on the quantity and distribution of microplastics in beach sediments of Granada (Spain). Sci Total Environ 746:142023. https://doi.org/10.1016/j.scitotenv.2020.142023
189. Dalmau-Soler J, Ballesteros-Cano R, Boleda MR, Paraira M, Ferrer N, Lacorte S (2021) Microplastics from headwaters to tap water: occurrence and removal in a drinking water treatment plant in Barcelona Metropolitan area (Catalonia, NE Spain). Environ Sci Pollut Res:1–11. https://doi.org/10.1007/s11356-021-13220-1
190. Simon-Sánchez L, Grelaud M, Garcia-Orellana J, Ziveri P (2019) River Deltas as hotspots of microplastic accumulation: the case study of the Ebro River (NW Mediterranean). Sci Total Environ 687:1186–1196. https://doi.org/10.1016/j.scitotenv.2019.06.168
191. Carretero O, Gago J, Viñas L (2021) From the coast to the shelf: Microplastics in Rías Baixas and Miño River shelf sediments (NW Spain). Mar Pollut Bull 162:111814. https://doi.org/10.1016/j.marpolbul.2020.111814
192. Franco AA, Arellano JM, Albendín G, Rodríguez-Barroso R, Zahedi S, Quiroga JM, Coello MD (2020) Mapping microplastics in Cadiz (Spain): Occurrence of microplastics in municipal and industrial wastewaters. J Water Process Eng 38:101596. https://doi.org/10.1016/j.jwpe.2020.101596
193. Castillejo Martínez N (2019) Fuentes de microplásticos marinos en el entorno de Gandia. Universitat Politècnica de València
194. van Schothorst B, Beriot N, Huerta Lwanga E, Geissen V (2021) Sources of light density microplastic related to two agricultural practices: the use of compost and plastic mulch. Environments 8:36. https://doi.org/10.3390/environments8040036
195. Gil Meseguer E, Gómez Espín JM (2011) Cultivos bajo cubierta en el Sureste de España. Papeles Geogr:155–170
196. Duque-Acevedo M, Belmonte-Ureña LJ, Plaza-Úbeda JA, Camacho-Ferre F (2020) The management of agricultural waste biomass in the framework of circular economy and bioeconomy: an opportunity for greenhouse agriculture in Southeast Spain. Agronomy 10:489. https://doi.org/10.3390/agronomy10040489
197. Rodríguez-Seijo A, Pereira R (2019) Microplastics in agricultural soils. In: Bioremediation of agricultural soils. CRC Press, pp 45–60
198. Romero-Gámez M, Suárez-Rey EM (2020) Environmental footprint of cultivating strawberry in Spain. Int J Life Cycle Assess 25:719–732. https://doi.org/10.1007/s11367-020-01740-w
199. Gobierno de España. Ministerio de Agricultura Alimentación y Medio Ambiente (MAPAMA) (2019) Anuario de estadística, 2016. Madrid
200. Pereira R, Hernandez A, James B, Lemoine B (2021) EIP-AGRI Focus Group – reducing the plastic footprint of agriculture Minipaper A: the actual uses of plastics in agriculture across EU: an overview and the environmental problems
201. Beriot N, Peek J, Zornoza R, Geissen V, Huerta Lwanga E (2021) Low density-microplastics detected in sheep faeces and soil: a case study from the intensive vegetable farming in Southeast Spain. Sci Total Environ 755:142653. https://doi.org/10.1016/j.scitotenv.2020.142653

202. Otsyina HR, Nguhiu-Mwangi J, Mogoa EGM, Mbuthia PG, Ogara WO (2018) Knowledge, attitude, and practices on usage, disposal, and effect of plastic bags on sheep and goats. Tropl Anim Health Prod 50:997–1003. https://doi.org/10.1007/s11250-018-1523-9
203. Dahl M, Bergman S, Björk M, Diaz-Almela E, Granberg M, Gullström M, Leiva-Dueñas C, Magnusson K, Marco-Méndez C, Piñeiro-Juncal N, Mateo MÁ (2021) A temporal record of microplastic pollution in Mediterranean seagrass soils. Environ Pollut 273:116451. https://doi.org/10.1016/j.envpol.2021.116451
204. Rodriguez-Seijo A, Lourenço J, Rocha-Santos TAP, da Costa J, Duarte AC, Vala H, Pereira R (2017) Histopathological and molecular effects of microplastics in Eisenia andrei Bouché. Environ Pollut 220:495–503. https://doi.org/10.1016/j.envpol.2016.09.092
205. Rodríguez-Seijo A, da Costa JP, Rocha-Santos T, Duarte AC, Pereira R (2018) Oxidative stress, energy metabolism and molecular responses of earthworms (Eisenia fetida) exposed to low-density polyethylene microplastics. Environ Sci Pollut Res 25:33599–33610. https://doi.org/10.1007/s11356-018-3317-z

Agricultural Land Degradation in Sweden

Ana Barreiro and Linda-Maria Dimitrova Mårtensson

Contents

1 Introduction to Agriculture in Sweden 300
2 Soil Compaction and Erosion (Water and Wind) 305
3 Overgrazing 309
4 Slash-and-Burn Agriculture 310
5 Soil Contamination 311
6 Salinity 313
7 Agrochemicals Use 313
8 Microplastics 315
9 Conclusions 316
References 316

Abstract The current status of agricultural soils in Sweden is dominated by a tendency of land abandonment in some areas vs. intensification which leads to soil degradation in other areas, like the south of the country. Overgrazing, slash-and-burn agricultural practices or salinization is currently not a problem for Swedish agriculture. On the other hand, soil compaction and erosion represent problems, mostly in some areas of Sweden. Agricultural soils are also exposed to different soil contaminants, pesticides and microplastics which can have negative or unclear effects in the soil ecosystem.

A. Barreiro (✉)
Department of Biosystems and Technology, Swedish University of Agricultural Sciences, Lomma, Sweden

Department of Soil Science and Agricultural Chemistry, Engineering Polytechnic School, University of Santiago de Compostela, Lugo, Spain
e-mail: ana.barreiro.bujan@usc.es

L.-M. D. Mårtensson
Department of Biosystems and Technology, Swedish University of Agricultural Sciences, Lomma, Sweden
e-mail: linda.maria.martensson@slu.se

Paulo Pereira, Miriam Muñoz-Rojas, Igor Bogunovic, and Wenwu Zhao (eds.),
Impact of Agriculture on Soil Degradation II: A European Perspective,
Hdb Env Chem (2023) 121: 299–324, DOI 10.1007/698_2022_916,

Published online: 14 December 2022

Keywords Agriculture, Land abandonment, Soil degradation, Sweden

1 Introduction to Agriculture in Sweden

Sweden is located in the north of Europe and is part of the Scandinavian peninsula. The elongated form of the country provides quite some variety on the growing conditions (Fig. 1). The climate varies from subarctic in the north to maritime and continental in the south, where the mean annual temperature ranges from −3°C in the north and 10°C in the south and the accumulated annual precipitation ranges from 400 mm in the northeast and 1,000 mm in the southwest [1].

The vegetation period is about 100 days in the north and 210 days in the south. The main soil type is podzol, which is dominated by forest production (Fig. 2). There

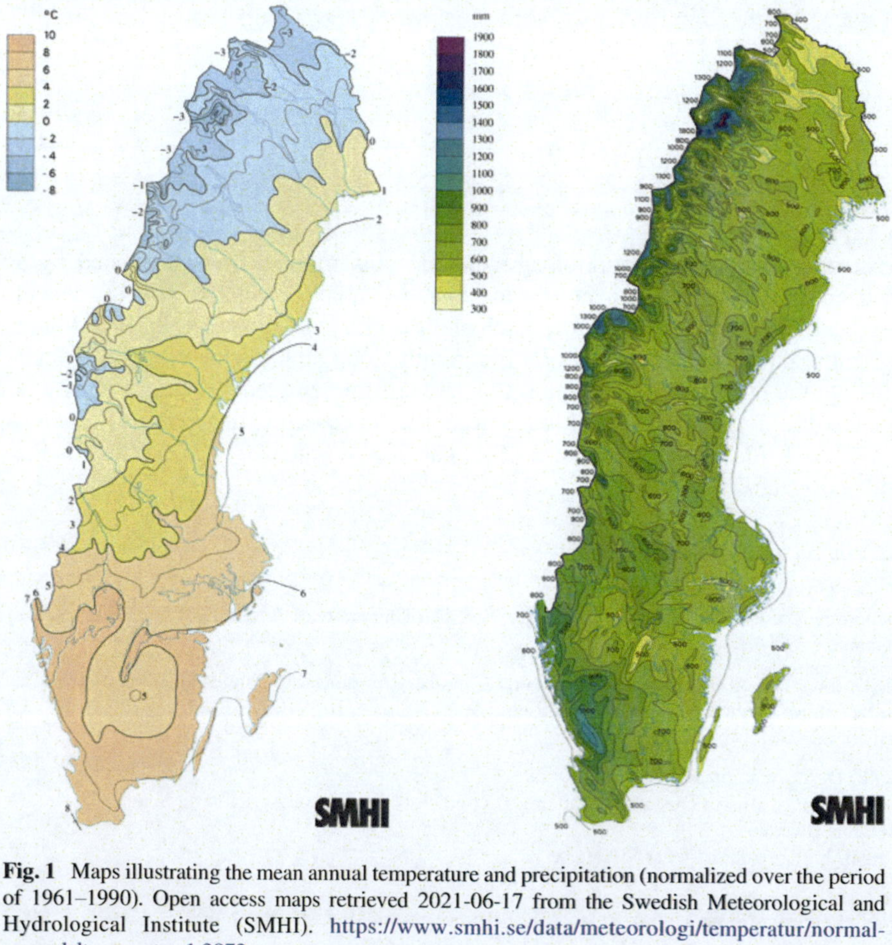

Fig. 1 Maps illustrating the mean annual temperature and precipitation (normalized over the period of 1961–1990). Open access maps retrieved 2021-06-17 from the Swedish Meteorological and Hydrological Institute (SMHI). https://www.smhi.se/data/meteorologi/temperatur/normal-arsmedeltemperatur-1.3973

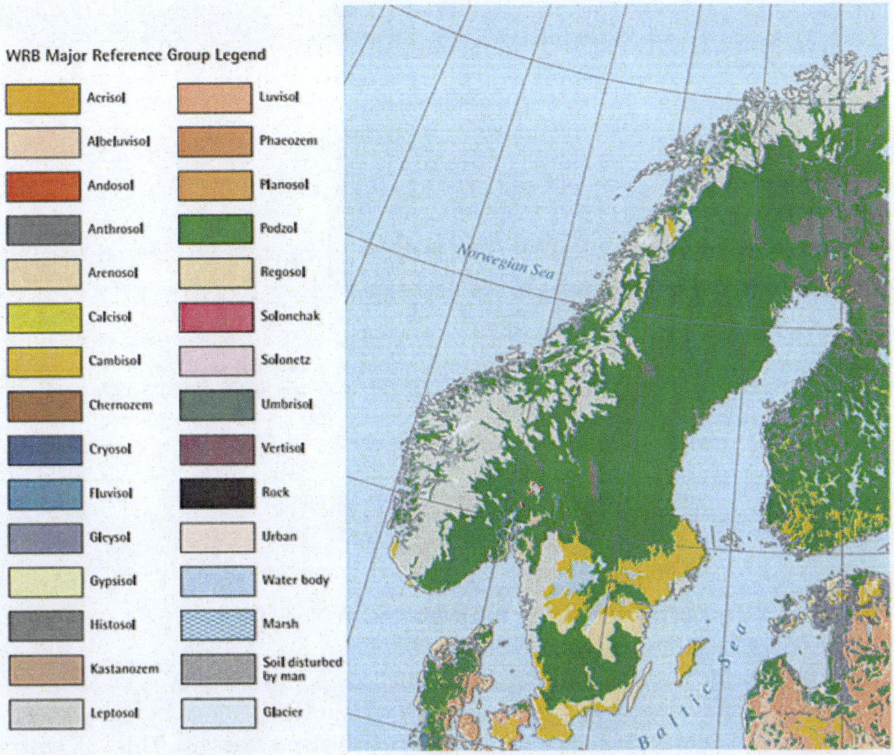

Fig. 2 Map illustrating the soil type according to the WRB classification. Soil Atlas of Europe, JRC European Commission

are large contributions of clay, silt and clayey moraine in some regions, and this is where agricultural activities are allocated, with Cambisols as the dominant soil type (Fig. 2).

The arable land in Sweden covers an area of around 2.6 million ha plus 0.4 million ha of pastures, corresponding to about 7% of the total land area, with animal husbandry being the dominant line of production for meat (mainly pork and beef) and dairy products [2]. The growing conditions form the demographics of the agricultural activities within the country, with smaller farms in the north and larger, aggregated farming enterprises with intense crop production in the central and south parts, representing as much as 60% of the national arable land [2]. Furthermore, half of the animal husbandry facilities are located in the south of the country, mainly in the southern counties of Sweden [2]. The climatic and soil conditions are more favourable for forest land use in the northern and mountainous areas of the country, while the central and southern regions are the most agriculturally productive.

The amount of precipitation in Sweden provides a majority of rainfed agricultural production (90%), with only 10% of the production requiring irrigation. Currently most of the irrigation practices in Sweden take place in the southern counties on

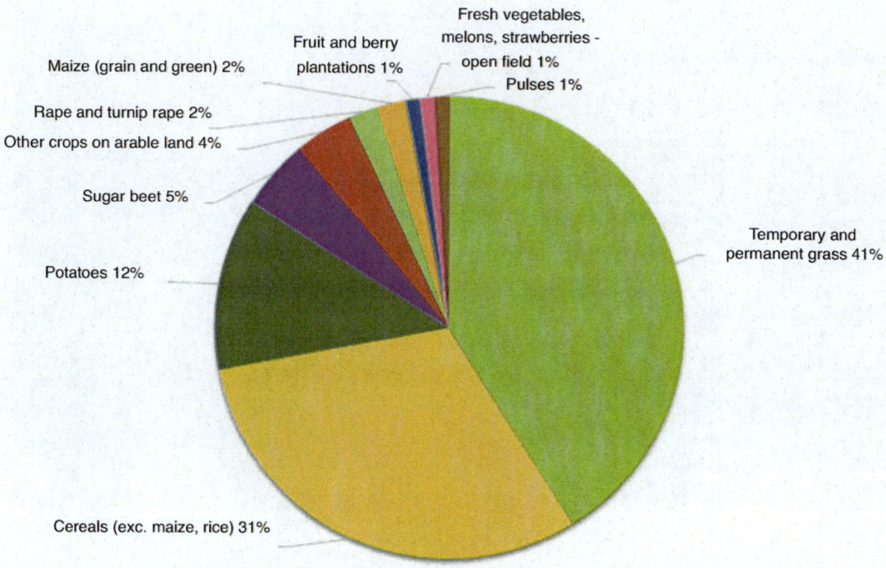

Fig. 3 Irrigated area by type of crops (%), Sweden, 2010. Source: Eurostat

sandy soils, mainly in grasslands, cereals and potatoes (Fig. 3). However, under most climate change scenarios the needs for irrigation will most likely increase early in the season for cereals and during summer for other crops like potatoes [3]. This will pose a huge future challenge for farmers in terms of building both knowledge and infrastructure for irrigation.

In Sweden, the crop production is strongly dominated by cereals and leys, with barley, oats and wheat as the most frequent cereals (40% of arable land; Table 1) used both for human consumption and as animal feed. The practice of crop rotations is well established and implemented for most of the farmers. One classic example of conventional agriculture in the south of the country is a 4-year rotation including winter wheat, winter oilseed rape, spring barley and sugar beet. In the southern parts of the country, the use of autumn-sown winter crops is possible, and they are beneficial for the recovery of nutrients and may both prevent leaching of nutrients and promote soil fertility [4]. The use of cover or intermediate crops, like grass-legume leys or oil radish, are other well-established practices to avoid bare soil and nutrient losses during winter among the Swedish farmers, both conventional and organic ones.

The organic share of total agricultural land in Sweden reached 19% in 2017 [5] and 17% of the milk production was under certified organic management [2]. Sweden is one of the countries within Europe with a bigger percentage of organic farming area, just behind Estonia and Austria, but with much bigger area than the neighbouring Nordic countries (Eurostat statistics).

The area of agricultural production has declined in the last decades in Sweden due to declined interest in the younger generations of farming as a career path, resulting in abandonment of land which successively became forests in most of the cases. In

Table 1 Use of arable land (hectares) during 1990–2019 period. Numbers refer to agricultural holdings with more than 2.0 ha of arable land or holdings with large animal stocks or holdings with at least 2,500 square metres outdoor horticultural cultivation or at least 200 square metres greenhouse area. Source: Swedish Board of Agriculture [2]

Crop	1990[a]	1995[a]	2000[a]	2005	2010	2018	2019
Cereals[b]	1,335,700	1,104,500	1,228,900	1,024,000	962,800	991,700	993,200
Thereof wheat	349,700	261,400	401,600	354,800	400,000	381,100	472,200
Thereof barley	492,000	453,400	411,200	378,600	318,800	397,200	299,900
Thereof oats	387,800	278,300	295,500	200,100	164,400	163,900	147,900
Leguminous plants	–	21,200	37,300	40,900	46,100	56,600	44,200
Grass and green fodder[b]	918,100	1,058,900	920,800	1,067,000	1,194,700	1,121,200	1,163,700
Potatoes	36,200	35,000	32,900	30,500	27,200	23,900	23,600
Sugar beets	50,000	57,500	55,500	49,200	37,900	30,700	27,300
Rape and turnip rape	167,900	104,600	48,200	82,200	110,200	99,400	105,600
Other crops[c]	–	46,400	55,000	55,000	67,200	54,400	51,300
Fallow	176,100	278,600	247,700	321,300	176,800	165,400	131,700
Not specified arable land[d]	–	–	79,700	31,700	10,500	11,000	10,900
Not utilized arable land	46,400	59,800	–	1,800	–	–	–
Total arable land	2,844,600	2,766,600	2,706,000	2,703,300	2,633,500	2,554,400	2,551,500

[a] For the years 1990, 1995 and 2000 the numbers refer to holdings with more than 2.0 ha of arable land
[b] From the year 2000, certain areas intended for green fodder and cereal/leguminous plant mixtures to be harvested as green fodder are included in the area of cereals
[c] Seed lay, Linseed, Energy forest, Horticultural plants and all other small crops on arable land are not included above. From the year 2005, the area of horticultural plants includes spice plants and seed vegetables
[d] Areas are not possible to divide amongst crops

Sweden, similarly to most European countries, this land abandonment is a great concern because of the impact in the rural communities and agricultural landscapes and the decrease in biodiversity that it causes [6]. The Swedish Board of Agriculture has a Rural Development Programme which is intended to promote growth, competitiveness, entrepreneurship and employment and slow down agricultural land abandonment. The number of farms thus has declined while the mean farm size has increased, the production has intensified with respect to external inputs and is nowadays more regionally specialized, with more intensive agricultural enterprises placed in the south of the country. From the organizational point of view, the Swedish farmers work mostly grouped in cooperatives [7], which are focused on a unique activity. Therefore, the same farmer can belong to more than one cooperative, for example a farmer with a mixed crop-livestock farm would be part of a dairy and a cereal production cooperative at the same time. Such culture allows for specialization at the individual farmer level but also for integration of different farming orientations at farm and landscape level.

The intensification caused well-known adverse impacts like the loss of soil biodiversity [8] or the decline in farmland bird population [9]. The agricultural intensification was implemented through the increase of mineral fertilizers and pesticides and the upgrade of machinery, usually with bigger and heavier equipment. This weight increment causes soil compaction and negatively affects the soil functioning [10]. In spite of this intensification, Sweden is not self-sufficient in agriculture, since the imports exceed the exports in this sector. The imports had a value of 164,144 million SEK while the exports had a value of 97,568 million SEK in 2019 [2]. Among the imports, meat and meat preparations, dairy products, vegetables, fruits, coffee and tea stand out. On the other hand, Swedish's most important exports are cereals.

In Sweden, the conception of agriculture as part of the natural environment is more settled than in other parts of the modern world and the agricultural landscape together with the values associated with them are protected [11]. For example, grazing activities have created semi-natural grasslands that are identified as some of the most species-rich biotopes in Sweden [12] and their preservation is encouraged, regulated and financially supported by the government. Regulations concerning livestock density, storage capacity and spreading of animal wastes were implemented already in 1995 [13]. Sweden has, compared with the rest of the European Union countries, a large number of environmental measures within the national application of the EU's Common Agricultural Policy [14], favouring biodiversity protection in agricultural contexts. The concept of sustainable economic development is emphasized by the Swedish government. The Swedish Board of Agriculture is the Government's expert and managing authority for agriculture, fishing and rural development and is the central and coordinating authority for farming subsidies and the organization of incentives for sustainable management in agriculture.

2 Soil Compaction and Erosion (Water and Wind)

Soil compaction and erosion have important negative impact on the soil productivity and the environment. In Sweden, soil compaction (Fig. 4) in agricultural land is mainly due to field traffic using heavy vehicles following the last decades of intensification. The need of a programme for controlling the machinery-induced compaction was highlighted decades ago [15], and several mechanisms were proposed: reduction of load on running gear, reduction of ground contact pressure and reduction in the area of fields subjected to traffic [16]. However, the weight of the

Fig. 4 Compacted soil under oilseed rape crop in the south of Sweden. Photo taken by Linda-Maria Dimitrova Mårtensson, in June 2020, on the Alnarp Estate surrounding the campus of the Swedish University of Agriculture in Alnarp, Sweden

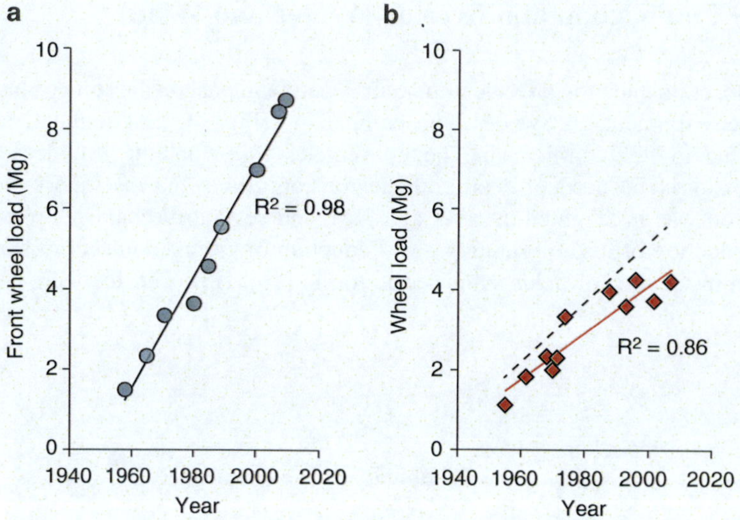

Fig. 5 Historical evolution of (**a**) front wheel loads of combine harvesters and (**b**) rear wheel loads of tractors. Figure from Keller et al. [10] (reproduced with permission of the publisher)

farm vehicles has continued to increase (Fig. 5), for example between 1989 and 2009 wheel loads of combine harvesters have increased 65% [17], causing an increment of the soil compaction levels corresponding to estimated costs of several hundred M€ year^{-1} for Sweden [10]. Swedish subsoils are highly susceptible to compaction because the soils are often wet during the field operations [18]. Despite the large contact area of the wheels distributing the pressure somewhat, the heavy weight on the soil surface distributes pressure into the subsoil layers. The degree of compaction is indeed correlated with the axle load and the compaction can be detected at 30 cm soil depth [19], which is right below the ploughing depth, where the ploughing itself also contributes to a subsoil compaction, i.e. the plough pan.

Different tillage systems have different impacts on the soil compaction. While the tillage system does not modify soil compaction significantly in the surface soil layer (0–10 cm), more superficial tillage strategies (for example, chiselling vs mouldboard) can cause an increment in the soil compaction in the deeper soil layers [20]. In soils with a documented plough pan at 25–30 cm, the application of inter-row subsoiling increases crop performance [21]. Most Swedish croplands are ploughed to a depth of at least 23 cm [22]; however, the current tendency in Sweden, similarly to the other Scandinavian countries, is towards reduced tillage strategies [23, 24]. On the other hand, the increase in the organic production in Sweden, which discards the use of herbicides for weed control, makes necessary the application of different tillage techniques (ploughing, cultivation, harrowing, inter-row cultivation) for weed control, even though non-mechanical strategies like cultivar choice, crop rotation, biological control or the use of cover crops are gaining more and more attention [25].

Soil compaction has a direct impact on the soil, increasing bulk density and decreasing water movement, which can negatively impact crop productivity [26]. An early study in Sweden showed an 11% yield decrease in barley in soils with low amount of organic matter [27], while a more recent one showed up to 72% loss in potato yield in the south of Sweden [21], as a consequence of soil compaction. In a controlled experiment in Sweden [28] where no traffic vs. three passes track-by-track was compared, decreases in yield differed depending on the crop: spring wheat (0.3%), winter rye (0.6%), winter wheat (3%), spring oilseed rape (3.6%), oats (8.7%), winter oilseed rape (8.8%), sugar beet (9.4%), potato (9.9%), pea (11.3%), and horse bean (21.7%). The type of crop in the rotation might as well influence the level of soil compaction. One example of this in Sweden is the sugar beet production, an important and profitable crop in the south region of the country. The introduction of six-row sugar beet harvesters, with total loads of around 35 Mg increases the risk of soil compaction up to 60% in the long term [29, 30].

Soil erosion from agricultural land is the main cause of soil loss and eutrophication processes. In Sweden, it was estimated that a maximum of 15% of the arable land is a source of surface erosion, directly via overland flow or indirectly via surface water inlets [31]. Soil erosion and consequent phosphorus losses are general problems in silt and clay agricultural soils in central and northern Sweden [32]. Most of the phosphorus losses (around 80%) are a consequence of soil water erosion events that occur in only 20% of catchment areas [33], known as critical source areas (CSA). It is difficult to identify these CSAs due to a wide range of factors governing phosphorous losses [34], but these authors identified two main causes of high phosphorus losses in Sweden: surface runoff on heavier soils and loss of dissolved phosphorus on sandy soils. Soils with values >35% of clay are associated with high suspended sediment concentrations in associated streams [31].

The soil management indeed has a high impact on the erosion rates and phosphorus losses, where reduced tillage and year-round vegetation reduce the losses [35]. Recent large-scale topographic data dictate the high relative erosion risk areas to be situated in the central and west part of the country, with soil loss rates of 2–5 T ha^{-1} $year^{-1}$ in arable lands [36]. Djodjic and Markensten [37] modelled the erosion risk in the central-south of Sweden, representing 90% of the arable land, and found that only 3% of the total area is included in the three highest erosion classes (>10 T km^{-2}). On the other hand, the model from Zhou et al. [38] only estimates erosion rates >5 T ha^{-1} in the very south of the country (Fig. 6). Projections of soil loss by water erosion in Europe by 2050 [39] predicts a drastic increase in the soil loss in agricultural soils in the central-south area of Sweden. A study in different agricultural streams (2004–2009) in Sweden measured between 0.01–1.75 T ha^{-1} $year^{-1}$ of suspended sediment and between 0.10–1.75 kg ha^{-1} $year^{-1}$ of total phosphorus [40].

Despite an improvement over the last decades, the phosphorus concentrations are more than twice the natural values in 75% of the Swedish lakes [41] and most of the eutrophicated lakes are situated in agricultural areas [42]. Engström et al. [43] point out eutrophication together with global warming and resource use as the most important sources of negative impacts on the surrounding ecosystems from Swedish

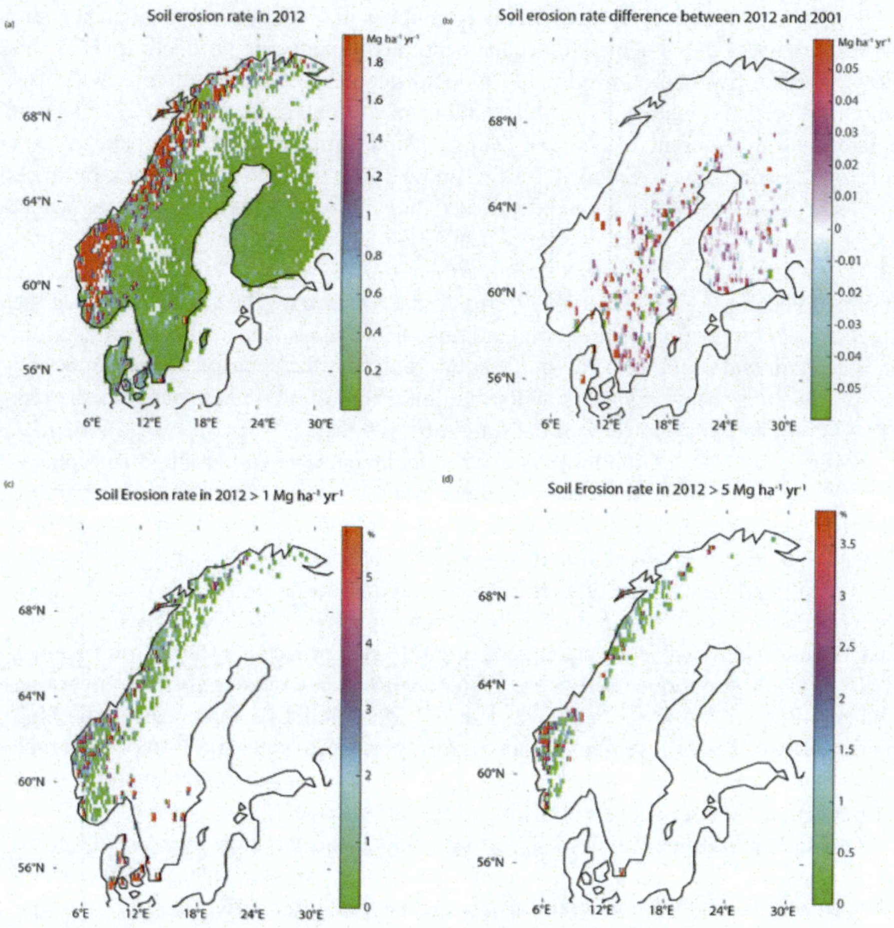

Fig. 6 Spatial distribution of the rate of soil erosion and the level of increased total soil erosion in agricultural land in the entire Scandinavia. The soil erosion rate in 2012 (**a**), soil erosion rate change between 2012 and 2001 (**b**), areas with soil erosion rate higher than 1 Mg ha^{-1} year^{-1} (**c**), areas with soil erosion rate higher than 5 Mg ha^{-1} year^{-1} (**d**). Maps in (**c**) and (**d**) are aggregated to 24 km spatial resolution to improve visualization, and the values show the percentages of area above the given thresholds in the aggregated grids. Agriculture land refers to the grid cells classified as agriculture land both in 2001 and 2012. Figure from Zhou et al. [38]

agriculture. Indeed, an increase in the concentration of suspended solids and total phosphorus from 0.4 to 0.7% over the period 1975–2004 in two large rivers in agricultural regions is assumed to be a consequence of climate change because different measures to reduce the phosphorus losses are successfully in place in other rivers [42].

Wind erosion is not a widespread problem in Europe, if compared with arid regions, however, it is a hazard in the northwest of the continent, in Lower Saxony (about 2 million ha of land), The Netherlands (97,000 ha), south-east of England

(260,000 ha), west of Denmark (about 1 million ha) and south of Sweden (170,000 ha) [44]. In Sweden, it occurs mostly in the southern county, Scania, which is exposed to periods of high wind erosion [45]. The total area of the arable land in Sweden exposed to wind erosion is estimated to be about 35,000 ha, of which about 20,000 ha is located in Scania [46]. The topography of the county is determinant for the sensitivity to wind erosion. Since Scania is rather flat with about 30% of the area below 30 m a.s.l., holding sandy and silty soils, around 23% of the arable land in the county has high potential wind erosion risk [47]. It is reported, in sugar beet fields, that the wind erosion in this area has increased since 1970 [48]. The reasons for this increment are increase in the frequency of strong winds, higher ratio of NE wind (more erosive), field enlargement, removal of wind breaking vegetation and new methods in beet cultivation. Areas heavily affected by wind erosion were transformed in the past from arable land by reforestation using pine plantations and there are dunes in the landscape consequence of past wind erosion events in agricultural fields. Despite this, in general, there is a lack of strategies to decrease wind erosion on arable land, being a problem for current agriculture in Scania [46].

3 Overgrazing

In general, overgrazing is not a problem in Sweden. On the contrary, a process of cessation of grazing and mowing species-rich semi-natural grasslands has taken place in the last years, which have had very negative consequences for grassland species diversity [49]. The process of management cessation in these species-rich semi-natural grasslands is resulting from the transition from the smaller scale, closed nutrient-loop-based agricultural practices with a grazing and mowing frequency allowing for a large range of grasses and forbs to coexist, to a more intense centralized livestock production relying on external energy and nutrient resources. The former semi-natural grasslands then suffer from encroachment, initially by shrubs and later by trees, leaving no room for the light-demanding species typical for, e.g., pastures and meadows. One particular example is the cessation of grazing which causes bush encroachment in the alvar vegetation, a unique type of grasslands with high richness of species and a great number of rare species, in the Swedish islands Öland and Gotland [50].

Governmental actions are taken to protect these semi-natural herbaceous habitats, which include protection or set-aside land with prescribed management. Such management often requires reintroduction of grazing. However, there is a risk that such reintroduced grazing causes local overgrazing, which in turn leads to a decrease in plant richness [50], thus missing the target of sustained plant species diversity and the protection of rare species. Another special fact in Sweden is that the number of horses for recreational use is bigger than the number of dairy cows [51], and the horse population has more than doubled since 1981 [22]. By law, the horses have to be outside at least, 1 h/day, every day, meanwhile the cows only have to be outside during the summer months. This regulation causes local overgrazing by the horses, while the meadows are overgrown due to the lack of grazing cattle [51]. There is a lack of studies regarding the impact of these local overgrazing events in the soil system.

On the other hand, overgrazing events due to reindeer husbandry have been highlighted as a problem in the north of Sweden, as well as in Norway and Finland [52]. The reindeer caused overgrazing, trampling, and the disappearance of bare soil and lichens, which is the main winter food for them [53]. To avoid these problems the authorities in these three countries regulated the number of reindeer in a reindeer village. Reindeer husbandry in Sweden is an exclusive right for the Saami, northern Scandinavia's indigenous people, and still in practice nowadays covering an area roughly 50% of Sweden [54]. Other authors stated that the overgrazing caused by the reindeer husbandry cannot be extrapolated, since strong grazing and trampling effects may be found just around enclosures and fences [55]. Johansen et al. [56] described change in plant species composition and soil erosion process in reindeer summer pastures in Norway. This activity is currently threatened by the loss of available grazing land due to forest and agricultural expansion, and to the difficulties in predicting the weather, due to climate change. Furthermore, a significant part of the previously species-rich semi-natural grasslands has been transformed to forests [57], and this agricultural land abandonment has been increasing since the 1940s in Sweden [58]. More research integrating the different socio-ecological challenges and aspects of the reindeer husbandry are needed [59].

Nowadays, domesticated animals are mainly kept on former arable land where a few high-yielding crop species are grown under high applications of fertilizers, instead of on less productive semi-natural grasslands [2]. Similarly, the ley production for winter and indoor feeding is done on fertile lands under high fertilization practices instead of in traditional hay meadows. Including such grass and grass-legume production for forage and fodder into ordinary crop rotations has several benefits like the increase in the soil organic carbon [60] with positive consequences for soil fertility and pest suppression, leading to improved yields in terms of both quantity and quality [61]. Here, a discussion on different goals and potential conflicts between goals may be advisable. However, we will leave such to another forum.

4 Slash-and-Burn Agriculture

Prohibition of slash-and-burn agriculture started in the 1570s in favour of the iron mining industry [62], but continued far into the twentieth century in long cultivation rotations, starting with the slash-and-burn clearing, continued with cultivation of more nutrient demanding annual crops and was later used for hay making providing winter fodder for the livestock, and at last, the area was used as pastures which slowly were left for the more natural succession into forested areas again [63]. This former use of slash-and-burn agriculture is recognized as one of the causes of the borealization of southern Sweden, where broadleaved forests were common before [64].

Today, the only remnant from this practice is prescribed burning, which is sometimes used in nature conservation areas where the grazing pressure or management intensity is insufficient to maintain high plant diversity and the targeted vegetation types. Prescribed burning can help avoid the competitive exclusion of rare plant species and encroachment from shrubs and trees in protected grassland and heather vegetation areas [65] otherwise occurring on the behalf of less competitive plant species.

5 Soil Contamination

Contaminants can enter the soil system in various ways: via atmospheric deposition transported from relatively near sources as well as from very distant ones; from the weathering of the underlying bedrock becoming the naturally occurring soil constituents; and through agricultural management. The accumulation of heavy metals in agricultural soils is considered an important and increasing environmental concern. Sweden has had policies in place since more than two decades ago towards zero accumulation of heavy metals in soils by reducing the input of heavy metals through atmospheric deposition, and fertilizer and sludge use [66]. Swedish boundary values for soil metal concentrations (Cd, Cu, Cr, Ni, Pb, Zn, Hg) are among the lowest in Europe, with the exception of Zn where Sweden allows higher concentrations in soils naturally rich in this metal [67]. The limits for cadmium represent the threshold values for human toxicity, but this might not be valid for soil microorganisms [67].

Cadmium is a naturally occurring heavy metal arising from the mother material from which soil mineral particles are weathered. Thus, the background levels of cadmium in agricultural soils depend on the underlying bedrock. At the European level, the background levels of cadmium are quite low in Sweden, with soil values generally between 0.02 and 0.09 mg kg^{-1} (below the average), with the exception of a non-agricultural area with higher values in the centre of the country [68]. The anthropogenic inflow of cadmium to arable soils in Sweden is dominated by atmospheric deposition and cadmium-containing phosphate fertilizers [69]. Airborne cadmium arises from smelter emissions, burning of fossil fuels and incineration of municipal waste. The resulting deposition of cadmium follows a gradient with the highest concentrations close to the continent and the lowest concentrations further away, in the north of Sweden [70].

Cadmium may also enter the agricultural land through the application of sewage sludge, which is an important strategy for closing nutrient loops in our current society. The Swedish boundary value on the concentration of metals in agricultural soil for the use of sewage sludge is 0.4 mg kg^{-1} dry matter [71]. Because of the large potential as agricultural fertilizer, the sewage sludge from two wastewater treatment plants in the south of Sweden was analysed on their content of heavy metals, including cadmium. The author reports that the sludge cadmium content has decreased from 3.5 mg kg^{-1} TS (in 1981) to 0.70 mg kg^{-1} TS (in 2017) [72]. The same study report on cadmium recovered in the crops after sludge application. The content averaged from 0.012 mg kg^{-1} TS in winter oilseed rape, 0.018 mg kg^{-1} TS in spring barley, 0.07 mg kg^{-1} TS in winter wheat, and 0.16 mg kg^{-1} TS in sugar beet. In a very few cases, the sludge application gave statistically supported higher cadmium content in the cereal grains or in sugar beet.

Furthermore, feed additives such as mineral feed and imported fodder and fodder additives further enrich soils with different nutrients when manure, slurry or urine are used as fertilizers, and so do liming activities [69]. The national mean of cadmium was estimated to 0.24 mg kg^{-1} dry soil [73], with 9% and 8% of evaluated agricultural soils reaching the categories of high and very high cadmium content [74] in relation to the Swedish boundary value of 0.4 mg kg^{-1} dry matter [71]. The solubility of cadmium in soils is related to the soil pH, with increasing solubility under pH 4.5, and thus influenced by anthropogenic acidification [75]. Cadmium present in agricultural soils and solved in the soil solution will partly end up in our products. Winter wheat, one of the widespread crops in Sweden, accumulates on average 0.05 mg kg^{-1} dry mass of grain [73]. However, according to a report from the Swedish Food Agency [76], cereals and potatoes contain rather low levels of cadmium, but since they are consumed at high levels, they still contribute largely to the exposure in both children and adults in the Swedish population. Other products, like mushrooms, crabs or moose kidney, hold higher levels of cadmium but are eaten less often or in lower amounts. A balanced diet is always recommended.

Sewage sludge application on arable land may also lead to other risks and potential problems. One important aspect is the spread of antibiotics and antibiotic resistant genes, and the resulting increasing occurrence of antibiotic resistance. However, a recent study on the long-term application of sewage sludge in arable land, with doses between 4–12 metric T ha^{-1} every 4 years, did not cause accumulation of antibiotics in soil [77]. The researchers found neither an increase in phenotypic resistance after sludge application nor an enrichment of resistance genes in soil. However, they did see subtle effects on microbial community composition and that the bioavailability of Cu was higher in long-term sludge-amended soil than in controls. However, other studies detected antibiotics in digestates coming from biogas production [78] that are used as bio-fertilizers on agricultural fields. The enrichment of harmful and potentially harmful organic substances, such as nonylphenol, did increase in soils amended by sewage sludge in the earlier days (in the 90s), but in later assessments, such enrichment could no longer be seen [79]. For most other organic substances, such as PAH (polycyclic aromatic hydrocarbon), PCB (polychlorinated biphenyls), DEHP (diethylhexyl phthalate), BBP (benzyl butyl phthalate), and DBP (dibutyl phthalate) the soils are still enriched when amended with sludge [79]. The DEP (diethyl phthalate) was below the detection limit and could not be quantified. It was concluded by Hörsing [79] that in the case of unhealthy organic substances, the concentrations are generally very low in the crops growing on land amended with sewage sludge and that large amounts of raw material need to be consumed to possibly reach boundary levels. Another type of substance that can end up in the agricultural soils and needs further attention are the synthetic nanoparticles which in some products are combined with elements such as titanium, zinc, silver or gold [67]. The knowledge about their toxicity and environmental impact is still limited.

6 Salinity

Sweden has more than 3,000 km of coastline, but salinity is not considered an agricultural problem. Salt intrusion does occur in the coastal areas, but here semi-natural pastures and meadows, i.e. continental salt marshes and salt meadows [80], are since long located and now under governmental protection for the sake of habitat and species protection. The coastal meadows have a natural distribution of 1,360 thousand ha, but only 3,600 ha is managed according to recommendations for habitat protection [81]. Furthermore, 72% of the boreal and 54% of the continental ones are Natura 2000 sites but despite this, the conservation of these ecosystems has been reported as unsuccessful, mainly due to abandonment [82]. Salt marshes and salt meadows are most often grazed by cattle due to their less precise preferences which can be seen in sheep and horses. The feed value differs with the dominating plant species and the local plant associations, with lower general quality and digestibility for *Phragmites* dominated swards and higher for *Juncus* dominated ones [83, 84].

However, local problems with soil salinity occur under specific conditions. For example, the use of salt for de-icing roads, a common practice during wintertime in Sweden can increase the salinity of streams or aquifers along the roads as well as soils [85]. On the other hand, the irrigation practices can be considered a risk to induce soil salinity, since almost all irrigation water contains some amount of dissolved salts. The relatively small percentage of land that is irrigated in Sweden (10% of the agricultural land, around 0.3 million ha) minimizes this risk, but in a future climate change scenario where irrigation practices would increase, it should be considered.

An interesting phenomenon is observed in the Baltic Sea, where Sweden has a long coast line. This sea has a natural low salinity of 0.3–0.9% due to the fact that it is an enclosed sea with a great input of freshwater from hundreds of rivers and with a single salt input point in the southwest. The Baltic has a problem of eutrophication due to the agricultural activities in the countries around the sea, Sweden being one of them. Indeed, the eutrophication process can decrease salinity, and this impact has been detected in the west coast of Sweden with a consequent negative impact in Charophytes algae [86].

7 Agrochemicals Use

The total hectare dose of agrochemicals used in Sweden generally lies within the range of 1.5–2 kg ha^{-1}, according to Statistics Sweden (2020) [87], with 58% use of herbicides, 26% fungicides, 13% insecticides and 3% growth regulators. One of the main environmental risks of pesticides is the contamination of water sources originated by leaching processes in the intensive agricultural areas, even though the impact on the stream's fauna and flora in intensive agricultural areas in Sweden has been currently described as limited [88]. The flow of pesticides is dominantly occurring through macroporous soils, but long-term storage along the transport pathways also occurs, presumably in subsoil horizons where degradation is slow [89]. The agricultural stream's pesticides concentration in Sweden is mostly below the European Uniform Principles, with only 2% of the samples above the limit, but this threshold might underestimate the long-term ecotoxicological potential of the pesticide's mixtures in the streams [90].

The presence of chemical pesticides in surface waters in several agricultural catchments in Sweden has been monitored since 2002 [91]. Pesticide concentrations measured in surface runoff from agricultural fields often exceed Swedish water quality standards [92]. Due to the recalcitrance of DDT (dichlorodiphenyltrichloroethane) and HCB (hexa-peri-hexabenzocoronene), residues are still found in Swedish agricultural soils even though they have been banned for decades. The metabolite of DDT, p,p-DDE and the HCB enters the human body primarily through animal products and has been found in the blood of almost all participants in a national monitoring programme [93]. The broad-spectrum herbicide glyphosate is now used instead. Glyphosate is found in freshwater [94, 95] and in urine from farmers after exposure [96], but the potential human health risks need further evaluation.

In Sweden, as well as in the EU, there is a vivid debate about the use of pesticides in agriculture, but regulation only defines maximum residue limits in food and feed, not in soils. Among the different pesticides, the use of glyphosate has caused a big controversy. Currently, the use of this herbicide is approved, but the approval expires in 2022. In Sweden, glyphosate represents 28% of the herbicide used for weed control, and the cessation of use would reduce between 5–8% of the income in the Swedish agriculture sector [97]. Glyphosate has been shown to have a negative impact on annelids (earthworms), arthropods (crustaceans and insects), molluscs, echinoderms, fish, reptiles, amphibians and birds [98] and mycorrhiza [99]. It should be noted that the monitoring of pesticide residues in foods of both plant and animal origin has revealed that the EU-harmonized maximum residue levels were not exceeded for any surveillance samples of Swedish origin in 2017 [100]. However, biomarkers of contemporary pesticides have been found in measurable concentrations in more than 50% of urine samples in Swedish adolescents, which indicates widespread exposure and a widespread pesticide exposure from diet in Swedish adolescents, but the concentrations are low and presumably below recommended acceptable daily intake values [101]. Indeed, a governmental inventory showed that 97% of the fruits, berries, vegetables and cereals had lower values than the statutory maximum residue level of pesticide residues [102].

A recent review on pesticide exposure on soil-dwelling, non-target invertebrates show high negative effects, quantified on a range of endpoints, from structural changes, such as visible, physical histological and/or morphological changes, to mortality, such as change in survival and/or lifespan [103]. Effects of pesticide exposure are dependent on type of pesticide, resulting in larger negative effects of insecticides on, e.g., beetles [103] and larger negative effects of fungicides on a variety of fungal taxa [104]. Furthermore, a differential response has been detected between the insecticide at hand and different invertebrate species [103]. Toxic effects of pesticides on soil microorganisms have been also described, regarding the inhibition of nitrification and denitrification mechanisms, in a study where 54 different pesticides were tested [105]. In terms of the less used pesticides, fungicides have been found toxic for bacterial communities in wetlands [106], while insecticides have been reported as a substantial risk to wild bees in agricultural landscapes [107] with the consequent negative impact on pollination. Furthermore, antimicrobial drugs can be used as growth regulators in animal production, known as antimicrobial growth promoters (AGP). In Sweden the use of AGP was banned as early as in 1986, being the first country setting up this ban allowing only the use of antimicrobials under veterinary prescription [108]. This legislation did not lead to economic repercussions in the animal husbandry sector, with the only exception of the porcine production where a slight increase in piglet mortality was observed.

In Sweden, 180,490 tonnes of inorganic nitrogen is applied annually to agricultural fields, which corresponds to 86% of the total 210,640 tonnes applied, where the total includes organic fertilizers [2]. The corresponding numbers for applied inorganic phosphorus and potassium are 13,460 (45%) and 24,190 (22%) tonnes, respectively. A meta-analysis shows that long-term inorganic fertilization in cropping systems increases microbial biomass, as a consequence of increased amount of soil organic matter [109]. However, inorganic fertilizers have proven to alter the soil bacterial community structure [110], down-regulate the activity of arbuscular mycorrhizal fungi [111] and nitrogen fixing bacteria [112] with consequences for plant nutrition and sustainable nutrient management in agroecosystems.

8 Microplastics

The presence of microplastics in the environment is a global concern. In Sweden, the main sources and routes of dispersal of microplastics are road traffic, followed by artificial sports grass lawns and households washing synthetic fabrics [113]. The microplastics are primarily entering the agricultural soils via stormwater, while the application of sewage water or sewage sludge from wastewater plants has been documented to add insignificant levels of microplastics. The use of sewage sludge in agricultural lands is currently under debate in Sweden, and the remains of microplastics is one of the main concerns, mainly because of the possible "cocktail effects" that can have when interacting with other substances present in the sludge [114]. Water treatment plants have shown to reduce the majority (98–99%) of the microplastics [115, 116]. Notwithstanding the water treatment plants are highly efficient in reducing microplastics, microplastics do occur in sewage sludge [117, 118], where a number of 720 particles kg^{-1} of sludge have been detected [117].

It cannot be excluded that cover materials, irrigation, flooding and atmospheric deposition also contribute to the entrance of microplastics into the agricultural soils [119]. In Sweden the largest greenhouse crops are cucumbers, lettuce, herbs and tomatoes, covering an area of around 125 ha in 2019 [2] and the significant use of plastics in horticulture through greenhouse covers, low and high tunnels, mulching films, etc., increases the risk of microplastic contamination.

9 Conclusions

In the 2.6 million ha arable land in Sweden, soil compaction and erosion represent problems, where high compaction due to the intensive management practices concurs with the natural susceptibility to wind erosion of agricultural soils. However, in the 0.4 million ha of pastures, overgrazing is not considered a problem and is thus not a driver behind soil erosion. Contrarily, the cessation of grazing and traditional hay making is deleterious for biodiversity in general. Furthermore, slash-and-burn agricultural practices do not currently suppose a risk for land degradation in Sweden. Swedish agricultural soils are exposed to many different contaminants, but the current regulations have resulted in low levels of such, resulting in a minor influence on human health. Salinity is currently not a problem for Swedish agriculture. The scientific knowledge on pesticide effects is not consistent and we may face threshold scenarios parallel to those discussed in Steffen et al. [120], who points out novel entities for which global-level boundaries cannot yet be quantified. The data on microplastics occurrence in soils are growing, while their impact on crops, humans, domesticated and wild organisms is still largely unknown.

References

1. Swedish Meteorological and Hydrological Institute. Open access data. https://www.smhi.se/en/services/professional-services/data-and-statistics. Retrieved 17 Jun 2021
2. Swedish Board of Agriculture and Statistics Sweden 2020. Open access data. http://www.scb.se/jo1901-en. Retrieved 17 Jun 2021 from Statistics Sweden
3. Grusson Y, Wesström I, Joel A (2021) Impact of climate change on Swedish agriculture: growing season rain deficit and irrigation need. Agric Water Manag 251:106858. https://doi.org/10.1016/j.agwat.2021.106858
4. Torstensson G, Aronsson H (2000) Nitrogen leaching and crop availability in manured catch crop systems in Sweden. Nutr Cycl Agroecosyst 56:139–152. https://doi.org/10.1023/A:1009821519042
5. Willer H, Schaack D, Lernoud J (2019) Organic farming and market development in Europe and the European Union. In: The world of organic agriculture. Statistics and Emerging Trends 2019. Research Institute of Organic Agriculture FiBL and IFOAM-Organics International, pp 217–254

6. Queiroz C, Beilin R, Folke C, Lindborg R (2014) Farmland abandonment: threat or opportunity for biodiversity conservation? A global review. Front Ecol Environ 12:288–296. https://doi.org/10.1890/120348
7. Moraru RA (2018) The cooperative system from Sweden agriculture: main features and evolution. Agronomy Ser 61:43–48
8. Tsiafouli MA, Thébault E, Sgardelis SP, De Ruiter PC, Van Der Putten WH, Birkhofer K, Hemerik L, de Vries FT, Bardgett RD, Brady MV, Bjornlund L, Bracht Jørgensen H, Christensen S, D' Hertefeldt T, Hotes S, Hol GWH, Frouz J, Liiri M, Mortimer SR, Setälä H, Tzanopoulos J, Uteseny K, Pižl V, Stary J, Wolters V, Hedlund K (2015) Intensive agriculture reduces soil biodiversity across Europe. Glob Chang Biol 21:973–985. https://doi.org/10.1111/gcb.12752
9. Wretenberg J, Lindström Å, Svensson S, Thierfelder T, Pärt T (2006) Population trends of farmland birds in Sweden and England: similar trends but different patterns of agricultural intensification. J Appl Ecol 43:1110–1120. https://doi.org/10.1111/j.1365-2664.2006.01216.x
10. Keller T, Sandin M, Colombi T, Horn R, Or D (2019) Historical increase in agricultural machinery weights enhanced soil stress levels and adversely affected soil functioning. Soil Tillage Res 194:104293. https://doi.org/10.1016/j.still.2019.104293
11. Saltzman K, Head L, Stenseke M (2011) Do cows belong in nature? The cultural basis of agriculture in Sweden and Australia. J Rural Stud 27:54–62. https://doi.org/10.1016/j.jrurstud.2010.09.001
12. Lindborg R, Bengtsson J, Berg Å, Cousins SAO, Eriksson O, Gustafsson T, Hasund K-P, Lenoir L, Pihlgren A, Sjödin E, Stenseke M (2008) A landscape perspective on conservation of semi-natural grasslands. Agric Ecosyst Environ 125:213–222. https://doi.org/10.1016/j.agee.2008.01.006
13. Lantbruksstyrelsen (1990) Lantbruksstyrelsens föreskrifter om begränsning av antalet djur i ett jordbruk. Lantbruksstyrelsens författningssamling 1988:44. Jönköping (in Swedish)
14. European Commission (2005) Agri-environment measures. Overview on general principles, types of measures and application. Unit G-4 e Evaluation of measures applied to agriculture studies. Directorate General for Agricultura and Rural Development: ec.europa.eu/agriculture/publi/reports/agrienv/rep_en.pdf
15. Håkansson I (1990) Soil compaction control—objectives, possibilities and prospects. Soil Technol 3:231–239. https://doi.org/10.1016/0933-3630(90)90003-L
16. Oskoui KE, Campbell DJ, Soane BD, McGregor MJ (1994) Economics of modifying conventional vehicles and running gear to minimize soil compaction. In: Developments in agricultural engineering, vol 11. Elsevier, pp 539–567
17. Schjonning P, van den Akker JJH, Keller MH, Greve MH, Lamandé M, Simojoki A, Stettler M, Ardvidsson J, Breuning-Madsen H (2015) Driver-Pressure-state-Impact-Response (DPSIR) analysis and risk assessment for soil compaction – a European perspective. Adv Agron 133:183–123. https://doi.org/10.1016/bs.agron.2015.06.001
18. Keller T, Arvidsson J (2006) Prevention of traffic-induced subsoil compaction in Sweden: experiences from wheeling experiments (Vermeidung von Unterbodenverdichtungen durch Landwirtschaftsmaschinen in Schweden: Erfahrungen aus Befahrungsversuchen). Arch Agron Soil Sci 52:207–222. https://doi.org/10.1080/03650340600631540
19. Trautner A, Arvidsson J (2003) Subsoil compaction caused by machinery traffic on a Swedish Eutric Cambisol at different soil water contents. Soil Tillage Res 73:107–118. https://doi.org/10.1016/S0167-1987(03)00104-1
20. Arvidsson J (1998) Effects of cultivation depth in reduced tillage on soil physical properties, crop yield and plant pathogens. Eur J Agron 9:79–85. https://doi.org/10.1016/S1161-0301(98)00026-4
21. Ekelöf J, Guamán V, Jensen ES, Persson P (2015) Inter-row subsoiling and irrigation increase starch potato yield, phosphorus use efficiency and quality parameters. Potato Res 58:15–27. https://doi.org/10.1007/s11540-014-9261-5

22. Poeplau C, Bolinder MA, Eriksson J, Lundblad M, Kätterer T (2015) Positive trends in organic carbon storage in Swedish agricultural soils due to unexpected socio-economic drivers. Biogeosciences 12:3241–3251. https://doi.org/10.5194/bg-12-3241-2015
23. Riley H, Borresen T, Ekeberg E, Rydberg T (2017) Trends in reduced tillage research and practice in Scandinavia. In: Conservation tillage in temperate agroecosystems. CRC Press, pp 23–45
24. Hydbom S, Olsson JA, Olsson PA (2020) The use of conservation tillage in an agro-intensive region: results from a survey of farmers in Scania, Sweden. Renew Agric Food Syst 35:59–68. https://doi.org/10.1017/S174217051800025X
25. Lundkvist A, Verwijst T (2011) Weed biology and weed management in organic farming. Res Org Farm:10–41
26. Arvidsson J, Håkansson I (1996) Do effects of soil compaction persist after ploughing? Results from 21 long-term field experiments in Sweden. Soil Tillage Res 39:175–197. https://doi.org/10.1016/S0167-1987(96)01060-4
27. Arvidsson J (1998) Influence of soil texture and organic matter content on bulk density, air content, compression index and crop yield in field and laboratory compression experiments. Soil Tillage Res 49:159–170. https://doi.org/10.1016/S0167-1987(98)00164-0
28. Arvidsson J, Håkansson I (2014) Response of different crops to soil compaction – Short-term effects in Swedish field experiments. Soil Tillage Res 138:56–63. https://doi.org/10.1016/j.still.2013.12.006
29. Arvidsson J, Trautner A, van den Akker JJH (2000) Subsoil compaction – risk assessment and economic consequences. In: Horn R, van den Akker JJH, Ardvisson J (eds) Subsoil compaction: preface advances in geoecology, vol 32, pp 3–12
30. Arvidsson J (2001) Subsoil compaction caused by heavy sugar beet harvesters in southern Sweden: I. Soil physical properties and crop yield in six field experiments. Soil Tillage Res 60: 67–78. https://doi.org/10.1016/S0167-1987(01)00169-6
31. Ulén B (2006) Sweden. In: Boardman J, Poesen J (eds) Soil erosion in Europe. Wiley
32. Ulén B, Jakobsson C (2005) Critical evaluation of measures to reduce phosphorus losses from arable land in Sweden. Sci Total Environ 344:37–50. https://doi.org/10.1016/j.scitotenv.2005.02.004
33. Sharpley AN, Bergström L, Aronsson H, Bechmann M, Bolster CH, Börling K, Djodjic F, Jarvie HP, Schoumans OF, Stamm C, Tonderski KS, Ulén B, Uusitalo R, Withers PJA (2015) Future agriculture with minimized phosphorus losses to waters: research needs and direction. Ambio 44:163–179. https://doi.org/10.1007/s13280-014-0612-x
34. Djodjic F, Elmquist H, Collentine D (2018) Targeting critical source areas for phosphorus losses: evaluation with soil testing, farmers' assessment and modelling. Ambio 47:45–56. https://doi.org/10.1007/s13280-017-0935-5
35. Ulén B, Kalisky T (2005) Water erosion and phosphorus problems in an agricultural catchment – need for natural research for implementation of the EU Water Framework Directive. Environ Sci Policy 8:477–484. https://doi.org/10.1016/j.envsci.2005.06.005
36. Panagos P, Borrelli P, Poesen J, Ballabio C, Lugato E, Meusburger K, Alewell C (2015) The new assessment of soil loss by water erosion in Europe. Environ Sci Policy 54:438–447. https://doi.org/10.1016/j.envsci.2015.08.012
37. Djodjic F, Markensten H (2019) From single fields to river basins: identification of critical source areas for erosion and phosphorus losses at high resolution. Ambio 48:1129–1142. https://doi.org/10.1007/s13280-018-1134-8
38. Zhou N, Hu X, Byskov I, Sandstad Næss J, Wu Q, Zhao W, Cherubini F (2021) Overview of recent land cover changes, forest harvest areas, and soil erosion trends in Nordic countries. Geogr Sustain 2:163–174. https://doi.org/10.1016/j.geosus.2021.07.001
39. Panagos P, Ballabio C, Himics M, Scarpa S, Matthews F, Bogonos M, Poesen J, Borrelli P (2021) Projections of soil loss by water erosion in Europe by 2050. Environ Sci Policy 124: 380–392. https://doi.org/10.1016/j.envsci.2021.07.012

40. Ulén B, Bechmann M, Øygarden L, Kyllmar K (2012) Soil erosion in Nordic countries – future challenges and research needs. Acta Agric Scand B Soil Plant Sci 62:176–184. https://doi.org/10.1080/09064710.2012.712862
41. SEPA (1999) Swedish Environment Protection Agency (Naturvårdsverket). Bedömningsgrunder för Miljökvalitet. Sjöar och Vattendrag. Report 4913. SEPA, Stockholm
42. Ulén B, Bechmann M, Fölster J, Jarvie HP, Tunney H (2007) Agriculture as a phosphorus source for eutrophication in the north-west European countries, Norway, Sweden, United Kingdom and Ireland: a review. Soil Use Manag 23:5–15. https://doi.org/10.1111/j.1475-2743.2007.00115.x
43. Engström R, Wadeskog A, Finnveden G (2007) Environmental assessment of Swedish agriculture. Ecol Econ 60:550–563. https://doi.org/10.1016/j.ecolecon.2005.12.013
44. Riksen M, Brouwer F, de Graaff J (2003) Soil conservation policy measures to control wind erosion in northwestern Europe. Catena 52:309–326. https://doi.org/10.1016/S0341-8162(03)00020-1
45. Mattsson JO (1987) Vinderosion och klimatändringar. Kommentarer till 1700-talets ekologiska kris i Skåne. (English summary: Wind erosion and climatic changes. Comments on the ecological crisis of Skane during the 18th century). Sw Geogr Yearb 63:94–108
46. Bärring L, Jönsson P, Mattsson JO, Åhman R (2003) Wind erosion on arable land in Scania, Sweden and the relation to the wind climate—a review. Catena 52:173–190. https://doi.org/10.1016/S0341-8162(03)00013-4
47. Riksen MJPM, de Graaff J (2001) On-site and off-site effects of wind erosion on European light soils. Land Degrad Dev 12:1–11. https://doi.org/10.1002/ldr.423
48. Jönsson P (1992) Wind erosion on sugar beet fields in Scania, southern Sweden. Agric For Meteorol 62:141–157. https://doi.org/10.1016/0168-1923(92)90012-S
49. Tyler T, Herbertsson L, Olsson PA, Froberg L, Olsson KA, Svensson A, Olsson O (2017) Climate warming and land-use changes drive broad-scale floristic changes in Southern Sweden. Glob Chang Biol 24:2607–2621. https://doi.org/10.1111/gcb.14031
50. Rosén E, van der Maarel E (2000) Restoration of alvar vegetation on Öland, Sweden. Appl Veg Sci 3:65–72. https://doi.org/10.2307/1478919
51. Hammer M, Bonow M, Petersson M (2017) The role of horse keeping in transforming peri-urban landscapes: a case study from metropolitan Stockholm, Sweden. Norsk Geografisk Tidsskrift – Norwegian J Geogr 71:146–158. https://doi.org/10.1080/00291951.2017.1340334
52. Bostedt G, Parks PJ, Boman M (2003) Integrated natural resource management in northern Sweden: an application to forestry and reindeer husbandry. Land Econ 79:149–159. https://doi.org/10.2307/3146864
53. Evans R (1996) Some impacts of overgrazing by reindeer in Finnmark, Norway. Rangifer 16:3–19. https://doi.org/10.7557/2.16.1.1177
54. Axelsson-Linkowski W, Fjellström AM, Sandström C, Westin A, Östlund L, Moen J (2020) Shifting strategies between generations in Sami reindeer husbandry: the challenges of maintaining traditions while adapting to a changing context. Hum Ecol 48:481–490. https://doi.org/10.1007/s10745-020-00171-3
55. Moen J, Danell Ö (2003) Reindeer in the Swedish mountains: an assessment of grazing impacts. AMBIO: A Journal of the Human Environment 32:397–402. https://doi.org/10.1579/0044-7447-32.6.397
56. Johansen B, Karlsen SR, Uhli C (2007) Mapping of vegetation, reindeer pastures and erosion-Kvaløya in Hammerfest municipality. Northern Research Institute, Report 9/2007, Tromsø. 43 pp (in Norwegian)
57. Dahlström A, Cousins SA, Eriksson O (2006) The history (1620-2003) of land use, people and livestock, and the relationship to present plant species diversity in a rural landscape in Sweden. Environ Hist 12:191–212. https://doi.org/10.3197/096734006776680218
58. Drake L (1999) The Swedish agricultural landscape–economic characteristics, valuations and policy options. Int J Social Econ. ISSN 0306-8293

59. Pape R, Löffler J (2012) Climate change, land use conflicts, predation and ecological degradation as challenges for reindeer husbandry in northern Europe: what do we really know after half a century of research? Ambio 41:421–434. https://doi.org/10.1007/s13280-012-0257-6
60. Prade T, Katterer T, Bjornsson L (2017) Including a one-year grass ley increases organic carbon and decreases greenhouse emissions from cereal-dominated rotations – a Swedish farm case study. Biosyst Eng 164:200–212. https://doi.org/10.1016/j.biosystemseng.2017.10.016
61. Lemaire G, Gastal F, Franzluebbers A, Chabbi A (2015) Grassland–cropping rotations: an avenue for agricultural diversification to reconcile high production with environmental quality. Environ Manag 56:1065–1077. https://doi.org/10.1007/s00267-015-0561-6
62. Emanuelsson M, Segerström U (2002) Medieval slash-and-burn cultivation: strategic or adapted land use in the Swedish mining district? Environ Hist 8:173–196. https://doi.org/10.3197/096734002129342639
63. Dove MR (2015) Linnaeus' study of Swedish swidden cultivation: pioneering ethnographic work on the 'economy of nature'. Ambio 44:239–248. https://doi.org/10.1007/s13280-014-0543-6
64. Lindbladh M, Axelsson AL, Hultberg T, Brunet J, Felton A (2014) From broadleaves to spruce-the borealization of southern Sweden. Scand J For Res 29:686–696. https://doi.org/10.1080/02827581.2014.960893
65. Valkó O, Deák B, Magura T, Török P, Kelemen A, Tóth K, Horváth R, Nagy DD, Debnár Z, Zsigrai G, Kapocsi I, Tóthmérész B (2016) Supporting biodiversity by prescribed burning in grasslands – a multi-taxa approach. Sci Total Environ 572:1377–1384. https://doi.org/10.1016/j.scitotenv.2016.01.184
66. Witter E (1995) Towards zero accumulation of heavy metals in soils. Fertilizer Res 43:225–233
67. Witter E (2009) Agricultural use of sewage sludge – is there a need to revise the Swedish regulations pertaining to heavy metals? Naturvardsverket
68. Salminen, R., Batista, M.J., Bidovec, M., Demetriades, A., De Vivo, B., De Vos, W., Duris, M., Gilucis, A., Gregorauskiene, V., Halamic, J., Heitzmann, P., Lima, A., Jordan, G., Klaver, G., Klein, P., Lis, J., Locutura, J., Marsina, K., Mazreku, A., O'Connor, P.J., Olsson, S.Å., Ottesen, R.-T., Petersell, V., Plant, J.A., Reeder, S., Salpeteur, I., Sandström, H., Siewers, U., Steenfelt, A., Tarvainen, T. 2005. Geochemical Atlas of Europe. Part 1 – Background information, methodology and maps. ISSN 951-690-913-2.
69. Fredrikson F, Holmberg J, Karlsson S (2002) Resource efficiency and toxic accumulation. II. Arable soils and cadmium stock dynamics. In: Fredrikson F (ed) Resource efficiency and toxic accumulation. Licentiate Thesis, Department of Physical Resource Theory, Chalmers and Göteborg University, Göteborg
70. Parkman H, Iverfeldt Å, Borg H, Lithner G (1998) Cadmium exposure in the Swedish environment Part I: Cadmium in Sweden – environmental risks. Kemi 1:98. The Swedish National Chemicals Inspectorate
71. SNFS (1998) Swedish Nature Protection Regulations on changes in the official announcement (SNFS 1994:2) on environmental protection, especially the soil, when sewage sludge is used for fertilisation in agricultural production – The Swedish Constitution on Nature Protection, SNFS 1998:4. Original title: Statens naturvårdsverks föreskrifter om ändring i kungörelsen (SNFS 1994:2) med föreskrifter om skydd för miljön, särskilt marken, när avloppsslam används i jordbruket – Statens naturvårdsverks författningssamling, SNFS 1998:4.
72. Dyrlund Martinsson U (2020) Slamtillförsel på åkermark. Slamrapport 2015-2018 (in Swedish)
73. Eriksson J, Mattsson L, Söderström M (2010) Current status of Swedish arable soils and cereal crops. Data from the period 2001–2007. Swedish Environmental Protection Agency Report 6349. Original title: Tillståndet i svensk åkermark och gröda. Data från 2001–2007. Naturvårdsverkets rapport 6349

74. Berndes G, Fredrikson F, Börjesson P (2004) Cadmium accumulation and salix-based phytoextraction on arable land in Sweden. Agric Ecosyst Environ 103:207–223. https://doi.org/10.1016/j.agee.2003.09.013
75. Bergkvist B, Folkesson L, Berggren D (1989) Fluxes of Cu, Zn, Pb, Cd, Cr, and Ni in temperate forest ecosystems. A literature review. Water Air Soil Pollut 47:198–216
76. Kristersson M (2017) Kadmium i livsmedel - Riskhanteringsrapport. Rapport 15 del 1. Swedish Food Agency. ISSN 1104-7089 (in Swedish)
77. Rutgersson C, Ebmeyer S, Lassen SB, Karkmana A, Fick J, Kristiansson E, Brandt KK, Flach C-F, Larsson DGJ (2020) Long-term application of Swedish sewage sludge on farmland does not cause clear changes in the soil bacterial resistome. Environ Int 137:105339. https://doi.org/10.1016/j.envint.2019.105339
78. Lehmann L, Bloem E (2021) Antibiotic residues in substrates and output materials from biogas plants – implications for agriculture. Chemosphere 278:130425. https://doi.org/10.1016/j.chemosphere.2021.130425
79. Hörsing M (2018) Sewage sludge on arable land – what do we need to know about organic substances? Svensk Vatten AB report 2018:4
80. EU (1992) European Union Council Directive 92/43/EEC of 21 May 1992 on the conservation of natural habitats and of wild fauna and flora
81. Swedish EPA (2011) Salta strandängar Atlantiska havsstrandängar (Galuco-Puccinellietalia maritimae) Atlantic salt meadows (Glauco-Puccinellietalia maritimae) Vägledning för svenska naturtyper i habitatdirektivets bilaga 1 NV-04493-11 (in Swedish)
82. European Environment Agency (2013) Habitats Directive Article 17 Reporting
83. Köster T, Viiralt R, Geherman V, Selge A (2004) Yield, quality and ecologically balanced utilisation of semi-natural grasslands on coastal areas of Estonia. Conference paper in: Land use systems in grassland dominated regions. Proceedings of the 20th General Meeting of the European Grassland Federation, Luzern, Switzerland, 21–24 June 2004. pp 906–908
84. Köster T, Kauer K, Kõlli R (2005) The grazing of livestock on coastal grasslands in Estonia. Conference paper in: Integrating efficient grassland farming and biodiversity. Proceedings of the 13th International Occasional Symposium of the European Grassland Federation, Tartu, Estonia, 29–31 August 2005. pp 376–379
85. Löfgren S (2001) The chemical effects of deicing salt on soil and stream water of five catchments in southeast Sweden. Water Air Soil Pollut 130:863–868
86. Blindow I (2000) Distribution of charophytes along the Swedish coast in relation to salinity and eutrophication. Int Rev Hydrobiol 85:707–717. https://doi.org/10.1002/1522-2632(200011)85:5/6<707::AID-IROH707>3.0.CO;2-W
87. Statistics Sweden (2020) Plant protection products in Swedish agriculture. Number of hectare-doses in 2019. ISSN 1654-3939
88. Bighiu MA, Hoss S, Traunspurger W, Kahlert M, Goedkoop W (2020) Limited effects of pesticides on stream macroinvertebrates, biofilm nematodes, and algae in intensive agricultural landscapes in Sweden. Water Res 174:115640. https://doi.org/10.1016/j.watres.2020.115640
89. Sandin M, Piikki K, Jarvis N, Larsbo M, Bishop K, Kreuger J (2018) Spatial and temporal patterns of pesticide concentrations in streamflow, drainage and runoff in a small Swedish agricultural catchment. Sci Total Environ 610–611:623–634. https://doi.org/10.1016/j.scitotenv.2017.08.068
90. Bundschuh M, Goedkoop W, Kreuger J (2014) Evaluation of pesticide monitoring strategies in agricultural streams based on the toxic-unit concept – experiences from long-term measurements. Sci Total Environ 484:84–91. https://doi.org/10.1016/j.scitotenv.2014.03.015
91. Boye K, Lindström B, Boström G, Kreuger J (2019) Long-term data from the Swedish national environmental monitoring program of pesticides in surface waters. J Environ Qual 48:1109–1119. https://doi.org/10.2134/jeq2019.02.0056
92. Larsbo M, Sandin M, Jarvis N, Etana A, Kreuger J (2016) Surface runoff of pesticides from a clay loam field in Sweden. J Environ Qual 45:1367–1374. https://doi.org/10.2134/jeq2015.10.0528

93. Swedish Food Agency and Swedish Environmental Protection Agency (2020) Contaminants in blood and urine from adolescents in Sweden – results from the national dietary survey Riksmaten Adolescents 2016–2017. Livsmedelsverkets samarbetsrapport S 2020:01. Uppsala
94. Jonsson O, Berggren K, Boström G, Gönczi M, Kreuger J (2019) Screening av bekämpningsmedel i dagvatten från bostadsområden – med fokus på glyfosat. CKB rapport 2019:2. Sveriges lantbruksuniversitet (in Swedish)
95. Nanos T, Kreuger J (2019) Resultat från miljöövervakningen av bekämpningsmedel (växtskyddsmedel). Årssammanställning 2017. Sveriges lantbruksuniversitet, Institutionen för vatten och miljö. Rapport 2019:1 (in Swedish)
96. Acquavella JF, Alexander BH, Mandel JS, Gustin C, Baker B, Chapman P, Bleeke M (2004) Glyphosate biomonitoring for farmers and their families: results from the Farm Family Exposure Study. Environ Health Perspect 112:321–326. https://doi.org/10.1289/ehp.6667
97. Kudsk P, Mathiassen SK (2019) Pesticide regulation in the European Union and the glyphosate controversy. Weed Sci 68:214–222. https://doi.org/10.1017/wsc.2019.59
98. Gill JPK, Sethi N, Mohan A, Datta S, Girdhar M (2018) Glyphosate toxicity for animals. Environ Chem Lett 16:401–426. https://doi.org/10.1007/s10311-017-0689-0
99. Zaller JG, Heigl F, Ruess L, Grabmaier A (2014) Glyphosate herbicide affects belowground interactions between earthworms and symbiotic mycorrhizal fungi in a model ecosystem. Sci Rep 4:5634. https://doi.org/10.1038/srep05634
100. Johansson A, Ahmed TM (2019) Monitoring of pesticide residues in Swedish Food. Swedish Food Agency Reports 2019:16. Original title: L 2019 nr 16: Kontroll av bekämpningsmedelsrester i livsmedel 2017. Livsmedelsverkets rapportserie. Uppsala. ISSN 1104-7089
101. Norén E, Lindh C, Rylander L, Glynn A, Axelsson J, Littorin M, Faniband M, Larsson E, Nielsen C (2020) Concentrations and temporal trends in pesticide biomarkers in urine of Swedish adolescents, 2000–2017. J Expo Sci Environ Epidemiol 30:756–767. https://doi.org/10.1038/s41370-020-0212-8
102. Jansson A, Fogelberg P (2018) Control of pesticide residues in food 2016. Swedish Food Agency. ISSN 1104-7089
103. Gunstone T, Cornelisse T, Klein K, Dubey A, Donley N (2021) Pesticides and soil invertebrates: a hazard assessment. Frontiers in Environmental Science 9:122. https://doi.org/10.3389/fenvs.2021.643847
104. Bunemann EK, Schwenke GD, Van Zwieten L (2006) Impact of agricultural inputs on soil organisms—a review. Aust J Soil Res 44:379–406. https://www.publish.csiro.au/sr/pdf/SR05125
105. Pell M, Stenberg B, Torstensson L (1998) Potential denitrification and nitrification tests for evaluation of pesticide effects in soil. Ambio 27:24–28
106. Milenkovski S, Bååth E, Lindgren PE, Berglund O (2010) Toxicity of fungicides to natural bacterial communities in wetland water and sediment measured using leucine incorporation and potential denitrification. Ecotoxicology 19:285–294. https://doi.org/10.1007/s10646-009-0411-5
107. Rundlof M, Andersson GKS, Bommarco R, Fries I, Hederstrom V, Herbertsson L, Jonsson O, Klatt BK, Pedersen TR, Yourstone J, Smith HG (2015) Seed coating with a neonicotinoid insecticide negatively affects wild bees. Nature 521:77. https://doi.org/10.1038/nature14420
108. Maron DF, Smith TJS, Nachman KE (2013) Restrictions on antimicrobial use in food animal production: an international regulatory and economic survey. Glob Health 9:48
109. Geisseler D, Scow KM (2014) Long-term effects of mineral fertilizers on soil microorganisms – a review. Soil Biol Biochem 75:54–63. https://doi.org/10.1016/j.soilbio.2014.03.023
110. Marschner P, Kandeler E, Marschner B (2003) Structure and function of the soil microbial community in a long-term fertilizer experiment. Soil Biol Biochem 35:453–461. https://doi.org/10.1016/S0038-0717(02)00297-3

111. Hammer EC, Nasr H, Wallander H (2011) Effects of different organic materials and mineral nutrients on arbuscular mycorrhizal fungal growth in a Mediterranean saline dryland. Soil Biol Biochem 43:2332–2337. https://doi.org/10.1016/j.soilbio.2011.07.004
112. Schipanski ME, Drinkwater LE, Russelle MP (2010) Understanding the variability in soybean nitrogen fixation across agroecosystems. Plant Soil 329:379–397. https://doi.org/10.1007/s11104-009-0165-0
113. Magnusson K, Eliasson K, Fråne A, Haikonen K, Hultén J, Olshammar M, Voisin A (2016) Swedish sources and pathways for microplastics to the marine environment – a review of existing data. (C 138). IVL Swedish Environmental Research Institute, Stockholm
114. Ekane N, Barquet K, Rosemarin A (2021) Resources and risks: perceptions on the application of sewage sludge on agricultural land in Sweden, a case study. Front Sustain Food Syst 5:94. https://doi.org/10.3389/fsufs.2021.647780
115. Simon M, van Alst N, Vollertsen J (2018) Quantification of microplastic mass and removal rates at wastewater treatment plants applying focal plane array (FPA)-based Fourier transform Infrared (FT-IR) imaging. Water Res 142:1–9. https://doi.org/10.1016/j.watres.2018.05.019
116. Ljung E, Borg Olesen K, Andersson P-G, Fältström E, Vollertsen J, Wittgren HB, Hagman M (2018) Microplastics in the water and nutrient-cycle. Svenskt Vatten Utveckling Rapport 2018-13
117. Magnusson K, Norén F (2014) Screening of microplastic particles in and down-stream a wastewater treatment plant. VL Swedish Environmental Research Institute
118. Norén K, Magnusson K, Westling K, Olshammar M (2016) Report concerning techniques to reduce litter in waste water and storm water. Swedish Meteorological and Hydrological Institute, Norrköping
119. Bläsing M, Amelung W (2018) Plastics in soil: analytical methods and possible sources. Sci Total Environ 612:422–435. https://doi.org/10.1016/j.scitotenv.2017.08.086
120. Steffen W, Richardson K, Rockström J, Cornell SE, Fetzer I, Bennett EM, Biggs R, Carpenter SR, de Vries W, de Wit C, Folke C, Gerten D, Heinke J, Mace GM, Persson LM, Ramanahatan V, Reyers B, Sörlin S (2015) Planetary boundaries: guiding human development on a changing planet. Science 347:6223. https://doi.org/10.1126/science.1259855

Agricultural Soil Degradation in Ukraine

Oleksandr Menshov and Oleksandr Kruglov

Contents

1 Introduction 326
1.1 Soil Compaction 330
1.2 Water and Wind Erosion 332
1.3 Soil Contamination 335
1.4 Salinity 340
1.5 Agrochemicals Use 341
1.6 Overgrazing, Slash-and-Burn Agriculture, Microplastics and Other Forms of Soil Degradation 343
2 Conclusions 344
References 345

Abstract The degradation of the arable soils is one of the critical issues for Ukraine's agriculture production. The agricultural area covers 19 million hectares within the Steppe zone, 16.9 million hectares within the Forest-Steppe zone and 5.6 million hectares in the Forest zone. The most important soils in Ukraine are chernozems, phaeozems and albeluvisols. This chapter aims to study the factors affecting Ukraine's agricultural soil degradation. The physical degradation of the Ukraine lands is related to excessive ploughing, poor balance of nutrients, insufficient application of organic matter, mineral fertilisers, ameliorants and pollution. Many economically valuable crops are growing without reliable erosion protection measurements application. The area affected by the radioactive elements in Ukraine is about 461.7 thousand hectares. Overgrazing, slash-and-burn agriculture are not

O. Menshov (✉)
Taras Shevchenko National University of Kyiv, Kyiv, Ukraine

O. Kruglov
NSC "Institute for Soil Science and Agrochemistry Research n. a. O.N. Sokolovskiy", Kharkiv, Ukraine

Paulo Pereira, Miriam Muñoz-Rojas, Igor Bogunovic, and Wenwu Zhao (eds.), 325
Impact of Agriculture on Soil Degradation II: A European Perspective,
Hdb Env Chem (2023) 121: 325–348, DOI 10.1007/698_2022_951,

Published online: 29 January 2023

typical in Ukraine. The urgent needs for Ukraine are innovation and investments in strengthening the material and technical base of the agricultural sector.

Keywords Erosion, Pollution, Soil degradation

1 Introduction

The Ministry of Agrarian Policy and Food of Ukraine (MAPFU) is responsible for forming and implementing the Agrarian Policy of Ukraine. The Department of Engineering and Technical Support and Agricultural Engineering of MAPFU is a subdivision of the Ministry. The department's main tasks are implementing state policy on engineering and technical support and developing national agricultural machinery production. Recently, the Ministry and its departments emphasised soil fertility preservation in Ukraine. Among the urgent needs are innovation and investments in strengthening the material and technical base of the agricultural sector, implementing environmentally friendly, resource and energy-saving technologies and conservation. Ukraine's State Agency of Land Resources is the central executive authority on land resources activity. The Cabinet of Ministers of Ukraine coordinates the Agency through the MAPFU [1].

The total area of Ukraine is 60.35 thousand km^2. Ukraine includes three agro-ecological zones and two mountain regions: The forest zone or Polissya with 19% of total land, the Forest-Steppe zone with 35%, Steppe zone with 40%, and the Carpathian and Crimean mountains, which occupy, respectively, the west and the southern parts of the territory [1]. The Steppe zone covers 19 million hectares of agricultural land, the Forest-Steppe zone includes 16.9 million hectares and the Forest zone is 5.6 million hectares (Fig. 1).

The most essential for agriculture and well-known fertile soil in the world is Ukraine Chernozems, black soils rich in humus. Chernozem soil covers about half of the country (68% of the arable land). Chernozems are some of the most common soils in Europe. There is still no consensus on the factors controlling Central European Chernozems' formation, conservation and degradation [2]. However, the on-going human-induced chernozem degradation is still underestimated and poorly studied in Ukraine. Other fertile and important agricultural soils in Ukraine are Phaeozems and Albeluvisols [3]. Physical, chemical and biological nominal data of Ukrainian soils and their classification were studied in the mid-twentieth century. Since then, no all-inclusive soil data update has been done [1, 4]. The soil organic matter content of chernozems ranges from 5.2% in wet Forest-Steppe, 5.7% in Forest-Steppe, 6.2% in the steppe and 3.4% or less in South Steppe. Fertility follows a similar pattern, decreasing from Forest Steppe to the Southern Steppe [1].

As of 01.01.2016, the farmlands in Ukraine cover 41,477.2 thousand hectares or 68.72% of the territory. The agricultural lands occupied 42,726.4 thousand hectares or 70.8%. They consist of arable land – 32,528.9 thousand hectares (78.3%),

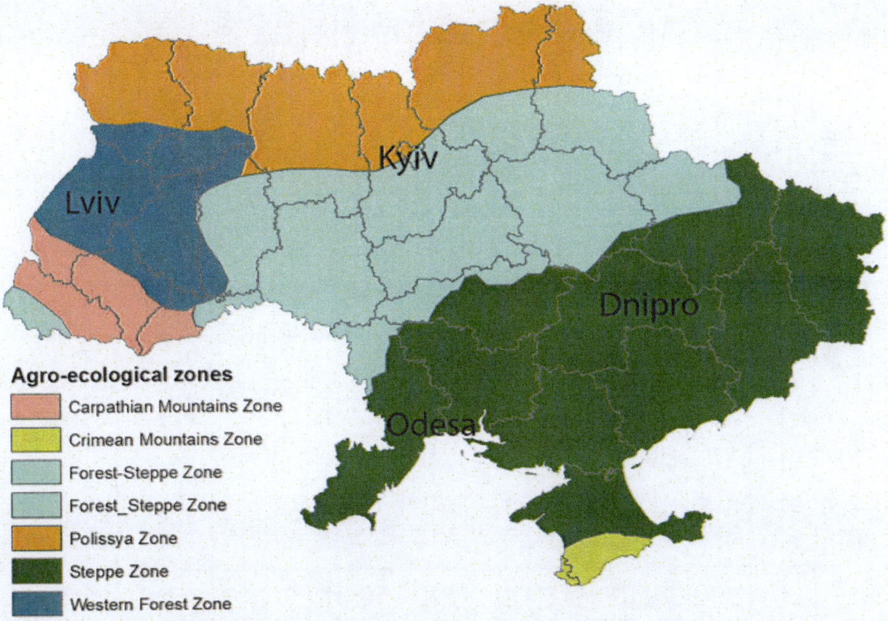

Fig. 1 Agro-ecological zones of Ukraine

perennial plantations – 879.4 thousand hectares (2.1%), meadows and pastures – 5,429.9 thousand hectares (13.1%), hayfields – 2,405.3 thousand hectares (5.8%) and fallow lands – 233.7 thousand hectares (0.6%). Detailed information about agricultural land design in Ukraine is described in Table 1. Ukraine is in the group of countries with a high level of ploughing, significantly exceeding the regional average level for Eastern Europe (61.7%). The part of the arable land compared to the total land area is 56%.

Since 2011, the area of arable land has increased by 52.4 thousand hectares. At the same time, the area of several other types of categories of agricultural land was decreased: perennial plantations by 17 thousand hectares, pastures by 52 thousand hectares, hayfields by 5 thousand hectares and fallow lands by 75.5 thousand hectares.

The highest degree of ploughing of the agricultural lands is observed in the Kherson and Cherkasy regions. The lowest ploughing of the arable land is typical for the Zakarpattia and Lviv regions [5].

The structure of agricultural land generally does not always reflect the actual state of land use. More reliability is the characteristic of economically active pharm enterprises. As of 2016, according to the State Statistics Service [5], there were 47,697 agricultural enterprises in Ukraine, which operated on 19,821.2 thousand hectares of agricultural land, which included 19,010 thousand hectares of arable land. Thus, the amount of all other types of land, except arable land, is less than 4.1%. This situation does not allow us to discuss modern land use's ecological balance.

Table 1 Agriculture lands of Ukraine [6]

Region	Agriculture land, thousand ha	Arable land		Perennial plantation		Meadows and pastures		Hayfields		Fallow lands	
		thousand ha	%	thousand ha	%	thousand ha	%	thousand ha	%	thousand ha	%
Polissya											
Volyn	1047.6	672.6	64.2	11.7	1.1	201.4	19.2	161.9	15.5	–	–
Zhytomyr	1510.1	1112.7	73.7	23.4	1.5	185.0	12.3	126.9	8.4	62.1	4.1
Transcarpathian	451.0	200.2	44.4	27.3	6.1	129.2	28.6	94.3	20.9	–	–
Ivano-Frankivsk	630.5	397.2	63.0	16.3	2.6	126.3	20.0	83.9	13.3	6.8	1.1
Lviv	1261.5	794.1	62.9	23.2	1.8	255.8	20.3	187.7	14.9	0.7	0.1
Rivne	926.2	656.8	70.9	11.7	1.3	127.7	13.8	126.5	13.7	3.5	0.4
Chernihiv	2067.5	1419.2	68.6	24.5	1.2	282.9	13.7	306.1	14.8	34.8	1.7
Total regions	**7894.4**	**5252.8**	**66.5**	**138.1**	**1.7**	**1308.3**	**16.6**	**1087.3**	**13.8**	**107.9**	**1.4**
Forest-Steppe											
Vinnytsia	2014.2	1725.5	85.7	51.4	2.6	186.1	9.2	50.2	2.5	1.0	0.0
Kyiv	1664.2	1355.5	81.5	44.7	2.7	134.9	8.1	116.7	7.0	12.4	0.7
Poltava	2165.5	1774.7	82.0	28.7	1.3	199.3	9.2	160.3	7.4	2.5	0.1
Sumy	1698.0	1226.3	72.2	24.4	1.4	166.8	9.8	280.4	16.5	0.1	0.0
Ternopil	1046.2	856.4	81.9	15.7	1.5	144.1	13.8	26.6	2.5	3.4	0.3
Kharkiv	2411.5	1933.2	80.2	48.9	2.0	304.9	12.6	117.0	4.9	7.5	0.3
Khmelnytsky	1566.2	1252.7	80.0	41.6	2.7	135.4	8.6	135.3	8.6	1.2	0.1
Cherkasy	1451.0	1272.0	87.7	27.3	1.9	78.4	5.4	64.8	4.5	8.5	0.6
Chernivtsi	469.7	330.8	70.4	30.2	6.4	67.8	14.4	40.9	8.7	–	–
Total regions	**14486.5**	**11727.1**	**81.0**	**312.9**	**2.2**	**1417.7**	**9.8**	**992.2**	**6.8**	**36.6**	**0.2**
Steppe											
Crimea	1792.5	1271.5	70.9	75.7	4.2	432.7	24.1	2.0	0.1	10.6	0.6
Dnepropetrovsk	2513.0	2127.4	84.7	53.1	2.1	315.1	12.5	17.4	0.7	–	–
Donetsk	2041.1	1652.7	81.0	57.9	2.8	287.2	14.1	42.6	2.1	0.7	–

Zaporizhye	2241.7	1903.6	84.9	38.7	1.7	216.1	9.6	83.3	3.7	–	–
Kirovograd	2032.2	1764.6	86.8	25.4	1.2	218.5	10.8	23.7	1.2	–	–
Luhansk	1908.6	1276.6	66.9	29.5	1.5	463.3	24.3	91.7	4.8	47.5	2.5
Mykolayivska	2006.0	1699.2	84.7	35.7	1.8	264.1	13.2	3.9	0.2	3.1	0.2
Odessa	2591.8	2075.5	80.1	86.5	3.3	351.9	13.6	50.6	2.0	27.3	1.1
Kherson	1969.4	1777.9	90.3	25.9	1.3	155.0	7.9	10.6	0.5	–	–
Total regions	**19096.3**	**15549.0**	**81.4**	**428.4**	**2.2**	**2703.9**	**14.2**	**325.8**	**1.7**	**89.2**	**0.5**
Total in Ukraine	**41477.2**	**32528.9**	**78.4**	**879.4**	**2.1**	**5429.9**	**13.1**	**2405.3**	**5.8**	**233.7**	**0.6**

In addition, one of the main factors in the development of soil degradation processes (primarily water and wind erosion) is the changes in the structure of sown areas of crops [5]. The areas under economically valuable crops are growing, but these practices increase erosion. As a result, compared to 2010, the area under sunflower increased to 5,107 thousand hectares (11.6%) in 2015, soybeans – increased to 2,158 thousand hectares (100%) and corn for grain increased to 4,123 thousand hectares (52.2%). At the same time, over 5 years, the area of perennial grasses decreased to 1,027 thousand hectares (20.5%).

1.1 Soil Compaction

Soil compaction is a part of the physical degradation of the land in Ukraine [7]. Nowadays, the processes of physical degradation are dangerous and common for Ukraine lands according to excessive ploughing, the poor balance of nutrients, insufficient application of organic matter, mineral fertilisers, ameliorants or pollution. Physical degradation is the effect of intensive agricultural land use, particularly excessive ploughing of soils, intensive mechanical cultivation and reduced soil organic matter content. These destructive processes covered the entire arable land of Ukraine. The result is the destruction of the upper layer (topsoil), lumps after ploughing, inundation and crusting, the presence of a plough sole, and finally, the compaction of the soil subsoil layers (horizon B and C) and deeper layers. The physically degraded soils are sensitive to erosion and characterised by decreased absorption and retention of atmospheric moisture, limiting plant root systems' development.

Soil compaction is the changes in the soil density related to anthropogenic activity resulting in decreased growth of cultivated plants. It is high in soils with high humidity (Voronic Chernozems Pachic) [8]. Conversely, the danger of soil compaction development is much less for soils with a light particle size structure (Luvic Phaeozems Albic), which are more common for Ukraine Polissya [9]. Thus, at present, the threat of over-compaction is significant for 2/3 (66%) of the arable land in Ukraine [7].

Soil compaction appears at more than 20% of the area when growing cereals. At the same time, the increase in the soil density for the row crops (sugar beets and corn for grain) is up to 45% [7]. 39% of Ukraine's arable land is characterised by the risk of soil compaction development and achievement of the total 72% of the destruction [10]. The potential financial losses were predicted in 2000 [11] at 500 million USD. The excess soil density negatively influences the overall productivity of the agrocenosis [12].

Over-compaction occurs under the influence of two factors: natural and anthropogenic. In the first case, it is called self-sealing. The main factors are climatic phenomena and gravity. None of the soil types in Ukraine can counteract the influence of these factors in the state of the arable land [12]. The most resistant to self-compaction are the sandy soils of Polissya, and the least resistant is the chernozem soil of Steppe and Forest-Steppe [7]. At the same time, the observation results detected a negative correlation between the total humus content and compaction for the soil of one type (R = −0.64) [13]. They did not observe the soil compaction in protected conditions (national parks). There is evidence that soil compaction is much more related to the arable land in Ukraine.

The anthropogenic changes are related to the combination of two processes. The first one is the agricultural machinery undercarriage (i.e. wheels). The second one is the working process and impact of the agricultural machinery working units (i.e. plough, harrow) [14]. The result of the agricultural machinery influence is the linear structures of the over-compaction of productive soil. Among these linear structures, the most important are tracks and continuous formation (plough sole). In the case of the tracks, soil compaction can reach 70–80 cm [15, 16]. In the second case of the continuous formation (plough sole), the effect is visible according to the depth of the soil tillage, which is 5–35 cm, depending on the tool type. The deepest layer of the impact is 50 cm.

The level and intensity of the soil compaction depend on the type of running system of machines and working units of the agricultural tools. Several attempts were made to regulate the allowable pressure values on soil [17–19]. Soil compaction is not only one type of loss and damage generated by the agricultural machinery undercarriage (i.e. wheels). Among other negative impacts are soil structure damage, resulting in a decrease in harvest [12].

In Ukraine, the agronomic classification of soil density is [13]:

- loose soil (<1.1 g/cm^3),
- optimal density (1.1–1.3 g/cm^3),
- dense soil (1.3–1.4 g/cm^3),
- compacted soil (>1.4 g/cm^3).

The primary measure of the compaction prevention of the arable horizon of agricultural soils in Ukraine is their cultivation. The studies in the research field demonstrated the effect of different measures of cultivation of typical chernozem on its density. At the same time, the optimal physical values of the lower part of the arable layer were increased when attracting the shallow cultivation with disk tools [12, 20]. These results confirm the need to attract the algorithm with the periodic deep cultivation (40–50 cm) with the combination of many different tools and machinery. For example, replacing traditional ploughing with tillage increases soil compaction in Ukraine (0.03 g/cm^3 on average) [21].

A significant increase in the density of the arable soil layer was detected in the first year if the tillage was not applied compared with the traditional ploughing. The increase of 0.07 g/cm^3 is 6%, and the average density increase under the studying was up to 7.9%. Generally, the relaxation time (lea) leads to the compaction increase throughout all arable layers of soil [22].

1.2 Water and Wind Erosion

Soil erosion is a critical factor in soil degradation in Ukraine [7] and all over the world [23].

From the beginning of the twenty-first century, the environmental and economic losses of agricultural production in Ukraine have reached the dramatic proportions, which were related to the anthropogenic impact and soil erosion. The main reason is excessive soil ploughing [7].

Water erosion in Ukraine involves up to 15 million hectares of agricultural land when the wind erosion affects up to six million hectares [24]. More than 20 million hectares are affected by wind erosion [8]. In several regions of Ukraine, the percentage of eroded lands is much higher than the national average [7] (see Table 2).

According to expert estimation, Ukraine lost up to 500 million tons of soil annually because of erosion. Accordingly, soil loss includes:

- 24 million tons of humus,
- 0.96 million tons of nitrogen,
- 0.68 million tons of phosphorus,
- 9.4 million tons of potassium.

The amount described above lost is comparable with the values of reverse mineral fertilisation [8]. The annual growth of eroded lands reaches 80 thousand hectares [25]. The largest increase is related to the Steppe zone of Ukraine [7]. The eroded soil is characterised by low fertility. The losses of the agricultural sector of Ukraine related to the shortage of yield reach annually (in equivalent): 9–12 million tons of grain, or in 2010 prices up to six billion USD [4]. The report's results confirm that arable land's erosion reaches 40% [7]. The largest areas of eroded soil in Ukraine were detected in the area of Vinnytsia, Luhansk, Donetsk, Odessa, Chernivtsi and Ternopil regions.[1] These regions include 4.5 million hectares with moderately and heavily eroded soils, and 68 thousand hectares have entirely lost the humus horizon [7].

The main reason for a significant increase in the soil erosion processes in Ukraine is the high level of ploughing in Ukraine. The ploughed land in Ukraine is 53.9% of the total area and 78.1% of the agricultural land [7]. The lowest part of the rill is in Polissya of Ukraine, the medium is in Forest Steppe and the highest is in the steppe areas [25]. The mentioned above three climatic zones of Ukraine (Polissya, Forest Steppe and Steppe) have differentiated in their types of erosion development [4].

The main factor of the erosion process occurrence in Ukraine is the relief conditions. The slopes of agricultural lands of Ukraine have a differentiation [25]:

[1] https://superagronom.com/news/200-naybilshi-problemi-z-eroziyeyu-gruntiv-v-shesti-oblastyah-ukrayini.

Table 2 Eroded lands in Ukraine: regions and agro-climatic zones (thousand hectares) [7]

Region/zone	Agricultural land	Arable land	Eroded land Total	Eroded land Agricultural land (%)	Arable land Total	Arable land Arable land (%)
Volyn	1051.4	674.3	362.4	34.5	225.4	33.4
Zhytomyr	1526.9	1092.8	87.8	5.8	60.7	5.6
Transcarpathian	453.2	200.6	39.6	8.7	35.5	17.7
Ivano-Frankivsk	631.9	381.6	133.7	21.2	98.4	25.8
Lviv	1267.8	797.2	525.0	41.4	380.1	47.7
Rivne	933.9	658	323.3	34.6	224.2	34.1
Chernihiv	2076.7	1396.1	81.0	3.9	53.3	3.8
Polissya	7941.8	5200.6	1552.8	19.6	1077.6	20.7
Vinnytsia	2017.1	1729	687.5	34.1	593.1	34.3
Kyiv	1668.4	1360.6	157.9	9.5	128.8	9.5
Poltava	2175.7	1768.8	517.7	23.8	420.3	23.8
Sumy	1701.6	1232.8	305.1	17.9	176.3	14.3
Ternopil	1049.7	854.0	244.0	23.2	239.7	28.1
Kharkiv	2418.7	1926.6	996.3	41.2	791.2	41.1
Khmelnytsky	1568.4	1254.8	628.4	40.1	501.9	40.0
Cherkasy	1451.4	1271.6	326.6	22.5	286.1	22.5
Chernivtsi	471.2	333.9	124.2	26.4	88.5	26.5
Forest-Steppe	14522.2	11732.1	3987.7	27.5	3225.9	27.5
Crimea	1798.4	1265.6	999.3	55.6	919.3	72.6
Dnepropetrovsk	2514.3	2125.0	1104.8	43.9	914.7	43.0
Donetsk	2045.2	1656.0	1757.4	85.9	1080.0	65.2
Zaporizhye	2247.7	1906.7	1212.5	53.9	640.8	33.6
Kirovograd	2039.9	1762.4	1102.4	54.0	886.7	50.3
Luhansk	1911.1	1269.7	1372.3	71.8	1237.9	97.5
Mykolayivska	2010.0	1698.1	964.5	48.0	914.8	53.9
Odessa	2593.4	2067.6	1214.0	46.8	1081.6	52.3
Kherson	1971.1	1777.6	686.2	34.8	961.0	54.1
Steppe	19131.1	15528.7	10413.4	54.4	8636.8	55.6
Total in Ukraine	***41595.1***	***32461.4***	***15953.9***	***38.4***	***12940.3***	***39.9***

– from 0° to 1.3° – 78%,
– from 1.3° to 3° – 17%,
– from 3° to 6° – 0.9%,
– from 6° to 12° – 2.1%,
– from 12° to 20° – 1.8%,
– ≥20° – 0.2%.

To predict soil erosion, the proposal was developed to transfer eight million hectares of arable land to less erosive dangerous territories [4]. However, the land reform in Ukraine is not completed now, and the present land legislation support does not allow the full implementation of the arable land transfer.

Among the measures to protect soils from erosion in Ukraine [4]:

- soil management,
- agroforestry,
- agronomic,
- special.

Among the soil management measures regulating land use is the implementation of the "Guidelines for the development of land management projects that provide environmental and economic justification of crop rotation and land management" [26].

The main criterion for the organisation of the crop rotation fields [26] is the inclination of the slope. There are three technological groups of arable land according to the inclination of the slope:

- Group I – up to 3°,
- Group II – 3-7°,
- Group III – more than 7°.

A definition of soil-protective technologies for growing crops is developed for each group. In group I, the recommendation is to cultivate special-zoned crops with intensive tillage, including row-crop technology. The first group is divided into two technological subgroups. Subgroup Ia is the plain land with an inclination of the slope up to 1°, where there are no restrictions in the choice of direction of cultivation and sowing. Subgroup Ib is the area between 1–3° inclination. Here, obligatory cultivation and sowing are permitted orthogonally to the slope direction or across the acceptable angle to the slope. In the areas of group II, the recommended is to apply grain-grass and soil-protective crop rotations except for the black steam field, row crops and other erosion susceptible crops. Group II is divided into two technological subgroups to differentiate the protective erosion measures' level. Subgroup IIa is the slope with an inclination of 3–5° without depressions. The subgroup IIb considers the slopes with 3–7° and 3–5°, the latter located in complex morphology. In the areas of the technological subgroup IIa, the recommendation is to perform grain-grass crop rotations. In the areas of subgroup IIb, it is required to organise the grassland soil-protective crop rotations.

Implementing the approaches and recommendations described above is limited by the high degree of inconsistency of the land market and land fragmentation. Up to seven million private land plots cover an area of about 28 million hectares in Ukraine. This is not favourable for implementing optimal and cooperative measures at the local and territorial levels.

The agrotechnical actions require considering the territory's soil, landscape and economic conditions. The set of erosion protection actions to reduce the erosion risk include [4]:

- tillage orthogonally the slope, contour tillage and tillage with the formation of inclined furrows;
- deep ploughing and ploughing with soil deepening, step ploughing, selfless tillage with stubble preservation;
- attraction of the flat cutter, chisel and minimum zero tillage;
- stubble peeling and soil disking;
- combined shelf and selfless ploughing;
- splitting and taming of the soil;
- rolling of soil with simultaneous splitting, and with simultaneous rolling and splitting;
- sowing with stubble drills with simultaneous formation of intermittent furrows;
- sowing of crops with simultaneous formation of furrows;
- intermittent furrowing and slitting of the soil during cultivation between rows of row crops;
- application of early steam, snow retention and regulation of snow melting.

In addition, in Ukraine, more than six million hectares of land are systematically affected by wind erosion, and dust storms influence up to 20 million hectares in years. The Southern Steppe is the most impacted zone in Ukraine by wind erosion and dust storms. Hence, the number of days per year with dust storms in the Southern Steppe is 159, in the Northern and Central Steppe – 188 days/per year, and in the Forest-Steppe and Polissya – about 33 days/per year.

1.3 Soil Contamination

The content of Co, Ni and Cr in the Ukraine soil depends on the bedrock [7]. Among the most important soil properties related to natural zonation are organic matter (humus) content, genetic type of the soil and horizons and particle size distribution [27]. Generally, the reference heavy metals content does not exceed the acceptable limits. The exception was registered for the territories of geochemical anomalies located in the Ivano-Frankivsk, Zakarpattia regions and the Autonomous Republic of Crimea [7, 28].

The Ukraine legislation requires the determination of mobile heavy metals, i.e. zinc, copper, lead, cadmium and mercury [29].

The soil contamination with heavy metals is typical for all regions of Ukraine. Very high content of mobile forms of copper is specific to 14.6% of Ukraine's soils. The peaks were registered in Kharkiv, Lviv, Zakarpattia and Chernivtsi regions. High and very high values of lead in soils occur in 1.6% of Ukraine's soils. In particular, the value reaches 2.8% in Polissya soil-climatic zone. High concentrations of mobile forms of zinc were found in 71.08 thousand hectares, and about 53.1 thousand hectares are in Polissya soil-climatic zone. An increased, high and very high content of mobile forms of cadmium was found in the territory of 7.6 thousand hectares or 0.047% of the Ukraine land. The predominance part is also located in the

Polissya area. The exceeding of the limit values was also registered for the Steppe zone of Ukraine (Zaporizhia, Luhansk, Mykolaiv, Kherson regions and Crimea). Soils with high mercury content were not detected [29].

At the same time, numerous cases of crop production with a permissible content of heavy metals were registered in the areas with high content of heavy metals in soil. Conversely, the excess of heavy metals in harvest happens in areas with relatively unpolluted soil. Such cases are common, mainly for sunflowers [7].

In Ukraine, according to the Institute of Soil Protection data, the maximum allowable concentration of lead is exceeded in 0.15% of the studied areas, cadmium – 0.14%, and zinc – 0.06%. Significant concentrations of copper are found in 0.3% of the area. These locations are concentrated in vineyards, berries and orchards in Odessa and Kyiv regions.[2]

Currently, the study of the pesticide content in Ukraine soil is of increased attention. Pesticides are harmful to crop growth and have been forbidden from being used at the national level of Ukraine. The level of soil contamination with forbidden pesticides gradually decreases due to their stopping use. In 1980, the area of the arable soils with the identified residues of 4.4-dichlorodiphenyltrichloroethane (DDT) and hexachlorocyclohexane (HCG) was up 72.7%. However, in 2010, these lands were only 21.0%.[3] Subsequently, in 2015, HCG pollution was recorded at an area of 2.52 thousand hectares (only 0.014% of the surveyed lands), DDT was observed at an area of about 6.69 thousand hectares (0.039%) and dimethylamine salt of the 2.4-dichlorophenoxyacetic acid (2.4-D) was detected for the 2.5 thousand ha (0.027%) [6].

The contamination of Ukraine's soil with radioactive substances is closely related to the accident at the Chernobyl nuclear power plant in 1986. The most affected are Vinnytsia, Volyn, Zhytomyr, Ivano-Frankivsk, Kyiv, Rivne, Sumy, Ternopil, Khmelnytsky, Cherkasy, Chernivtsi and Chernihiv regions. The monitoring of the radionuclides is organised constantly. The contaminated areas remain in 12 regions of Ukraine, where 8.8 million hectares have been surveyed. The contamination with radionuclides of caesium and strontium agricultural soils was 01.01.2010, as shown in Table 3 [7].

During 2011–2015, about 19 million hectares of land were surveyed for radioactive isotope contamination. Caesium-137 contamination in the range up to 5 Ci/km^2 is observed at the area of 99.97%, more specifically, up to 1 Ci/km^2–94.8%, 1–5 Ci/km^2–4.9% [6]. Ukraine legislation accepts this level as admissible for economic activity. The contamination of the lands with an intensity of more than 15.5 Ci/km^2 is subject to conservation [30]. The relevant legislation determines the legal regime of the contaminated areas [31].

The caesium-137 contamination in agriculture soils is over 37 kBq/m^2 and is spread over 461.7 thousand hectares. The arable land is 345.9 thousand hectares (Zhytomyr region – 156 thousand hectares, Cherkasy – 76 thousand hectares,

[2] http://www.iogu.gov.ua/monitorynh-objektiv-dovkillya/vazhki-metaly/.

[3] http://www.iogu.gov.ua/monitorynh-objektiv-dovkillya/vazhki-metaly.

Table 3 The density of contamination with radionuclides of caesium and strontium of soils of agricultural lands as of 01.01.2010 [7]

			Contamination Density. kBq/m^2							
			Caesium-137				Strontium-90			
Region	Land/arable	Square thousand hectares	<37	37–185	186–555	>555	< 0.74	0.74–5.55	5.56–111	>111
Vinnytsia	Total	1241.6	1192.1	49.4	0.1	–	–	1241.6	–	–
	Arable	1223.7	1176.5	47.1	0.1	–	–	1223.7	–	–
Volyn	Total	547.4	546.0	1.4	–	–	547.4	–	–	–
	Arable	346.4	346.3	0.1	–	–	346.4	–	–	–
Zhytomyr	Total	1150.3	994.4	145.6	10.3	–	354.3	757.5	38.5	–
	Arable	967.2	856.6	104.6	6.0	–	307.2	633.9	26.1	–
Ivano-Frankivsk	Total	276.1	267.1	9.0	–	–	87.4	186.2	2.5	–
	Arable	223.1	216.1	7.0	–	–	72.4	149.0	1.7	–
Kyiv	Total	525.0	491.4	33.5	0.1	–	517.7	6.6	0.7	–
	Arable	523.2	489.6	33.5	0.1	–	515.9	6.6	0.7	–
Rivne	Total	420.7	369.0	51.4	0.3	–	392.5	27.9	0.3	–
	Arable	310.4	281.0	29.3	0.1	–	293.2	16.9	0.3	–
Sumy	Total	1153.9	1143.6	10.3	–	–	1133.0	20.9	–	–
	Arable	1079.2	1070.0	9.2	–	–	1060.1	19.1	–	–
Ternopil	Total	127.5	126.8	0.7	–	–	93.3	34.2	–	–
	Arable	125.7	125.7	–	–	–	92.0	33.7	–	–
Khmelnytsky	Total	950.3	947.0	3.3	–	–	949.2	1.1	–	–
	Arable	941.3	938.4	2.9	–	–	940.2	1.1	–	–
Cherkasy	Total	330.6	254.5	75.3	0.8	–	25.1	286.9	18.6	–
	Arable	296.1	233.1	62.4	0.6	–	24.6	257.4	14.2	–

(continued)

Table 3 (continued)

Region	Land/arable	Square thousand hectares	Contamination Density. kBq/m^2							
			Caesium-137				Strontium-90			
			<37	37–185	186–555	>555	< 0.74	0.74–5.55	5.56–111	>111
Chernivtsi	Total	228.8	210.2	18.4	0.2	–	–	228.8	–	–
	Arable	189.3	173.7	15.4	0.2	–	–	189.3	–	–
Chernihiv	Total	1836.7	1785.1	48.4	3.0	0.2	85.6	1699.0	51.6	0.5
	Arable	1393.7	1366.4	26.4	0.9	–	65.3	1294.8	33.6	–
Total	***Total***	***8788.9***	***8327.2***	***446.7***	***14.8***	***0.2***	***4185.4***	***4490.8***	***112.2***	***0.5***
	Arable	***7619.3***	***7273.4***	***337.9***	***8.0***	–	***3717.3***	***3825.4***	***76.6***	–

Rivne – 52 thousand hectares, Chernihiv – 52 thousand hectares, Vinnytsia – 50 thousand hectares, Kyiv – 34 thousand hectares).[4]

The special attention in Ukraine related to the presence in the Rivne region of 18.6 thousand hectares of contaminated peatlands, which are a source of migration of radionuclides because of wind erosion (National Report, 2010). The density of the strontium-90 contamination has the following distribution: up to 0.02 Ci/km^2–76.3% of the surveyed areas, 0.02–0.15 Ci/km^2–22.6% and 0.15–3 Ci/km^2–1,1%.[5] A comparatively unpolluted area as of 2015 is 99.7% of the surveyed area, where agricultural production is possible [6]. At the same time, strontium-90 pollution depends on emissions from the Chernobyl nuclear power plant. Strontium-90 contaminated 4.6 million hectares in the range of 0.74–5.55 kBq/m^2, 52% of the surveyed area. Such distribution of this radionuclide in the territory of Ukraine is caused, first of all, by global emissions of strontium-90 [7].

As mentioned above, the use of radiation-contaminated areas is determined by some regulations. At the same time, there is the danger of obtaining radiation-contaminated products within lands with low radionuclide content in soils. The increased danger of radioactively contaminated products remains at the pastures and hayfields of contaminated zones located on meadow-swamp and peat-swamp wetlands, which are characterised by high coefficients of biological assimilation of caesium-137 in plants. Milk production, especially in private farms, is dangerous concerning the radioactive feed [6]. Critical zones outside the exclusion territory are widely registered, and this crop contains a high amount of radionuclides, predominantly in soils of a light mechanical composition. Thus, more than 90% of the surveyed lands by the density of radionuclide contamination are suitable for economic activity without any restrictions. The zone of guaranteed voluntary resettlement includes 5 thousand hectares (0.1%) of lands contaminated with caesium-137 and more than 50 thousand hectares (0.3%) of lands contaminated with strontium-90 [6].

At the same time, significant progress was observed compared to the situation immediately after the Chernobyl accident. This progress is due to the natural rehabilitation (i.e. radioactive decay, fixation and redistribution of radionuclides) and the established agronomic measures. The measures are the reduction of food contamination, the introduction of enhanced radioecological monitoring of agricultural products, precision soil pollution mapping. [6]

The soil protection measures from contamination have several peculiar characteristics, primarily related to the characteristics of the soil as a natural object. This is because soil and soil physical and chemical properties are less dynamic and more inertial than water and atmospheric air. When water and air can be easily cleaned of contamination, the soil (and even the pedosphere) are sometimes inaccessible to recovery.

[4] http://www.iogu.gov.ua/monitorynh-objektiv-dovkillya/radionuklidy.

[5] http://www.iogu.gov.ua/monitorynh-objektiv-dovkillya/radionuklidy.

[6] http://www.iogu.gov.ua/monitorynh-objektiv-dovkillya/radionuklidy.

1.4 Salinity

Saline soils are characterised by the presence within the soil horizons (all or partly) of the soil profile, the easily soluble salts in the amount that inhibits the growth and development of soil (pedogenesis). The crucial value of the inhibition is more than 0.1–0.3%. In Ukraine, there are chloride, carbonate and sulphate salinisation according to the chemical composition of salts. The predominant territories with saline lands in Ukraine are located in the central and southern parts of the Dnieper and in the Azov Sea area.

Saline soils in Ukraine cover an area of about 1.92 million hectares. The 1.71 million hectares are for agricultural use. The lightly saline soil covers up to 1336.6 thousand hectares, medium saline soil – 224.3 thousand hectares and highly saline soil – 116.3 thousand hectares. The most saline soil in Ukraine is solonchak (salt marshes, salt-affected soil) which includes 32.8 thousand hectares [8].

According to the genesis, soil salinisation is divided into primary and secondary. Primary salinisation is associated with natural processes, while secondary is related to human activities. The soil degradation in Ukraine related to salinisation is associated with reducing the irrigated land [32]. The irrigated land in Ukraine decreased from 2,624 thousand hectares in 1992 to 551.4 thousand hectares in 2020. [7]

There is another critical factor for soil salinity formation. The waters in Ukraine used for irrigation have increased rigidity, especially in the southern regions. About 70% of the irrigated soils attract water sources with a salinity of less than 1 g/dm^3. In all other areas – more than 1 g/dm^3 [33]. The relevant standards regulate the quality of the irrigation water. In the case of natural water use, the regulation is involved from the [34]. For wastewater, the regulation is involved from the [33]. Moreover, in 2020 the Cabinet of Ministers of Ukraine adopted the resolution, which formalised all requirements together [35].

Using water with high mineralisation causes changes in the ratio between sodium (increase) and calcium (trend or steady decrease). The irrigation leads to the increase of the absorbed sodium content of the number of exchangeable cations. When using the irrigation water of the first class (according to DSTU 2730–94), the increase is from 0.6–1.0 to 1.5–2.0%. When second or third class water class is applied, there is an increase of up to 3–10% [36].

The irrigated lands in Ukraine consist of 350 thousand hectares of saline soil, including 70–100 thousand hectares of secondary saline soils. The area of the solonetzes soil (Haplic Solonetz in WRB) is 2.8 million hectares, predominantly within the Steppe zone of Ukraine. About 2/3 of them are ploughed, and about 0.8 million hectares are irrigated [8]. The annual chemical reclamation applies to solonetzes soil, although their amount is insignificant within 6–7 thousand hectares

[7] https://superagronom.com/news/12128-ploschi-zroshuvanih-zemel-v-ukrayini-zbilshuyutsya-v-2020-polito-ponad-550-tis-ga.

(during 2011–2015) [6]. In 2014–2015, 78% of the reclaimed saline soils were located in Ukraine's Kherson and Mykolaiv regions.

As of 1990, plastering was carried out at an area of about 305 thousand hectares. The annual requirement for soil plastering is estimated at 160–200 thousand hectares [7].

Hence, the salinisation of the Ukraine lands demonstrated the risk of soil loss. According to the agricultural survey [6], more than 19% of soils are acidic, and the saline soil includes more than four million hectares. The liming of the acidic soil and plastering of the solonetz soil is required. These procedures remain one of the main components of agricultural measures to improve saline soil's physical and chemical properties in Ukraine.

1.5 Agrochemicals Use

The results of the soil surveys indicate a low supply of soils in Ukraine with easily hydrolysed nitrogen. The low and very low content is typical for 93.1% of the surveyed areas. At the same time, the situation is the opposite for the content of phosphorus and potassium mobile compounds. The low or very low content of mobile phosphorus compounds was observed only in 10.4% of soils. The soils with a low phosphorus and potassium content are located mainly in the Polissya area and the Chernivtsi region.

In general, the low content of mobile potassium compounds was detected only in 8.4% of soils. Such lands are located in Polissya oblasts, Chernivtsi, Kyiv and Kharkiv regions [6]. Such kind of distribution of the elements shapes the strategy of applying mineral fertilisation in Ukraine. Nitrogen fertilisers predominate, and much less potassium is applied (Fig. 2).

The balance of mineral fertilisers (the difference between applying them and their removal from the crop) has been negative in the last 15 years [29]. The negative balance ranges from minus 108 kg/ha in 2011 to minus 67 kg/ha in 2015. Regarding macronutrients, almost half of the total amount is formed due to potassium deficiency.

Another problem related to Ukraine agrochemistry is the declining quality of soils associated with the unbalanced use of mineral fertilisers. To achieve the optimal ratio between nutrients, the amount of phosphorus and potassium fertilisers should be increased at least three times [6].

There is a gradual increase in mineral fertilisation – the minimum application registered in the late '90s (Fig. 3).

The data presented in Fig. 3 illustrate the general trends of the application of mineral fertilisers in Ukraine. The increase in the number of nutrients is provided mainly by nitrogen fertilisers. The priority crops (in 2019) that are the primary

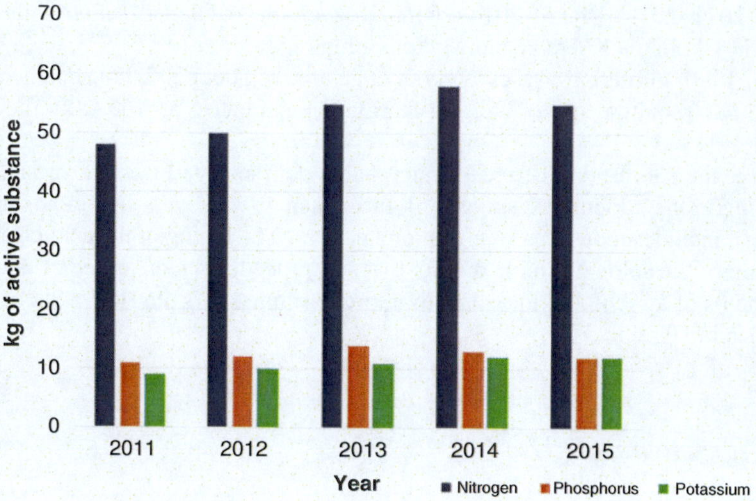

Fig. 2 The application of mineral fertilisation in Ukraine in 2011–2015

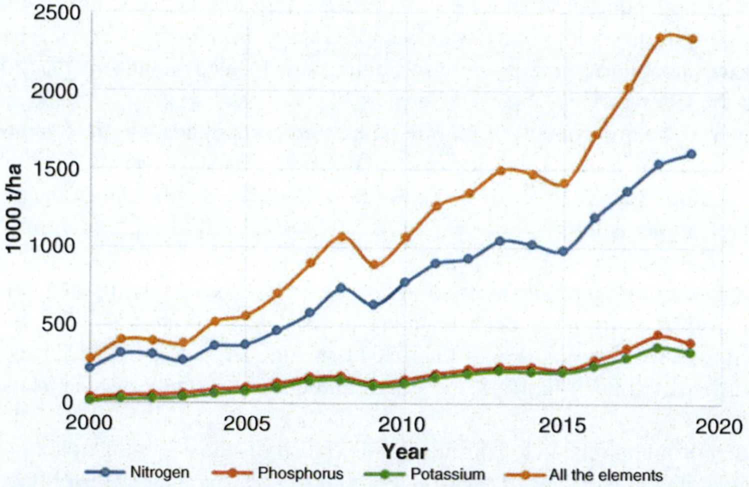

Fig. 3 The fertilisation in Ukraine with the different elements http://www.ukrstat.gov.ua/operativ/operativ2018/sg/vmod/vmod1990-2019_u.xls

consumers of mineral fertilisers in Ukraine are vegetables (299 kg d.r./ha of specified sown area), sugar beet (240 kg d.r./ha) and rapeseed (166 kg d.r./ha).[8]

The application of organic fertilisers in the last 30 years has decreased from the level of about 260 million tons in the early 1990s to 10–11 million tons in

[8] http://www.ukrstat.gov.ua/operativ/operativ2018/sg/vmod/vmodsg2019.xls.

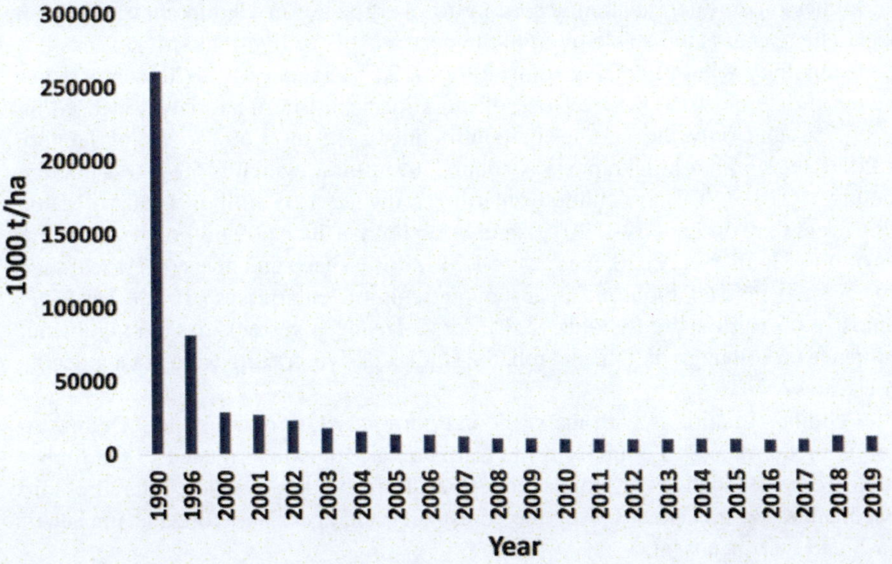

Fig. 4 Organic fertilisation in Ukraine

2010–2015 (see Fig. 4). This is related to the sharp decrease in the number of cattle and pigs.[9]

The sharp decrease in the organic fertiliser application in Ukraine led to the gap in the organic matter (humus) content in arable soils. The negative balance in 2011 was minus 0.37 t/ha, in 2012 – minus 0.36 t/ha, in 2013 – minus 0.13 t/ha, in 2014 – minus 0.20 t/ha and in 2015 – minus 0.13 t/ha. The regions located in the south part of Ukraine are characterised by the biggest lack of organic matter content in arable soil [6].

The humus balance improvement in the ploughing of the non-marketable part of the crop (in 2015–5073.9 thousand hectares), as well as the green manure crops (in 2015–233.4 thousand hectares), increased [6]. However, these measures do not have a decisive impact because of the limited areas of the application.

1.6 Overgrazing, Slash-and-Burn Agriculture, Microplastics and Other Forms of Soil Degradation

Some varieties of the soil destruction, that are common all over the world, are not typical for Ukraine.

[9] http://www.ukrstat.gov.ua/operativ/operativ2021/sg/ksgt/arh_ksgt2021_u.html.

Studies on soil degradation according to overgrazing in Ukraine have not been held. The reason is the stable trend to the decrease of the livestock. Most farms use the technology without grazing (on-site food). As of January 1, 2020, compared to the exact date of 2019, the livestock of most species of farm animals decreased by 4.7–5.7%, and only the number of poultry increased by 3.6%. [10] As of January 1, 2020, the number in Ukraine was about 3.14 million, which is 5.7% less than on January 1, 2019. The agricultural enterprises owned 1.05 million heads of cattle (7.5% less than on January 1, 2019), and households – had 2.09 million heads (4.7% less). As of January 1, 2020, there were 1.21 million sheep and goats (4.7% less than a year earlier): 164 thousand heads of agricultural enterprises (10.0% less) and almost 1.05 million households (3.8% less). Hence, livestock has no significant influence on soil degradation, and reliable studies of overgrazing were not applicable in Ukraine.

A similar situation is with the slash-and-burn agriculture in Ukraine. Deforestation is predominantly common in the Carpathian Mountains, related to the timber trade. 99% of changes in forest cover result from deforestation, and fires are responsible for less than 1% of the forest area change. Therefore the impact of slash and burn is absent.

Until recently, the study of microplastics was not provided in the Ukrainian guidelines for soil protection monitoring [37, 38]. The integration into the European research community required the starting of microplastics monitoring.

One of the first studies of the microplastic content in the natural sites and objects in Ukraine was launched as part of the Joint Danube Research #4 in 2019. The results demonstrated that the amount of microplastics varies from 2 g/kg of the suspended matter to 10 g/kg. The value of the microplastic content is 2.19 in Kiliya at the Danube area and 2.42 g/kg in the Tisza of Ukraine. [11]

Special legislative acts are being developed in Ukraine to reduce the accumulation of plastic waste in the environment [39].

2 Conclusions

Ukraine consists of a Forest zone or Polissya with 19% of total land, a Forest-Steppe zone with 35%, a Steppe zone with 40%, and the Carpathian and Crimean Mountains. The agricultural area covers 19 million hectares within the Steppe zone, 16.9 million hectares within the Forest-Steppe zone and 5.6 million hectares in the Forest zone. The important soils responsible for Ukraine's highly intensive agricultural production are the chernozems, which cover about half of the country (68% of the arable land). Other fertile soils in Ukraine are phaeozems and albeluvisols. One of the biggest challenges for Ukrainian farming is stabilisation of the situation with the

[10] http://ukrstat.gov.ua/.

[11] https://www.davr.gov.ua/news/pro-mikroplastik-u-nashomu-povsyakdennomu-zhitti.

on-going human-induced degradation. The farmlands in Ukraine cover 41,477.2 thousand hectares or 68.72% of the territory. One of the critical factors in the development of soil degradation processes is the changes in the structure of sown areas of crops. The number of economically valuable crops is growing without reliable erosion protection measurements application. The physical degradation of the Ukraine lands is related to excessive ploughing, the deficient balance of nutrients, insufficient application of organic matter, mineral fertilisers, ameliorants and pollution. The soil compaction appears at more than 20% of the area with the grain crops. More than six million hectares of land are systematically affected by wind erosion in Ukraine, and dust storms influence up to 20 million hectares in years. Wind erosion and storms are common in the Southern Ukraine. The soil contamination with heavy metals is typical for all regions of Ukraine. The radioactive contamination of the agricultural land of Ukraine with caesium-137 is over 37 kBq/m^2. The affected area is about 461.7 thousand hectares. At the same time, significant progress was observed compared to the situation immediately after the Chernobyl accident. Ukraine's predominance territories with saline soil are located in the central and southern parts of the Dnieper and in the Azov Sea area. The Ukraine soils are characterised by a low supply of easily hydrolysed nitrogen and high content of phosphorus and potassium mobile compounds. The problematic issue is the unbalanced use of mineral fertilisers. Overgrazing is not typical for Ukraine because of the decrease in livestock grazing. A similar situation is with the slash-and-burn agriculture in Ukraine.

Among the urgent needs for Ukraine are innovation and investments in strengthening the material and technical base of the agricultural sector, implementing environmentally friendly, resource and energy-saving technologies and conservation.

References

1. Fileccia T, Guadagni M, Hovhera V, Bernoux M (2014) Ukraine: soil fertility to strengthen climate resilience
2. Ibáñez JJ, Pérez-Gómez R, Martínez FSJ (2009) The spatial distribution of soils across Europe: A fractal approach. Ecol Complex 6(3):294–301
3. WRB, I. W. G. (2014) World reference base for soil resources 2014. International soil classification system for naming soils and creating legends for soil maps. World Soil Resources Report 106
4. Baliuk SA (2010) Scientific and applied bases of soil protection against erosion in Ukraine. Monography. Kharkiv, 538 p
5. Nizalov D, Dankevych V, Ivinska K (2018) Monitoring of land relations in Ukraine 2016-2017. Statistical yearbook, 168 p
6. Yatsuk IP (2017) About the condition of soils of agricultural lands of Ukraine
7. National Report on the State of Soil Fertility in Ukraine (2010) Kyiv, 112 p
8. Balyuk SA, Medvedev VV, Miroshnychenko MM, Skrylnyk EV, Timchenko DO, Fateev AI, Khristenko AO, Tsapko YL (2012) Ukr Geogr J 2:38–42
9. Krupskiy NK, Polupan NI (1979) Atlas of Ukrainian SSR soils. Urojay, Kyiv, p 160

10. Medvedev VV (2013) Physical degradation of chernozems. Diagnosis, causes, consequence, warnings Kharkiv, 326 p
11. Medvedev VV, Laktionova TM (2002) Estimation of crop losses in Ukraine from soil compaction. Bull Agric Sci 3:53–59
12. Shevchenko MV (2019) Scientific bases of tillage systems in conditions of unstable and insufficient moisture. KhNAU, Kharkiv, p 210
13. Medvedev VV, Lyndina TE, Laktionova TN (2004) Soil density (genetic, ecological and agrotechnical aspects). KP "Drukarnya № 13", Kharkiv. 244 p
14. Dyakov VP (2014) About the soil as a material of the impact of the working bodies of the machines of the technological complex. Agriculture 8:11–12
15. Bakhtin PU (1969) Studies of physical-mechanical and technological properties of the main types of soils of the USSR. Kolos, Moscow. 272 p
16. Rabochev IS, Bakhtin PU (1978) Industrialisation of agriculture and soil fertility. Problems of agriculture, pp 157–160
17. Medvedev VV, Laktionova TM (2008) Soil-technological requirements for tillage implements and running systems of machine-tractor units. KP "Drukarnya № 13", Kharkiv. 68 p
18. Sdobnikov SS (1976) Ways to increase the efficiency of tillage. Agriculture 1:30–31
19. Kuznetsova IV (1978) Sealing effect of the Belarus tractor on the chernozems of the Kursk region. Soil Sci 10:53–58
20. Shevchenko MV (2008) Tillage systems, agriculture Kyiv, Ekmo, vol 80, pp 33–39
21. Shevchenko NV, Lebed EM, Pivovar NI (2015) Comparative assessment of minimum tillage technologies in the cultivation of winter wheat in the northern steppe of Ukraine. Agriculture 2: 20–21
22. Shevchenko MV (2009) Results of application of "No-till" technologies in the conditions of the left-bank Forest-steppe, vol 1. Visnyk of Dnipropetrovsk State Agrarian University, pp 32–35
23. Borrelli P, Robinson DA, Panagos P, Lugato E, Yang JE, Alewell C et al (2020) Land use and climate change impacts on global soil erosion by water (2015-2070). Proc Natl Acad Sci 117(36):21994–22001
24. Tararico OG, Kuchma TL, Ilyenko TV, Demyanyuk OS (2017) Erosion degradation of Ukraine soils according to climate change. Agroecol J 1:7–15
25. Bulygin SY, Antonyuk D (2016) Soil erosion in Ukraine. Crop Soil Sci 235
26. (2013) Methodical recommendations for the development of land management projects that provide ecological and economic justification for crop rotation and land management. Land Manag Bull 10:52–63
27. Samokhvalova V, Fateev A, Luchnikova E, Lykova O (2012) Ecological and geochemical studies of the content of different forms of Co, Ni, Cr in soils of different genesis in Ukraine. Bull Lviv Univ Ser Biol 60
28. Fateev AI, Pashchenko YV (2003) Background content of microelements in the soils of Ukraine. Kharkiv, 112 p
29. Yatsuk IP (2015) About the condition of soils on agricultural lands of Ukraine. Kyiv
30. Order of the Ministry of Agrarian Policy, April 26, № 283 (2013) About approval of the procedure for land conservation. https://zakon.rada.gov.ua/laws/show/z0810-13#Text
31. Law of Ukraine (1991) About the legal regime of the territory that was exposed to radioactive contamination as a result of the chornobyl catastrophe. Information of the Verkhovna Rada of the USSR (VVR), vol 16, p 198
32. Stashuk VA, Balyuk SA, Romashchenko MI (2010) Scientific bases of protection and rational use of irrigated lands of Ukraine. Agrarna nauka, 624 p
33. DSTU 7369: 2013 (2014) Sewage. Requirements for wastewater and its sediments for irrigation and fertilisation. K.: Ministry of Economic Development of Ukraine. 8 p
34. DSTU 2730-94. (1994) System of standards in the field of environmental protection and rational use of resources. Quality of natural water for irrigation. Agronomic criteria. State Standard of Ukraine

35. Resolution of the Cabinet of Ministers of September 2, 2020 N 766 (2020) About norms of ecologically safe irrigation, drainage, management of irrigations and drainage. https://zakon.isu.net.ua/sites/default/files/pdf/pro_normativi_ekologichno_bezpec-3-477346.pdf
36. Baliuk SA, Medvedev VV, Nosok BS (2018) Adaptation of agrotechnologies to climate change: soil and agronomic aspects. Stylish Printing House, Kharkiv. 364 p
37. Medvedev VV (2012) Monitoring of soils of Ukraine. Concept. Results. Tasks. KP "City Printing House", Kharkiv. 536 p
38. Miroshnichenko MM (2016) Theory and practice of soil protection monitoring. FOP Brovin O. V., Kharkiv. 384 p
39. Law Draft of Ukraine "About Restrictions on the Circulation of Plastic Bags at the Territory of Ukraine". http://search.ligazakon.ua/l_doc2.nsf/link1/JI00504W.html

Correction to: Agricultural Land Degradation in the Czech Republic

David Zumr

Correction to:
Chapter "Agricultural Land Degradation in the Czech Republic" in: Paulo Pereira, Miriam Muñoz-Rojas, Igor Bogunovic, and Wenwu Zhao (eds.), *Impact of Agriculture on Soil Degradation II: A European Perspective*, Hdb Env Chem, https://doi.org/10.1007/698_2022_928

Chapter, "Agricultural Land Degradation in the Czech Republic" was previously published as non-open access. It has now been changed to open access under a CC BY 4.0 license and the copyright holder updated to "The Author(s)." The book has also been updated with this change.

The updated original version for this chapter can be found at https://doi.org/10.1007/698_2022_928

Paulo Pereira, Miriam Muñoz-Rojas, Igor Bogunovic, and Wenwu Zhao (eds.), *Impact of Agriculture on Soil Degradation II: A European Perspective*, Hdb Env Chem (2023) 121: C1–C2, DOI 10.1007/978-3-031-32052-1_1016,